# Information Technology and Military Power

A VOLUME IN THE SERIES

*Cornell Studies in Security Affairs*

Edited by Robert J. Art, Robert Jervis, and Stephen M. Walt

A list of titles in this series is available at cornellpress.cornell.edu.

# Information Technology and Military Power

JON R. LINDSAY

Cornell University Press

*Ithaca and London*

First published 2020 by Cornell University Press

Library of Congress Cataloging-in-Publication Data

Names: Lindsay, Jon R., author.
Title: Information Technology and Military Power / Jon R. Lindsay.
Description: Ithaca, New York : Cornell University Press, 2020. |
    Series: Cornell studies in security affairs | Includes bibliographical
    references and index.
Identifiers: LCCN 2019044585 (print) | LCCN 2019044586 (ebook) |
    ISBN 9781501749568 (hardcover) | ISBN 9781501749582 (pdf) |
    ISBN 9781501749575 (epub)
Subjects: LCSH: Military art and science—Information technology—
    United States. | Military art and science—Technological
    innovations—United States. | Information technology—Military
    aspects—United States. | Military art and science—Automation. |
    War—Technological innovations.
Classification: LCC UG478 .L56 2020 (print) | LCC UG478 (ebook) |
    DDC 355.00285—dc23
LC record available at https://lccn.loc.gov/2019044585
LC ebook record available at https://lccn.loc.gov/2019044586

# Contents

*Acknowledgments*                                                    *vii*

*List of Abbreviations*                                              *ix*

Introduction: Shifting the Fog of War                               *1*

1.  The Technology Theory of Victory                                *13*

2.  A Framework for Understanding Information Practice               *32*

3.  Strategic and Organizational Conditions for Success:
    The Battle of Britain                                           *71*

4.  User Innovation and System Management: Aviation
    Mission Planning Software                                       *109*

5.  Irregular Problems and Biased Solutions: Special
    Operations in Iraq                                              *136*

6.  Increasing Complexity and Uneven Results:
    Drone Campaigns                                                 *180*

7.  Practical Implications of Information Practice                  *212*

*Appendix: Methodology*                                             *243*

*Notes*                                                             *249*

*Index*                                                             *283*

# Acknowledgments

This book is inspired by my experience in war, but writing it has been a more difficult battle in many ways. Barry Posen, Wanda Orlikowski, Kenneth Oye, and Merritt Roe Smith provided encouragement and criticism during the earliest incarnation of this project at the Massachusetts Institute of Technology. Risa Brooks, Michael Horowitz, Joshua Rovner, and Janice Stein offered sage advice at a 2016 book workshop at the University of Toronto. Many other colleagues were kind enough to read drafts of chapters and provide helpful comments along the way, including Carmen Cheung, Bridget Coggins, Ron Deibert, Michael Desch, Jesse Driscoll, Kenneth Geers, Eugene Gholz, Ryder McKeown, Neil Narang, Jacquelyn Schneider, Rebecca Slayton, Lucy Suchman, and William Wohlforth. Others provided wise counsel at critical junctures, including Emanuel Adler, Jean-Pierre Dupuy, Erik Gartzke, Piet Hut, Brian Cantwell Smith, and Peter Godfrey Smith.

The citations in the book do not adequately express my intellectual debts to a number of pioneering thinkers, such as Terrence Deacon, Edwin Hutchins, Robert Jervis, Bruno Latour, Douglass North, James C. Scott, and Francisco Varela. I have also learned a lot from comrades in the MIT Security Studies Program who made this lonely pursuit bearable and occasionally even enjoyable. In addition to those already mentioned, I would like to thank Hanna Breetz, Keren Fraiman, Benjamin Friedman, Brendan Green, Phil Haun, Llewelyn Hughes, Shirley Hung, Colin Jackson, Stephanie Kaplan, Austin Long, William Norris, Andrew Radin, Joshua Shifrinson, Paul Staniland, and Caitlin Talmadge. Many people provided support and mentorship in connection to my mobilization to Iraq, which forms the basis of chapter 5. I would especially like to acknowledge Jeff Cadman, Lance

Dettmann, Mary Ann Dorsey, James P. Ford, John Jacobs, Dan Johaneckt, D. John Robinson, Michael Sayer, Dane Thorleifson, Shane Voodrin, and Pete "Bullfrog" Wikul. I owe many debts to others who cannot be named. Michael Perkinson deserves special credit, or blame, for inspiring my interest in intelligence and war.

This project has received institutional and financial support from various sources. I am grateful to the MIT Security Studies Program and the Program on Emerging Technology, the University of California Institute on Global Conflict and Cooperation, the Department of Defense Minerva Initiative (Office of Naval Research Grant N00014-14-1-0071), and the Munk School of Global Affairs and Public Policy at the University of Toronto. Jason Lopez at UC San Diego and Ivy Lo and Elisa Lee at the Munk School worked administrative wonders on my behalf. I received able research assistance from Creed Atkinson, Kirpanoor Badwal, Jinwei Jiao, Lennart Maschmeyer, Atul Menon, Mingda Qiu, and Benjamin Smalley. Jasmine Chorley Foster and Alexander Moulton assisted with copyediting. I am extremely grateful to Roger Haydon at Cornell University Press for believing in this project when others did not and for providing the right balance of encouragement and correction throughout several years of revisions. The anonymous reviewers for the most part provided very helpful advice. I would also like to thank Eric Levy for his meticulous editing, Susan Specter for shepherding the book through production, and Ken Bolton for indexing.

Some material in this book draws from or expands on articles that have appeared previously: chapter 4 expands on "'War upon the Map': User Innovation in American Military Software," *Technology and Culture* 51, no. 3 (2010): 619–51; chapters 1 and 5 overlap with "Reinventing the Revolution: Technological Visions, Counterinsurgent Criticism, and the Rise of Special Operations," *Journal of Strategic Studies* 36, no. 3 (2013): 422–53; chapter 5 overlaps with "Target Practice: Counterterrorism and the Amplification of Data Friction," *Science, Technology, & Human Values* 42, no. 6 (2017): 1061–99. I am obliged to say that the opinions expressed in this book are mine alone and do not represent those of the U.S. government, but my critical comments should make that obvious.

I am grateful most of all to my family for putting up with me during a very long writing process. Eleanor was born and learned to walk and talk during the years this book was under review. Lily kept asking me if I had finished the book yet. I can finally answer yes, but I have also started another one! I am grateful to my parents, who have been role models for connecting science and books to the public, and their desert casita was a wonderful writing retreat. Most importantly, words cannot express my gratitude to Heather Silverberg, the love of my life. She was there for me at the beginning of this journey and at the end, and at many points in between. This book, and so much more, is dedicated to her.

# Abbreviations

| | |
|---|---|
| A2/AD | antiaccess/area denial |
| ADGB | Air Defence of Great Britain |
| AFMSS | Air Force Mission Support System |
| AI | airborne intercept |
| AQAP | al-Qaeda in the Arabian Peninsula |
| AQI | al-Qaeda in Iraq |
| C2 | command and control |
| C4ISR | command, control, communications, computers, intelligence, surveillance, and reconnaissance |
| CAOC | Combined Air Operations Center |
| CFPS | Combat Flight Planning Software |
| CIA | Central Intelligence Agency |
| CJSOTF | Combined Joint Special Operations Task Force |
| COIN | counterinsurgency |
| CONOP | concept of operations |
| CONUS | continental United States |
| CRT | cathode ray tube |
| CS-2 | Cromemco System 2 |
| DCGS | Distributed Common Ground System |
| DEVGRU | NSW Development Group (SEAL Team 6) |
| DF | (radio) direction finding |
| F2T2EA | find, fix, track, target, engage, assess |
| F3EA | find, fix, finish, exploit, analyze |
| FATA | Federally Administered Tribal Areas |
| FDSP | Federal Directorate of Supply and Procurement |

| | |
|---|---|
| FOB | forward operating base |
| FPLAN | Flight Planner |
| GCI | ground-controlled intercept |
| GPS | Global Positioning System |
| GTRI | Georgia Tech Research Institute |
| HQ | headquarters |
| HUMINT | human intelligence |
| HVI | high-value individual |
| IED | improvised explosive device |
| IFF | identification friend-or-foe |
| ISAF | International Security Assistance Force |
| ISIS | Islamic State in Iraq and Syria |
| ISR | intelligence, surveillance, and reconnaissance |
| JFCOM | (U.S.) Joint Forces Command |
| JMPS | Joint Mission Planning System |
| JOC | joint operations center |
| JSOC | Joint Special Operations Command |
| LADA | London Air Defence Area |
| MEF | Marine Expeditionary Force |
| MIDB | Modernized Integrated Database |
| MSS | Mission Support System |
| NSA | National Security Agency |
| NSW | Naval Special Warfare |
| ODD | Ogden Data Device |
| OIC | Operational Intelligence Center |
| ONI | Office of Naval Intelligence |
| OODA | observe, orient, decide, act |
| OR | operational research |
| PC | personal computer |
| PERT | Program Evaluation and Review Technique |
| PFPS | Portable Flight Planning Software |
| PLA | People's Liberation Army |
| PPG | Presidential Policy Guidance |
| PPI | plan position indicator |
| PSP | Principles, Standards, and Procedures |
| PUC | person under control |
| RAF | Royal Air Force |
| RDF | range and direction finding (Radar) |
| RDT&E | research, development, test, and evaluation |
| RMA | revolution in military affairs |
| RPA | remotely piloted aircraft |
| SCIF | sensitive compartmented information facility |
| SEAL | Sea, Air, and Land |
| SIGACTS | significant activities |

| | |
|---|---|
| SIGINT | signals intelligence |
| SIOP | Single Integrated Operational Plan |
| SIPRNET | Secure Internet Protocol Router Network |
| SLAM | Standoff Land Attack Missile |
| SOCOM | (U.S.) Special Operations Command |
| SOF | special operations forces |
| SOTF | Special Operations Task Force |
| SPAWAR | Space and Naval Warfare Systems Command |
| SSE | sensitive site exploitation |
| STRATCOM | (U.S.) Strategic Command |
| STS | science, technology, and society |
| TACLAN | tactical local area network |
| TAMPS | Tactical Air Mission Planning System |
| TIP | target intelligence package |
| UAV | unmanned aerial vehicle |
| WERV | Western Euphrates River Valley |
| WILD | Wartime Integrated Laser Demonstration |

# Introduction

## Shifting the Fog of War

Below, where the fight was beginning, there was still thick fog; on the higher ground it was clearing, but nothing could be seen of what was going on in front. Whether all the enemy forces were, as we supposed, six miles away, or whether they were nearby in that sea of mist, no one knew.

Leo Tolstoy, *War and Peace*

Does information technology lift the fog of war or thicken it? Advances in computing and communications during the Cold War encouraged speculation about a revolution in military affairs (RMA). The idea continues to capture imaginations today, albeit under different names, such as "cyberwar," "the third offset," and Chinese "informatization." Skeptics argue that networked forces often fail to live up to the hype, or worse. This book charts a middle way between enthusiasm and doubt by recognizing both advantages and disadvantages of technology. It explains why organizational and strategic context is the key to understanding the performance of information systems.

The ideas developed in this book help to explain why the U.S. military, despite all of its advantages in information technology, has struggled for so long in unconventional conflicts against weaker adversaries. The same perspective suggests that the United States retains important advantages against advanced competitors like China that, despite major investments in information technology, are less prepared to cope with system complexity in wartime conditions. A better understanding of technology in practice can inform the design of command and control systems, improve the net assessment of military power, and promote reforms to improve organizational performance. This book also contributes to the growing scholarly literature on military innovation and effectiveness, which emphasizes the

1

importance of operational doctrine and administrative institutions, not just technological quality or force quantity, for success in war.

To understand why networked forces succeed or fail, it is important to understand how organizations actually use their computers. Digital systems now mediate almost every effort to gather, store, display, analyze, and communicate information. As a result, military personnel now have to struggle with their own information systems as much as with the enemy. Local representations of the world (charts, maps, models, slides, spreadsheets, documents, databases) must be coordinated with whatever distant reality they represent (friendly troops, enemy forces, civilian actors, weather, topography, battles, political events). When personnel can perceive things that are relevant to their mission, distinguish friend from foe, predict the effects of their operations, and get reliable feedback on the results, then they can fight more effectively. When they cannot do these things, however, then tragedies like friendly fire, civilian deaths, missed opportunities, and other counterproductive actions become more likely. If military organizations are unable to coordinate their representations with reality, then all of their advantages in weaponry or manpower will count for little.

I describe the organizational effort to coordinate knowledge and control as *information practice*. I argue that the quality of practice, and thus military performance, depends on the interaction between strategic problems and organizational solutions. Some geopolitical situations are easier to represent than others because they have a well-defined and unchanging structure. Other problems are more ambiguous or fluid. Centralized command and formalized processes may work well in the former case, but decentralized arrangements and informal adaptation work better for the latter. Mismatches between problems and solutions create *information friction* that heightens the risk of accident, targeting error, myopia, and mission failure. Friction becomes more likely, moreover, as operations and organizations become more complex. Warfighting problems and technical solutions keep on changing, but information practice is always stuck in between them.

## The Rumpus Room

I was a junior officer in the U.S. Navy just two years out of college when war broke out in Kosovo. A Serbian crackdown on Kosovar Albanians had raised fears that the Balkans were about to relapse into ethnic cleansing. As NATO began developing military options, I volunteered for an assignment with the U.S. Joint Task Force headquarters in Naples, Italy. The intelligence targeting division consisted of a handful of officers working in a cleared-out storage closet that we called "the Rumpus Room." We read analytical reports, attended staff meetings, updated a Microsoft Excel target spreadsheet, made PowerPoint slides of bombing targets, and briefed senior officers. We carried no weapons but instead managed data about where weapons should go. Our knowledge of Yugoslavia came to us through glowing rectangles—computer monitors, cable television, and video teleconferences. I spent my nights in a comfortable hotel built on Roman ruins and enjoyed mozzarella di bufala and espresso, and the only danger to me personally was the Neapolitan traffic. Not a single NATO service member died in Kosovo, and Slobodan Milosevic capitulated without any NATO boots on the ground. To many it seemed like the United States had reaped the rewards of the RMA.

But mistakes were made. The planned forty-eight-hour air campaign dragged on for two and a half months. More than two-thirds of the thousands of Albanians killed by Serbian forces died during the war that NATO had launched to protect them. Sometimes NATO hit the wrong targets, and sometimes it missed the right ones. One airstrike overshot a military compound and tore through a neighborhood of civilian houses, apartments, and medical clinics. In another, the video feed from a Standoff Land-Attack Missile revealed a passenger train crossing a bridge just milliseconds before the missile slammed into it. Another bomb hit another bridge where a group of people had gathered to celebrate an Orthodox holiday; the next bomb on the same bridge hit the people who had rushed over to help the first group. Some military targets turned out to be empty or full of cleverly disguised decoys, so civilian casualties caused by those strikes seemed especially pointless. More than five hundred civilians died as a result of NATO airstrikes during the campaign. Serbia capitalized on the mistakes to promote a media narrative about NATO human rights atrocities. After each blunder, NATO staff officers reviewed and adapted their processes to avoid making the same mistakes. Then they made new ones.

Our most spectacular failure was the accidental bombing of the Chinese embassy in Belgrade. General Wesley Clark's impetuous demand for "two thousand targets" made it easy to justify any target's connection to the Milosevic regime, however tenuous. This gave the Central Intelligence Agency (CIA) an opportunity to undermine Libyan weapons deals with Belgrade under the pretext of disrupting arms supplies to the Yugoslav army. The CIA thus assembled a target package on the Federal Directorate of Supply and Procurement (FDSP), its only target nomination of the war. CIA analysts knew the street address of the FDSP headquarters from prior reporting, but to find its coordinates they relied on paper maps and land navigation procedures that were ill suited for picking aerial bombing targets. They were also unaware of irregularities in the street layout of that part of Belgrade. As a result, they mistakenly picked a different building located four hundred meters north of the intended target. It turned out that

Beijing, improbably enough, had recently relocated its diplomatic mission in Yugoslavia to the very same building. This structure was known simply as "Belgrade Warehouse 1" in the Modernized Integrated Database (MIDB), the authoritative Department of Defense catalog of facilities worldwide. Although U.S. State Department officials were aware of the Chinese move, that information had not yet percolated into the MIDB in the course of routine updates. Target reviews then failed to notice the Chinese embassy, because no one was looking for it. As the original error percolated across different intelligence agencies, and as supporting documents and imagery accumulated in target folders, circular reporting created the illusion of multiple validations of the target, but it really just amplified the error. The PowerPoint slide that summarized the target appeared just as well vetted as hundreds of others that depicted targets that NATO bombed without incident. The target was approved during a regular briefing session that was supposed to provide high-level quality control, in which President Bill Clinton personally reviewed targets with significant collateral damage concerns.[1]

On the night of the strike, an intelligence analyst phoned the military headquarters in Naples to express doubts about the importance of the FDSP target. When queried explicitly about collateral damage, he did not raise any concerns, so the strike was allowed to proceed. B-2 stealth bombers delivered satellite-guided munitions precisely to the coordinates provided, but the coordinates were precisely wrong. The intended target, the actual FDSP headquarters just four hundred meters away, was unscathed. By a remarkable coincidence, the bombs hit the Chinese defense attaché's office and the embassy's intelligence cell, killing three Xinhua reporters and injuring twenty others. The number of casualties predicted on the PowerPoint slide was quite close to the actual number, ironically enough, but their expected identity was dead wrong.

The Chinese embassy bombing ignited street protests around the U.S. embassy in Beijing. Chinese citizens and officials alike refused to accept that so precise a strike by so advanced a military could have been an accident, speculating instead about veiled threats to a rising power. Disbelief colors Sino-U.S. relations to this day. The Chinese Central Military Commission met in the immediate aftermath of the bombing and made the decision to accelerate the development of cyber and space weapons for "seeing far, striking far, and striking accurately." Chinese writers likened their envisioned capabilities to an "assassin's mace" or trump card that would prevent the United States from ever coercing China in the same way. A glitch in the U.S. RMA thus encouraged China to accelerate its own military "informatization" (*xinxihua*), which would come to pose a serious challenge to U.S. military power in the following decades.[2]

Several formal investigations found no evidence that the United States had deliberately targeted China.[3] As a *New York Times* article described it, the embassy bombing was "an immense error, perfectly packaged."[4] Coincidentally I used the same phrase in a journal entry on the "human consequences of packaging information." Packaging information for people and weapons had become an essential part of the everyday conduct of war. Our slides and spreadsheets were just patches of light on a screen—neat, cheap, and changeable. Yet the strikes they enabled were violent, costly, and irreversible. The contrast sometimes seemed surreal. For the majority of targets we attacked, targeting information was packaged well enough to enable controllable operations with predictable effects. Yet when digital symbols became misaligned with distant situations, the results could be tragic. The possibility of inadvertent targeting error also raised the possibility that an adversary might deliberately alter our data through "information operations" (i.e., cyber attack). Either accident or subversion might thus transform a superpower's information system from an asset into a liability. Life inside the RMA was both quotidian and unsettling.

## Target Practice

Almost a decade after the Rumpus Room, I had another opportunity to observe the RMA in the wild. In 2007 I mobilized to active duty to join a Special Operations Task Force in western Iraq. Chapter 5 presents a participant-observer study of this deployment. The strategic and technological contexts of Kosovo and Iraq could not have been more different. Whereas Kosovo was a short air war against a unified state, Iraq was a long ground war against multiple insurgencies. The Rumpus Room had focused mainly on finding bombing targets for NATO aircraft, but the task force in Anbar Province had to "find, fix, and finish" insurgents as well as work "by, with, and through" Iraqi partners. The forward operating base (FOB) in Fallujah was decidedly more spartan than the Hotel Terme in Naples, and the dangers of mortar fire and improvised explosive devices (IEDs) surpassed the risks of the Autostrada. Nine years of progress in information technology, moreover, was an eternity by the standards of Silicon Valley.

Yet there were some remarkable similarities. All of the everyday struggles to produce reliable knowledge felt eerily familiar. Even on a FOB in a combat zone, or perhaps especially on a FOB, we experienced the battlefield through glowing rectangles. We cobbled together Excel lists of targets and missions, built PowerPoint decks about "high-value targets" and our operations to catch them, received and forwarded innumerable emails, endured endless video teleconferences, briefed leadership and troops, entertained a ceaseless flow of visitors, and argued with other units about who was allowed to access which data. Our operations on the FOB and in

5

the Naples headquarters were both pickup games of assorted personnel mobilized from civilian careers or detailed from other military units, thrown together to make sense of a war that developed beyond their control. The RMA unfolded through a combination of bureaucratic routine and jury-rigged improvisation. Personnel hacked and reconfigured their digital tools on the fly, adapting Microsoft Office and other commercial products rather than relying on systems procured from defense contractors. Sometimes this ferment of bottom-up adaptation improved mission performance. Sometimes, however, the hacks just created friction or reinforced cultural attitudes that were ill suited to the strategic environment. In both wars, all the mundane activities within expeditionary office spaces contrasted markedly with the lethal actions that they enabled.

Information pathologies on the FOB were less dramatic than the accidental bombing of the Chinese embassy, but in some ways they were more insidious. The RMA worldview expects the "persistent stare" of reconnaissance assets and a "common operational picture" to improve "situational awareness." I found instead that the task force's information systems amplified the biases of the Naval Special Warfare community and obfuscated the political situation in Anbar. An ingrained cultural preference for "direct action" (attacking insurgents) over "indirect action" (negotiating with locals) encouraged Navy SEALs to focus on hunting "bad guys" at the expense of other priorities. The organization's information technologies enabled them to do so, but in the process they elided the complex tribal politics of Anbar. Risks to the civilian population or political implications beyond the immediate scope of the deployment became harder for the unit to perceive given the ways in which it gathered and processed information. It is not possible to conclusively determine whether the unit helped or hindered the course of the counterinsurgency, but this itself is an important finding: the information system was not designed to ask or answer critical questions about its own performance. Some types of errors will make it obvious to practitioners that something is wrong, but other errors remain hidden in the epistemic infrastructure of an organization. Information systems in Kosovo were plagued by errors of omission, but in Iraq they were distorted by errors of commission.

This book turns the RMA inside out by examining how people conduct war in an information-intensive environment. What are military personnel actually doing behind all those glowing rectangles? How do they manage to construct reliable representations of the world that enable them to influence it? Under what conditions does information practice improve organizational performance, and when does it produce logistical delay, targeting error, unintended civilian deaths, tragic fratricides, or mission failure? When are informational advantages lasting or fleeting? When do well-meaning attempts at lifting the fog of war end up shifting it elsewhere? Why, furthermore, does technology keep inspiring such great hope and fear when actual systems have so often disappointed? And what, ultimately, are the implications of all this for defense strategy and policy? In short, what are the microfoundations of military power in the information age?

## The Argument

The relationship between information technology and military power is incredibly complicated. My goal is to bring some simplicity and clarity to these phenomena. It is important to understand the concept of information not simply in terms of bits and bytes but rather as a system of pragmatic relationships (between representations and referents, format and meaning, text and context, humans and computers, users and designers, allies and enemies, etc.). Information practice is preoccupied with the management and repair of these relationships. For explanatory simplicity, I will focus on an organization's information about operationally relevant entities on the battlefield. Information practice is like a bridge across a deep canyon that allows data to traffic back and forth between the organization and the environment. If either side of the bridge becomes unstable, then it becomes more difficult for traffic to move across it safely and reliably. The validity of operational information, likewise, depends on the structural stability of the warfighting situation on one side and of the military organization on the other. If a military is unable to connect its internal representations to external realities, then it cannot know where to aim, it cannot hit the things it aims at, and it receives poor feedback about what it hits and misses.

The variability of strategic problems and organizational solutions creates a difficult coordination challenge for military organizations. Simplifying again, the external problem can be described as *constrained* or *unconstrained*. Constrained problems are structured by geography, technology, and social institutions in such a way that the types of things that need to be represented are relatively clear and stable over time, even if the relevant entities stand in tightly coupled relationships. Unconstrained problems are more ambiguous, changeable, or loose. The internal solution, likewise, can be *institutionalized* or *organic*. Institutionalized problems are structured by formal rules, bureaucratic controls, and social identities in such a way that there is general agreement throughout the organization about what, how, and why to represent data. Organic solutions, by contrast, are more informal, self-organized, or heterogeneous. The interaction between problems and solutions produces four ideal types of information practice. Two can improve military performance, and the other two can undermine it.

Table I.1 summarizes the explanatory framework with illustrative examples from this book. Most of these are drawn from the U.S. experience after the Cold War. The United States has had more combat experience with digital technologies than any other country. It has also experienced a wide variety of successes and failures in its employment of those technologies, which enables some controlled comparison. My own experience in the U.S. military, furthermore, enables the use of ethnographic methods to document practice in the field that would otherwise remain inaccessible. The cases presented in chapter 4 on aviation mission planning and in chapter 5 on special operations in Anbar have not, to my knowledge, been examined by other scholars. The exception to my focus on the United States is the Battle of Britain, which has a particular importance in the history of command and control. Contemporaries explicitly described Royal Air Force (RAF) Fighter Command as an "information system" for air defense, and they analyzed it as such with "operational research" methods. Britain established a template for command and control that went on to have an important influence on U.S. Air Force doctrine. The United States, however, did not always fight with the same organizational and strategic advantages that Britain enjoyed in 1940.[5]

The interaction of an institutionalized solution and a constrained problem produces the pattern most conducive to improving military performance, which I call *managed practice*. The RAF fought with the geopolitical advantages of island defense and decades of prior preparation. The

**Table I.1 Theory of information practice, with examples**

| *Internal solution* | *External problem* | |
| --- | --- | --- |
| | *Constrained* | *Unconstrained* |
| Institutionalized | I. Managed practice (Performance improving) Chap. 3: RAF Fighter Command Chap. 4: Combined Air Ops Center Chap. 5: Joint Special Ops Command Chap. 6: Obama targeting policy | II. Insulated practice (Performance degrading) Chap. 3: Luftwaffe strategic bombing Chap. 4: Air Force Mission Support System Chap. 5: Special Ops Task Force in Anbar Chap. 6: Drone signature strikes |
| Organic | IV. Problematic practice (Performance degrading) Chap. 1: Chinese embassy bombing Chap. 2: USS Vincennes shoot-down Chap. 5: Improvised explosive devices Chap. 6: Drone targeting errors | III. Adaptive practice (Performance improving) Chap. 3: Operational research Chap. 4: FalconView Chap. 5: Marine Corps in Anbar Chap. 6: Big Safari |

Luftwaffe, by contrast, had a more difficult offensive campaign and suffered numerous bureaucratic pathologies. While the British planned to fight the last war, the Germans obligingly gave it to them. The information practice framework provides fresh insights into one of the most studied episodes in military history. The salutary performance of Churchill's "few" in the air was enabled by "many" on the ground who worked in a methodically organized and geographically anchored information system. Not only does this case illustrate the importance of organizational and strategic factors over technology, but it is also literally easier to see the computational interactions between people and machines in a simpler vintage of technology. It is more difficult, by contrast, to understand how data flow through modern, digital, classified, distributed command and control systems.

The interaction of an organic solution and an unconstrained problem produces the second-best outcome: *adaptive practice*. Throughout the four major U.S. air campaigns from 1991 to 2003, computer users had to cope with incompatible systems and changing missions after the Cold War. During this same time, bottom-up initiatives flourished in a manner reminiscent of Silicon Valley start-ups, but in the unlikely milieu of a military bureaucracy. Tech-savvy officers responded to changing needs and computational possibilities by writing software prototypes. The official system procurement offices, meanwhile, struggled to meet formal requirements that were outdated by the time systems were delivered. The emergence of the application known as FalconView is an exceptional case of sustained and successful user innovation, but it is hardly unique. Indeed, customized hardware and software can be found everywhere in modern military units, even as network managers and procurement offices struggle to standardize systems. A major problem for contemporary militaries is figuring out how to encourage user innovation while minimizing its liabilities for interoperability, reliability, and cybersecurity.

A bad fit between external problems and internal solutions will complicate information practice and undermine performance. The interaction of an organic solution with a constrained problem results in *problematic practice*. The Chinese embassy was accidentally bombed because uncoordinated agencies interacted opportunistically in a constrained environment of distinct civilian and enemy facilities. The 1988 accidental shoot-down of an Iranian airliner by the USS Vincennes is another tragic example of an uncoordinated system in a tightly coupled environment. Combat interactions between belligerents can also be analyzed under the rubric of problematic practice. Anarchy, or the lack of any overarching institution whatsoever, is essentially an extremely organic system. Enemy attacks by one part of the system explicitly try to confuse or subvert information practice elsewhere in the system, as exemplified by insurgent IEDs that ambushed coalition forces in Iraq.

The interaction of an institutionalized solution and an unconstrained problem results in *insulated practice*. The Special Operations Task Force in Iraq mentioned above had an institutionalized preference for commando raids but was myopic regarding the more ambiguous politics of Anbar. Other examples include the Luftwaffe strategic bombing campaign in 1940, which persisted in a suboptimal targeting strategy and ignored feedback. FalconView emerged as a workaround to more insular and less responsive procurement programs like the Air Force Mission Support System. The distinguishing difference between problematic and insulated practice is that breakdowns become immediately apparent to practitioners in the former, which prompts investigation and repair, but frictions are covered over in the latter, which allows organizations to persist in their folly.

Real cases are almost always a mixture of these ideal types. The empirical chapters tackle this problem by exploring within-case variation. For example, whereas the Special Operations Task Force in Anbar exhibited insulated practice, Marine Corps units in the same province took a more adaptive approach to counterinsurgency. A Joint Special Operations Command (JSOC) task force performed the same counterterrorism mission as the Anbar unit but with the benefit of more resources and a more managed style of practice. Different patterns can be found at different levels of analysis too. During the Battle of Britain, civilian operational research scientists worked side by side with military radar operators to analyze and adjust machines and work processes. Adaptive practice at one level thus helped to stabilize managed practice at higher echelons.

Patterns of practice also tend to change over time because of an endogenous interaction between problems and solutions. In table I.1, this canonical sequence is labeled I, II, III, IV. Management drifts into insulation as adversaries alter the warfighting problem. Personnel defect from institutional standards to make local adjustments to the organizational solution. Adaptation drifts into problems as uncoordinated modifications interfere with other parts of the organization. Problems prompt management to intervene to overhaul the institutions of control. Adaptive exploitation destabilizes while managerial reform restabilizes. Repetition of this exploitation-reform cycle tends to cause information systems and practice to become more complex over time. Recent U.S. drone campaigns exemplify this increasing complexity. The emergence of the armed Predator was the result of rapid prototyping and warfighter innovation that worked around extant bureaucracy. Ad hoc processes resulted in several cases of civilian deaths, fratricide, and the inadvertent killing of U.S. citizens. In response the Obama administration piled on many layers of oversight, and error rates decreased. The Trump administration then cut back the red tape in an effort to reduce bureaucratic insulation, but this appears to have increased civilian casualty rates. Information systems develop through ongoing cycles of exploitation and reform. As a result, information technology becomes more complex and increasingly essential for military performance without, however, providing any lasting decisive advantage on the battlefield.

## Policy Implications

The technology theory of victory is exemplified by the RMA ideology of the 1990s, but there are many other historical manifestations. Scientific advances in artificial intelligence and quantum computing are already encouraging new variations on familiar themes. Chapter 1 explores the eternal return of technology theory and reviews the scholarship on military innovation and effectiveness that it has inspired. This book explains why it is wise to be skeptical of technology theory.

Nevertheless, military competitors will keep on looking for technological advantages and military personnel will still have to rely on computers. What is to be done? The Clausewitzian answer to fog and friction is "genius," or *"an intellect that, even in the darkest hour, retains some glimmerings of the inner light which leads to truth; and . . . the courage to follow this faint light wherever it may lead."*[6] Genius in information practice is expressed in small ways whenever personal hack, modify, repurpose, or improve their own information systems. Customized software and hardware solutions are prevalent in military units, yet the ferment of bottom-up innovation has been largely overlooked by RMA champions and critics alike. Top-down approaches to "systems integration" associated with the RMA worry that amateur interventions will create interoperability, reliability, and security problems. Indeed, geniuses who try to compensate for friction sometimes just end up creating more friction. Yet dangers are also created by formally integrated engineering programs, or centrally managed command centers, that are unable to coordinate their systems with the changing realities of war.

In chapter 7, I describe practical steps that an organization can take to combine the advantages of managed and adaptive practice while mitigating the risks of problematic and insulated practice. I call this approach *adaptive management*. The same principles that work for maneuver warfare—promoting a common understanding of the mission while enabling subordinates to take the initiative—can be leveraged to improve information practice. The challenge is to apply a combined-arms mindset to the design of information systems and not simply to their use. No solution is a panacea, however, because the performance of any information solution is partly a function of changing information problems. Different types of information systems are better or worse for different types of military tasks, and the fit between solutions and problems changes over time.

My account of the microfoundations of military power has macroimplications for national security strategy. Debates about U.S. grand strategy often make assumptions about the impact of advanced technologies on the military balance. Chinese military modernization poses real challenges to U.S. military power in East Asia. Yet performance assessments of the People's Liberation Army (PLA) should also take into account the challenges of information practice that the PLA will undoubtedly experience in war. The U.S. military still has important advantages over the PLA in this respect. Its institutions and personnel have a lot of experience working through friction in combat conditions. Unfortunately, the technology-centric "third offset strategy" tends to overlook or underemphasize the ability of U.S. personnel to respond productively to breakdowns. Adaptive practice is an undervalued asset that is at risk of being squandered by the U.S. military's increasingly top-down focus on procurement, systems integration, and cybersecurity.

Even in the best of circumstances, there are bounds on the ability of information practice to enhance performance. At the low end of the conflict spectrum, warfighter innovation can optimize operations for irregular wars that end up being more costly than victory is worth. Problematic and insulated practice become serious risks whenever unconventional adversaries are more resolved to survive and adapt than the intervening organization. At the other end of the spectrum, some strategists assume that the targeting advantages of the conventional RMA will transfer directly into the nuclear realm. Yet a highly institutionalized nuclear targeting culture carries dangerous risks of insulated practice in a dynamic warfighting scenario. Unlike drone counterterrorism campaigns, nuclear counterforce campaigns cannot afford to learn from their targeting mistakes.

Cybersecurity can be understood as a second-order problem of using information practice to exploit or protect information practice itself. Hackers can either create friction for or exploit friction in target systems, which can shift the balance of power in subtle ways. Yet they also have to cope with the friction created by the complexity and abstraction of a large-scale, human-built, collectively constituted, heterogeneous information infrastructure. This book focuses on the conduct of military combat operations, so I do not address cybersecurity in detail here. Yet any understanding of digital conflict should be grounded in an understanding of information practice for the simple reason that attackers and defenders alike are social organizations that rely on it for everything they do.

Fog and friction will remain vexing problems in future war. Yet it would be a mistake to see friction as only a source of resistance. Friction can also provide traction for creative insight and innovation. The notion of breakdown, likewise, can be understood as either an unpleasant surprise or a deliberate investigation. Friction and breakdown can reveal the limits of knowledge, but they can also inspire the search for ways to go beyond limits. Effective military operations, and political strategies that avoid war, will require a new level of sensitivity to both what information means and how information works. We should endeavor to improve them both.

# The Technology Theory of Victory

The war of today is being fought with new weapons, but so was the war of yesterday and the day before. Drastic change in weapons has been so persistent in the last hundred years that the presence of that factor might be considered one of the constants of strategy.

Bernard Brodie, *A Layman's Guide to Naval Strategy*

The American "electronic battlefield" of the 1970s, the Soviet "reconnaissance-strike complex" of the 1980s, the "revolution in military affairs" (RMA) of the 1990s, Chinese "informatization" in the 2000s, and the American "third offset strategy" in the 2010s all share a family resemblance. They all assume that accurate intelligence sensors connected to precision weapons via common networks and protocols—seeing all, striking all, all together—can improve military performance. Smaller yet smarter forces that can identify and prosecute targets more quickly can thereby defeat larger but slower enemies. The pessimistic converse of this story is that weaker adversaries can use cheap and effective cyber operations to cripple the vital networks of powerful nations. Either way, computational networks will determine success or failure. I call this idea the technology theory of victory.

A theory of victory is an intellectual concept that describes how military organizations can prevail in war.[1] It reflects intuitions or folk theories about what war will be like and how militaries will fight it. Concepts may be codified in doctrinal manuals and strategy documents, or they may just be shared beliefs in a military community of practice. Strategic bombing is a theory of victory that holds that air attacks on critical nodes of an enemy's organization or economy will force it to capitulate. Counterinsurgency is a theory of victory that holds that protecting civil society makes it possible to find and defeat insurgents. Both of these ideas have run into trouble historically because they underestimate the operational difficulties of the task or ignore the political counteractions that undermine it.[2] The technology theory of victory has also run afoul of operational and political problems. Decades of effort to substitute information for mass have culminated in cyberwarfare and drone operations, but decisive victory remains elusive.

Hopes and fears about better fighting through information technology are evergreen. New variants of technology theory are already emerging in response to advances in machine learning, embedded computing, and quantum computing. Skeptics have highlighted numerous practical and conceptual shortcomings of technology theory. Scholars have offered more nuanced perspectives on the social factors that shape the adoption and employment of material capabilities. In this chapter I will argue that all of these different strands of debate reflect different perspectives on the same underlying phenomenon: the historically increasing complexity of information practice.

## Better Fighting through Information Technology

An early anticipation of technology theory appeared in a 1909 essay on "war in the future" by Field Marshal Alfred von Schlieffen, the recently retired chief of the German general staff:

> An immediate consequence of the improved rifle is a greater expansion of the battlefront. . . . As big as these battlefields will be, there will be little to see. . . . The commander will be located further in the rear in a house with spacious offices. Wired and radio telegraph, telephone, and signaling equipment are at hand. Fleets of cars and motorbikes ready for the longest rides await orders. There, in a comfortable chair in front of a wide desk, the modern Alexander has a map of the whole battlefield in front of him. He telephones inspiring words and receives the reports from army and corps commanders, tethered balloons, and dirigible airships that observe the movements of enemy forces along the whole front and monitor their positions.[3]

Schlieffen assumes that the commander who can gather up information from remote sensors in a central representation (or "common operational picture" in modern parlance) will be able to fight more efficiently and more effectively. Efficiency refers to the ratio of inputs to outputs (doing more with less). Effectiveness refers to the likelihood of desired outputs (getting the job done). Militaries historically have prioritized effectiveness over efficiency. Masses of troops and massed fires may not have been cheap, but they were useful for offsetting the uncertainty of war. Larger forces were useful for searching a larger area for enemies that remained hidden, and they could tolerate losses when they made contact with them. When individual soldiers were poor marksmen, similarly, volleys of fire were more likely to hit something. Forces that had more troops and firepower also had important advantages in contests of attrition. Maneuver strategies, by contrast, could improve efficiency by concentrating a smaller number of forces at some decisive point, but they needed reliable intelligence about where that point might be. Mass remained an important insurance policy for times when maneuvers did not go according to plan. Reserve capacity could then compensate for the uncertainty of combat. Mass offset uncertainty. The technology theory of victory, by contrast, aims to use information to substitute for mass. Militaries that are smarter are assumed to be both smaller and more lethal.

Schlieffen, of course, is better remembered as the architect of the eponymous plan for German victory in 1914. That plan, or one inspired by it, resulted not in decisive maneuver, but rather in bloody stalemate and, ultimately, German humiliation. Generals in chateau headquarters had little ability to monitor or control their massive armies once they left the trenches. They developed detailed plans for artillery barrages and infantry advances, but they inevitably failed to push through enemy barbed wire and machine guns. The breakthrough battles of 1918 finally ended the stalemate as doctrinal innovation on both sides empowered small units to maneuver independently. Infantry units used colored flares to communicate their progress and coordinate fire support. Artillery units leveraged aerial photography and silent registration techniques. Schlieffen's vision of top-down management via information technology was superseded by the bottom-up adaptation of information practice.[4]

## FROM THE ELECTRONIC BATTLEFIELD TO
## NETWORK-CENTRIC WARFARE

The modern Alexander was reincarnated decades later. General William Westmoreland, while serving as chief of staff of the U.S. Army in 1969, claimed that "we are on the threshold of an entirely new battlefield concept" due to innovation in surveillance and electronics.[5] These technologies would enhance both efficiency and effectiveness by substituting information for mass:

> Inherent in the function of destroying the enemy is fixing the enemy. In the past, we have devoted sizeable portions of our forces to this requirement. In the future, however, fixing the enemy will become a problem primarily in time rather than space. More specifically, if one knows continually the location of his enemy and has the capability to mass fires instantly, he need not necessarily fix the enemy in one location with forces on the ground. On the battlefield of the future, enemy forces will be located, tracked, and targeted almost instantaneously through use of data links, computer assisted intelligence evaluation and automated fire control. With first round kill probabilities approaching certainty, and with surveillance devices that can continually track the enemy, the need for large forces to fix the opposition physically

will be less important. . . . I am confident the American people expect this country to take full advantage of its technology—to welcome and applaud the developments that will replace wherever possible the man with the machine.[6]

Much as Schlieffen foresaw "the whole battlefield in front of" the commander at his desk, Westmoreland described "an improved communicative system" that "would permit commanders to be continually aware of the entire battlefield panorama down to squad and platoon level."[7] Just as Schlieffen is remembered more for the tragedy of 1914, however, Westmoreland is better known for presiding over the United States' debacle in Vietnam. Technological prowess and firepower did not make up for political miscalculations about Communist resolve. At the same time, the Vietnam War did witness numerous experiments in remote sensors, data links, computer processing, and precision munitions that anticipated the RMA capabilities to come. As a chastened U.S. military returned its attention to Central Europe, it began fielding an elaborate arsenal of stealth aircraft, battlefield surveillance networks, and smart weapons that would make Westmoreland's vision of "a more automated battlefield" actually feasible.[8]

Soviet observers like Marshal Nikolai Ogarkov characterized these developments as nothing less than a "military-technical revolution." He advanced a Marxist argument that a transformation in the means of destruction must determine a transformation in military operations. The "reconnaissance-strike complex" was the natural product of the electronics age, just as blitzkrieg was the natural product of the industrial age. Soviet ideas entered U.S. strategic discourse through the efforts of Andrew Marshall in the Pentagon's Office of Net Assessment. Marshall, in turn, encouraged U.S. planners to reinforce Soviet fears by doubling down on the reconnaissance-strike complex, thereby forcing Moscow to run an economic race that it could not hope to win. The resulting "offset strategy"— operationalized in the U.S. Army's "AirLand Battle" and NATO's "Follow-On Forces Attack" doctrines—was supposed to substitute Western quality for Soviet quantity in Central Europe. These capabilities would ultimately have their wartime debut in the deserts of Iraq. The geographical conditions of Iraq were even more favorable than Europe for combined-arms warfare. Saddam Hussein's army was also a poor copy of the Red Army, against which the U.S. military had trained and equipped itself to fight. The result of this most favorable alignment of environment and organization was a dramatic rout, but many observers focused on the spectacle of technology instead.[9]

The 1990s were a perfect storm for military futurism. The sudden end of the Cold War, decisive victory in the Gulf War, the explosion of the internet economy, and an atmosphere of millennial hype encouraged breathless speculation about the transformation of war.[10] Visionaries argued that revolutionary technologies required revolutionary concepts and transformational leadership. The Soviet idea of a "military-technical revolution" was rebranded the "revolution in military affairs" to emphasize that radical innovation involved more than just technology. Yet RMA proponents understood this in terms of a technological imperative to alter doctrine and organizations before a strategic competitor like Russia or China made the jump to a higher level of progress. Studies funded by Marshall's office thus likened the 1990s to the 1930s as a critical period of interwar innovation that would separate winners from losers.[11] As Marshall later reflected, "The effort yielded what seemed to be a consensus that we were in a period of major change; in short, that the Russian theorists were right. We also concluded from military history that changes of the scale that we were talking about would involve new concepts of operation, and new organizational structures and processes to execute these concepts."[12]

John Boyd's concept of "the OODA loop" became particularly influential in RMA discourse.[13] The perspicacious pilot developed the concept in the late 1970s to describe the "observe, orient, decide, and act" phases of a cybernetic decision cycle. Boyd argued that victory went to the man-machine system that could outthink and outmaneuver an enemy system. His concept tended to conflate tactical and strategic decision making. While the OODA loop was centralized in the cockpit of a single-seat fighter, the ideal of "unity of command" remained elusive in distributed military operations. While speed of decision might be vital in a tactical dogfight, strategic patience could be as useful as tactical agility. Yet the OODA loop resonated strongly for a generation of officers who sought to explain the importance of information in war. In an influential article on "network-centric warfare," Vice Admiral Arthur Cebrowski and a coauthor argued that digital networks would enable military forces to cycle the OODA loop so rapidly that "it appears to disappear." Whereas "situational awareness" deteriorates for industrial-age militaries as they suffer damage in combat, they wrote, information-age militaries would be able to "self-synchronize" the "situational awareness" of the network in order to "lock out" opponents with slower OODA loops. Networked OODA loops would thus transform the episodic nature of war described by Clausewitz into a decisive knockout blow.[14]

Marshall and Cebrowski's approach stressed the technological imperative of institutionalizing the RMA. Another navy flag officer, vice chairman of the Joint Chiefs of Staff Admiral William Owens, was the intellectual force behind a 1996 white paper entitled *Joint Vision 2010*. This influential document stated, "Information superiority and advances in technology will enable us to achieve the desired effects through the tailored application of joint combat power."[15] Efforts to institutionalize this vision enjoyed bipartisan support. The Clinton administration created a new four-star unified command called Joint Forces Command (JFCOM) to develop new concepts and doctrine for RMA ideas such as "dominant battlefield awareness," "network-centric warfare," and "effects-based operations." In the George W. Bush administration, Secretary of Defense Donald Rumsfeld pushed an ambitious agenda of "defense transformation" that urged the services to embrace the RMA.[16] Rumsfeld also created a new Office of Force Transformation and invited Cebrowski to become its civilian director. At about the same time, Owens published a book entitled *Lifting the Fog of War*, which included a passage that seemed to channel the ghost of the modern Alexander:

> In a future conflict . . . an Army corps commander in his field headquarters will have instant access to a live, three-dimensional image of the entire battlefield displayed on a computer screen, an image generated by a network of sensors including satellites, unmanned aerial vehicles, reconnaissance aircraft, and special operations soldiers on the ground. The commander will know the precise location and activity of enemy units—even those attempting to cloak their movements by operating at night or in poor weather, or by hiding behind mountains or under trees.[17]

Owens's timing was about as unfortunate as Schlieffen's. Owens wrote in 2000 that "technology can give us the ability to see a 'battlefield' as large as Iraq . . . with unprecedented fidelity, comprehension, and timeliness; by night or day, in any kind of weather, all the time."[18] In 2003 his nation went to war on a battlefield exactly "as large as Iraq," but it experienced unprecedented confusion, misunderstanding, and protraction. The initial "Shock and Awe" campaign appeared to realize the promise of smaller, faster, smarter forces, as Baghdad fell in just three weeks. Yet the intelligence on Iraqi weapons of mass destruction that was used to justify the war turned out to be faulty, and the ensuing occupation was unprepared to combat several insurgencies that emerged, supported by foreign jihadists and Iranian operatives. Joint military operations in Iraq (as well as Afghanistan) were plagued by information overload, interoperability problems, targeting errors, security vulnerabilities, and human mistakes.[19]

Despite paying lip service to the RMA, none of the services seriously transformed their procurement priorities or processes following the Cold War.[20] Rumsfeld was replaced in the second Bush term, and the Office of Force Transformation was disestablished after Cebrowski succumbed to cancer in 2005. In 2008, JFCOM commander General James Mattis directed the organization to cease using jargon about "effects based operations." He stated, "War is not composed of the tactics of targetry or an algebraic approach to measuring effects resulting from our actions," and he quoted a dictum from General William T. Sherman: "Every attempt to make war easy and safe will result in humiliation and disaster."[21] JFCOM was shuttered in 2011 as a cost-saving measure.

Yet the RMA did not fade away. Network-centric concepts and effects-based jargon had by then become ingrained in the U.S. military and defense industry. Curiously enough, technology theory not only survived but thrived in the quagmires of Afghanistan and Iraq. Just as World War I armies eventually modified their doctrine and technology to overcome deadlock in the trenches, coalition forces in Iraq eventually began to adapt their technologies and processes to cope with the messy realities of counter-insurgency. The U.S. military eagerly adopted many new data systems, sensor networks, and autonomous weapons, even as combat revealed shortfalls in the network-centric way of war. While systems often fell short of ambitious goals, improvements always seemed possible for those who used them. The U.S. special operations community proved especially adept at repurposing network-centric concepts originally developed for air and naval warfare for hunting human targets instead.[22] While the RMA as such disappeared as an active conversation, the allure of networked technology remained strong. As Lieutenant General David Deptula proclaimed in 2010, "We stand at the cusp of a new era in military operations in which the speed of information, advancements in technology, networking of our organizations, and mind-set of our people will directly shape the success or failure of our future military activities. The foundations of our achievement will hinge on the ability to sense, know, decide, and act ahead of our adversaries on a global scale."[23]

Technology theory continues to find new expressions in strategic thought. Some argue that the RMA in conventional warfare has also brought about a "new era of counterforce" in nuclear warfare.[24] Burgeoning worry about "cyberwar," meanwhile, inverts classic RMA themes. The paranoid belief that weak hackers can paralyze strong states is the converse of the euphoric belief that networked forces can trounce outmoded foes. The empirical record of cyber conflict to date, by contrast, has been distinguished by low-intensity intelligence operations rather than high-intensity warfare.[25] Growing reliance on space-based information systems for communication and reconnaissance creates still more opportunities and threats. Beijing's 2015 defense white paper thus advances the technology theory of victory with Chinese characteristics: "The world revolution in military affairs (RMA) is proceeding to a new stage. Long-range, precise, smart, stealthy and unmanned weapons and equipment are becoming increasingly sophisticated. Outer space and cyberspace have become new commanding heights in strategic competition among all parties. The form of war is accelerating its evolution to informatization."[26]

Chinese strategists developed the concept of informatization through close study of the U.S. RMA.[27] In response to Chinese modernization, U.S. strategists then proposed a "third offset strategy" that hearkened back to the offset strategy of the late Cold War that produced the original RMA.[28]

They proposed countering China's networked "antiaccess/area denial" (A2/AD) capabilities with networks of autonomous weapons, artificial intelligence, quantum computers, and other exotic capabilities, all in partnership with Silicon Valley. The third offset again looked to technology to counter a Chinese imitation of the U.S. RMA, which itself was an appropriation of a Russian description of an American innovation, much of which was anticipated by a Prussian officer. The technology theory of victory gets amplified through repetition.

Every new generation of information technology inspires a new generation of military strategists to envision a new revolution in warfare. The report by Xi Jinping to the Nineteenth National Congress of the Communist Party of China on October 18, 2017, announced a goal of "speeding up the development of an intelligentized military," suggesting that the format of war has shifted again from "informatized" (*xinxihua*) to "intelligentized" (*zhinenghua*) warfare. An official interpretation of Xi's report in the *PLA Daily*, entitled "Intelligentization Will Revolutionize Modern Warfare," bears an uncanny resemblance to Schlieffen's vision from over a century before:

> In the future intelligentized battlefield, networks of robots on land, sea and air will be supported by big data and cloud computing. More sophisticated artificial intelligence technology will shift from assisting human operators to replacing them, especially in high-risk situations that are not suitable for people. . . . Once command activities are intelligentized, battlefield data will be acquired in a useful, orderly, prompt, and accurate manner. This will greatly condense the combat decision cycle of planning operations, allocating tasks, striking targets, and evaluating losses. . . . The intelligentized cluster of combat systems will be extremely survivable and effective on the battlefield.[29]

China would thus take Westmoreland's ambition to "replace wherever possible the man with the machine" to an extreme.[30] Information would substitute not only for mass but also for human beings completely. Whereas the modern Alexander had a large map on his desk to methodically direct remote operations, now he is replaced altogether by an autonomous robot with a vast neural network running the OODA loop with astonishing efficiency. Not to be outdone, the 2018 *Department of Defense Artificial Intelligence Strategy* also proclaimed the automation of the eternal return. Machine learning techniques, it declared, are "poised to change the character of the future battlefield and the pace of threats we must face. We will harness the potential of [artificial intelligence] to transform all functions of the Department positively, thereby . . . improving the affordability, effectiveness, and speed of our operations."[31]

Different historical manifestations of technology theory all share a family resemblance. They assume that the widespread adoption of computing systems, whether by a military force or by society at large (or both), will make conflict more offensive, rapid, efficient, or decisive. Technocentric visions of information dominance eventually run afoul of the timeless problems of unintended consequences and enemy counteractions. Lieutenant General H. R. McMaster thus describes the ideology of better fighting through information technology as a "vampire fallacy" that keeps coming back to life every time it gets killed off in war.[32] The RMA is dead, long live the RMA.

## Scholarship on Military Innovation and Effectiveness

Great expectations for the RMA prompted a skeptical backlash almost immediately. The debate also inspired a rich scholarly literature on military innovation and effectiveness.[33] Most scholarship rejects technological determinism as a simplistic reading of history.[34] There is less consensus about how and why different social, institutional, economic, and strategic factors matter instead. There are four broad counternarratives to technology theory.

### THE FOG OF WAR

The first argues that technology cannot overcome, and often exacerbates, the inevitable uncertainty of combat. A mantra of the strategic studies community is that the nature of war never changes but the conduct of war changes constantly. Clausewitz describes this nature as a "paradoxical trinity" of violence, chance, and politics in chaotic interaction.[35] Technology can never eliminate this chaos and can even make it worse. Digital networks just connect more people who panic under stress, couple more machines that malfunction in the field, and offer more ways for the enemy to counteract war plans.

Much of the criticism of the RMA from the practitioner community falls into this category.[36] Officers in the army and Marine Corps tend to be more skeptical of the RMA, while RMA enthusiasts tend to be overrepresented in the air force and navy. The ground services are more preoccupied with the confusion of close combat amid restive populations, where intangible factors like resolve, morale, and intuitive judgment are particularly important. The more technologically oriented services operate in less cluttered domains (sea, air, space) where technical reconnaissance is more reliable (indeed, essential) and the technical characteristics of weapons (range, targeting, lethality) have a profound impact on combat interactions.

Scholarship on the complexity of organizations complements practitioner concerns about the inevitability of friction. Charles Perrow's theory of "normal accidents" argues that complex, tightly coupled systems are prone to catastrophic breakdowns.[37] The incompatible sense-making processes of different military units can suddenly and unexpectedly come into contact in such systems.[38] Tragic accidents such as targeting errors and fratricides thus become more likely. The Chinese embassy bombing described in the introduction can be described as a normal accident. A complex distributed organization in a tightly coupled wartime environment was unable to identify and handle exceptional circumstances. Quality-control measures like senior target reviews failed to catch the fatal mistake and instead served to validate it. Another modern pathology is that attempts to computerize and automate operations often end up removing the slack capacity that organizations need for resilience.[39] Automated systems become too brittle to deal with unforeseen situations, yet human employees become disempowered to adapt. Complex weapon systems also tend to induce complementary growth in the logistical complexity of the organizations that field them, which heightens the risk of "rogue outcomes" that nobody foresees.[40]

The fog of war school is compelling, and the recent history of networked operations provides countless examples of friction and confusion. At the same time, it does not adequately explain how and why some types of technology sometimes seem to improve some types of operations. It also does not explain why military organizations remain so hungry for networked technology even when lasting improvement seems so elusive. Either military practitioners are just ignorant about their business, which is uncharitable at the least, or the likelihood of information friction is conditional on some other factors.

## CONTINGENT DOCTRINE

A second perspective accepts the inevitability of fog and friction, yet holds that military organizations can improve their ability to cope with uncertainty, so long as they adopt administrative institutions appropriate to the technological possibilities of the era. This view is consistent with what is known in organization science as "contingency theory": management practices are neither good nor bad in absolute terms but rather must be tailored to organizations' size, their infrastructure, the nature of the task, and so on.[41] Many thoughtful proponents of the RMA fall into this category. They argue that, historically, those militaries that have developed appropriate doctrinal concepts and organizational structures in peacetime have tended to outperform their competitors in wartime.[42] Thus militaries must be willing and able to invest in the organizations and doctrine that turn technological possibilities into usable military capabilities.[43] The U.S. Navy in the Cold War is often portrayed as an exemplary case of adapting doctrine to get the most out of underwater sensors, satellite reconnaissance and communications, shipboard computation, and data links that connect afloat units and shore-based intelligence fusion centers.[44]

Some scholars who are very critical of the RMA thesis per se still argue that particular organizational arrangements are conducive to improving military effectiveness in the grueling conditions of modern battle. Stephen Biddle asserts that the "modern system" of combined arms warfare (a set of offensive and defensive tactics for concealment, suppressive fire, independent maneuver, combined arms cooperation, differential concentration, and defense-in-depth) can protect soldiers from lethal exposure and enable them to maneuver, even if they are outnumbered by enemy forces or outclassed by better weapons.[45] The United States overwhelmed the Iraqi military in 1991 because it had mastered the art of combined arms warfare and Iraq had not. In the 2003 sequel, Iraqi insurgents had an underground organization that enabled them to survive and impose costs on American units that had not yet mastered the art of counterinsurgency. While Biddle is skeptical of RMA technology, it is important to recognize that any implementation of the modern system necessarily relies on communication technology to manage combined arms and joint operations, coordinate independently maneuvering units, pass targeting data to artillery and close air support, and manage a responsive logistics system. The modern system is not so much an alternative to the information revolution as an appropriation of it.

Ryan Grauer focuses specifically on the fundamental problem of uncertainty in war, which technology theory assumes can be alleviated.[46] Grauer draws explicitly on contingency theory to argue that uncertainty is a function of the size of a military organization, the sophistication of its technology, and the complexity of its operating environment. Under these conditions, organizations that adopt a differentiated and decentralized command structure are better able to cope with uncertainty. Flexible control arrangements empower local commanders to seize the initiative while relieving information overload at higher echelons. Grauer provides a theoretical explanation for longstanding practitioner intuitions about the advantages of decentralized maneuver warfare guided by "mission command" rather than detailed task instructions.[47] The former specify the commander's intentions but leave the implementation up to the discretion of subordinates; the latter specify exactly how subordinates should implement a task, leaving little room for interpretation. It is important to note here that the specialized and decentralized command system that Grauer argues is best suited for modern war is, in part, a reaction to the very complexity that modern information technology creates. Biddle and Grauer both highlight the virtues of tactical self-organization, which has a strong affinity with the user-innovation school discussed below. Yet combined arms warfare is also predicated on shared, standardized, institutionalized systems of tactical doctrine, administrative procedure, communications, and data-processing protocols. Self-organization on the battlefield is enabled by prior training and proficiency to ensure that units can shoot, move, and communicate within a framework of commonly understood tactics, techniques, and procedures. Centralized coordination of doctrine and commander's intent thus facilitates decentralized innovation in tactical execution.

It is ironic that the institutional factors that are usually offered as an alternative to technological determinism for explaining military innovation and performance—doctrine, training, procurement, logistics, intelligence, command, planning, administration, and so forth—are themselves thoroughly infused with information technology. Whenever network-centric doctrine improves situational awareness, the modern system overcomes entrenched lethal defenses, or decentralized command systems mitigate uncertainty in modern warfare, we find personnel interfacing through digital representations of the battlefield and communicating over electronic circuits. As mentioned above, the institutions that enable military operations now rely to an unprecedented degree on information technology. Indeed, it is the very ubiquity of computers in the vital institutions of command that makes cyberwarfare a frightening prospect, insofar as the disruption of computation might disrupt the processes that facilitate combat coordination. Nonmaterial institutions work, in part, because they rely on material infrastructure to improve information processing. Yet those same institutions can also lead an organization to use its equipment in ways that undermine its own performance. Information technology thus appears to be necessary but not sufficient for military performance.

ORGANIZATIONAL CULTURE

A third perspective, accordingly, holds that organizational culture leads militaries to use new technologies in pursuit of parochial goals that are often counterproductive in war.[48] Military organizations are bureaucratic agencies that attempt to enhance their autonomy and acquire resources in competition with other services and agencies, while at the same time reducing uncertainty through standard operating procedures and formal doctrine. Services also have strong identities and cultures shaped by their traditional operating environments and formative combat experiences.[49] Together these factors tend to give military organizations a preference for offensive doctrines that require expensive weaponry and operating budgets, enable the organization to impose its plan on the enemy, and reinforce a warrior ethos.[50] The culture school sees organizations less as a means to an end like battlefield victory, and more of an end unto themselves. When things go wrong in war, an armed bureaucracy may just "do its thing," even if this exacerbates a deteriorating strategic situation.[51] Civilian intervention can sometimes override problematic military preferences, but intervention can also make the situation worse. Civilians, after all, are just members of other bureaucratic organizations, subject to all of the same imperatives for political survival and enrichment.[52]

There is considerable affinity between what I call the organizational culture and contingent doctrine schools. Military performance can improve when an organization's culture matches its strategic problem. Yet the culture school further maintains that organizations will usually be more responsive to internal prerogatives when there is a mismatch. Even worse, they may actively try to change the strategic problem to match their ex ante preferences and legacy capabilities. Strong martial cultures that emphasize bold, offensive, or heroic operations can be especially problematic for unconventional missions like counterinsurgency that require subtle political acumen, detailed local intelligence, and military restraint.[53] Investing in the RMA, by this logic, is actively counterproductive for military performance in the kinds of wars the United States is most likely to fight.[54] There is also overlap between the organizational culture and fog of war schools in that bureaucratic culture can inadvertently cover up dangerous frictions. Diane Vaughan explains that the Challenger space shuttle disaster resulted from the "normalization of deviance" at NASA; engineers and astronauts adopted and reinforced idiosyncratic practices over time but failed to correct aberrations from established standards that increased the likelihood of disaster.[55] Lynn Eden argues that U.S. nuclear planners systematically ignored the risks of nuclear firestorms because they were seen as less controllable and did not accord with the dominant values and identities of the nuclear targeting culture.[56] The culture school further points out that information technology often increases the bureaucratic load on all types of military organizations by enabling senior officers to micromanage events. New technologies of control burden personnel with reporting requirements, resulting in a myopic fixation on bureaucratic processes.[57]

By and large, a top-down bias runs through many RMA narratives and institutional counternarratives.[58] Information technology is taken to be an external input from the commercial marketplace, which military leadership then figures out how to adopt and implement in the force. Major defense firms receive contracts to build complex command and control networks. Senior leaders draft vision statements and craft doctrine to shape new technologies. Bureaucratic culture, in turn, either impedes or facilitates adoption. Agency is exercised mainly at the level of the organization as a whole. The networked organization, therefore, operates on behalf of the commander. Studies of large distributed organizations, however, often find that their members take significant liberties in their interpretation of corporate policy as they tailor its implementation to local circumstances.[59] Military practitioners, likewise, often exercise agency in local circumstances to reinterpret and customize technology for their particular mission. A subcurrent in the military culture school does recognize the importance of bottom-up adaptation to emerging strategic problems; thus, "learning organizations" will support grassroots initiatives because the organization encourages creative experimentation.[60]

USER INNOVATION

The fourth counternarrative to technology theory explicitly recognizes the importance of bottom-up adaptation in military organizations. While military user innovation is prevalent in history, the phenomenon has only recently begun to receive attention from security scholars.[61] Field expedient adaptation is as old as the Trojan Horse, and accounts of jury-rigged solutions are a staple of most combat histories.[62] In 1928, navy lieutenant Charles Momsen developed a prototype submarine emergency-escape breathing apparatus and tested it himself in three hundred feet of water.[63] After the landings at Normandy in June 1944, the Allied advance slowed to a crawl in the dense hedgerows that hid German defenses; American GIs fashioned hedgerow cutters for Sherman tanks out of German beach obstacles, and infantrymen, tankers, and engineers pioneered cooperative tactics to break out of the Bocage.[64] On the other side of the war, the Nazi occupation of France co-opted a system of punch cards that the French government developed in the 1930s for military conscription, using it instead to identify French Jews for arrest.[65] The U.S. wars in Afghanistan and Iraq became contests of user innovation as insurgents employed the improvised explosive device (IED) as their weapon of choice and American personnel responded with improvised countermeasures.[66] Whenever a powerful state relies on centrally managed representations to know and control its territory and population, bottom-up innovation becomes a strategy for resistance that enables people to remain illegible to the state.[67]

The interdisciplinary field of science, technology, and society (STS) emphasizes the agency of individuals and groups in shaping technological systems.[68] Research in many different sectors has found that consumers and users are often agents of technological change.[69] For example, American farmers converted the Ford Model T into a general-purpose engine to run washing machines, water pumps, and machine tools.[70] Hot-rod enthusiasts in California invented numerous performance enhancements that major automobile makers later incorporated.[71] Individuals will use their discretionary resources and local know-how to invent things that they will benefit from using rather than selling. They often freely reveal them to other users who can make further improvements. Manufacturers may later incorporate user prototypes into new products once lucrative opportunities are thereby revealed.[72] Thus an avid mountain biker who happened to be an emergency medical technician rigged an intravenous drip bag, thereby creating the first prototype of the Camelbak hydration system.[73] Practitioners in information-intensive situations also tend to shape the tools that shape their knowledge of their world. Ethnographic studies of corporate information practice regularly reveal what Claudio Ciborra calls "the importance of the unfinished, the untidy, the irregular, and the hack as fundamental systems practices."[74] As Edwin Hutchins points out in a pathbreaking study of shipboard navigation, "We are all cognitive bricoleurs, opportunistic assemblers of functional systems composed of internal and external structures."[75]

The user innovation school offers an important corrective to the top-down bias in the military innovation literature. At the same time, it tends to neglect or downplay the security and safety risks of uncoordinated experimentation. The STS tradition provides a rich set of descriptive concepts that reveal the historical contingency and "sociotechnical" complexity of interactions between nonmaterial institutions and material infrastructures. Yet the historical and ethnographic approaches that predominate in STS often shy away from advancing explanatory theories of cause and effect. Postmodern perspectives on the technological mediation of military experience, moreover, tend to be more preoccupied with criticizing dominant power structures than with identifying ways to improve national security.[76] The user innovation school demonstrates that bottom-up adaptation in war is both possible and important, but it does not reveal the conditions under which warfighter initiative is more likely to help or hinder military performance.

## The Increasing Complexity of Information Practice

Each of the four alternatives to the technology theory of victory—fog of war, contingent doctrine, organizational culture, user innovation—offers genuine insight into how organizations use and misuse information technology. Together they offer overwhelming evidence against popular but simplistic narratives of technological determination and disruption. Unfortunately, the very diversity of these perspectives creates another problem. The theoretical treatments of information technology have become as complex as the networked systems they are supposed to explain. How do these different schools of theory interrelate? Do they really constitute four separate understandings? Or, as in the old story about the blind men and the elephant, are they just four different vantage points on the same phenomenon? If so, then under what conditions should we expect these different aspects of practice to be more pronounced? As circumstances change, moreover, what pathways lead through these different understandings?

TOWARD A SYNTHESIS

One reason that all four schools seem plausible is that organizations tend to cycle through the phenomena that they emphasize in a characteristic sequence. Combat operations between complex organizations inevitably produce friction and unintended consequences (fog of war). The experience or anticipation of operational failure, in turn, prompts leadership to impose systematic reforms to standardize best practices that improve performance (contingent doctrine). Yet as resources start flowing, reforms are often captured by bureaucratic interests that become insulated from battlefield pressures (organizational culture). Deployed personnel in those same organizations become exposed to emerging problems in combat, which encourages them to take things into their own hands (user innovation).

The interaction of top-down reforms, bottom-up adaptation, and enemy counteraction, furthermore, tends to increase the overall complexity of military information systems over time. The term of art for what Napoleon would have called "command" has picked up a lot of excess baggage in the intervening centuries. The growth of staffs and their communicative paraphernalia in the mid-twentieth century gave rise to C2 (command and control). The modern acronym is C4ISR (command, control, communications, computers, intelligence, surveillance, reconnaissance), which struggles to convey in a single term both the specialization of information functions and their integration in a common organization.[77] The terminological and practical results can be confusing. C4ISR becomes vital for military performance, even as military outcomes are never assured. Complexity just creates new potentials for interference, friction, and breakdown (yet again, the fog of war).

This indefinitely repeating spiral helps to explain why technology theory is evergreen. New scientific possibilities create new opportunities for bottom-up exploration and top-down management. These in turn introduce dangers of increased complexity and myopic bureaucracy. Sometimes information technology works well, but then warfighting conditions change. Sometimes information technology fails, but then practitioners change their organization. In the long run, the complexity of information practice tends to increase without ever creating lasting advantages or disadvantages for military performance.

## THE INFORMATIONAL TURN IN WAR

The real novelty of information-age warfare is not some sort of transformation of the nature of war, but rather the increasing mediation of warfighter experience by layers of technology. Military personnel are now more likely to perceive battlefield situations through symbolic displays than through personal observation.[78] They are also more likely to engage in intellectual work in office-like settings rather than physical fighting on the battlefield. This is not to imply that war has become safe or bloodless, which would be contrary to its nature. Several thousand U.S. troops have been killed in the wars launched after September 11, 2001, and hundreds of thousands more have been physically and mentally injured. At the same time, a growing proportion of the U.S. fighting force now manages violence from afar without physically laboring on the battlefield. U.S. military personnel now enjoy an unprecedented degree of protection from lethal exposure, to say little of an American public insulated by two oceans and largely unbothered by Washington's interminable wars. By contrast, the enemy targets and civilian populations on the receiving end of U.S. military power have suffered more. Remote drone operations and cyber operations take this trend to an extreme. Yet even in forward-deployed units, many warfighters now make their contributions to the fight mainly within expeditionary office spaces.

The historical inflection point of the informational turn is not the invention of the digital computer or any other information technology, but rather the rise of the general staff. The information revolution is as much a story about organization as it is about computing. Throughout the nineteenth century, as the functions of command became impossible for any one person to manage, a new society of staff officers emerged to manage the computational burdens of large-scale campaigns.[79] As the historian Dallas Irvine observes,

> The art of war therefore became, in a sense that it had not been before, an art to be pursued upon the map, and with an immensely greater number of permutations and combinations possible than ever before. Obviously, the conduct of war upon this level required a far different order of intelligence, knowledge, preparation, and skill than the command of a visible mass of men upon a visible terrain. . . . The new situation consequently required a specially competent and more or less considerable staff to assist the commander in the exercise of his functions, and such a staff to be very effective would need to be trained in time of peace.[80]

The "operational level of war" between fighting tactics and political-military strategy has continued to grow in complexity as staffs attempt to manage larger, more differentiated, and more distributed forces. To ease the computational burden, militaries historically have embraced new forms of information technology (a category that I use broadly to include filing cabinets, newspapers, charts, telephones, software platforms, smartphone apps, supercomputers, and the internet). They have also created new occupational specialties to deal with the growing computational load, such as intelligence, cryptology, electronic warfare, space operations, meteorology, operations research, computer network operations, psychological operations, and public affairs. Even officers in combat specialties such as aviation, infantry, armor, or surface warfare have come to spend much more of their time doing information work as planners, managers, analysts, and liaison officers.

In the U.S. military today, civilian contractors and civil servants work right alongside uniformed personnel, and they often perform the same type of intelligence, operations, and logistics tasks. These changes reflect a long-term shift in military labor patterns that deemphasizes physical close combat and instead emphasizes more cognitive tasks. The "tooth-to-tail" ratio, which measures the number of personnel in combat arms against the number in headquarters, intelligence, and logistics, has been steadily declining. Combat arms constituted nearly 80 percent of an infantry division in the American Expeditionary Force in 1918, but only 43 percent of a Modular Combined Arms Brigade in 2004. The American Expeditionary Force overall included 53 percent combat specialties (with 8 percent in headquarters and 39 percent in logistics), but only 25 percent of the U.S. Army in Iraq in 2005 consisted of combat specialties (with 18 percent in headquarters and 56 percent in logistics).[81] Throughout the twentieth century, the ratio of officers to enlisted personnel in the U.S. military also increased, from 5 percent in 1900 to over 20 percent in 2000, which can be interpreted as an indication of the growing importance of management in military organization.[82] Enlisted personnel in the U.S. military, moreover, are now more likely than ever to have college degrees and to be promoted quickly into noncommissioned officer ranks. Officers, meanwhile, spend more time in postgraduate and professional military education programs. The U.S. military has become smaller yet smarter.[83] Warfighting now emphasizes brain over brawn.

There are many reasons why the battle rhythms of military organizations have converged with the corporate routines of administrative offices. The availability of equipment for gathering, processing, and communicating data has transformed diplomatic, corporate, and regulatory practice over the past two centuries.[84] Given the general-purpose nature of much information technology, it is unsurprising that military bureaucracies would be eager to adopt it too. As a rich democracy, moreover, the United States has long substituted firepower for manpower wherever possible to exploit its industrial advantages and limit casualties; this substitution tends to increase the physical distance on the battlefield between American personnel and their enemies.[85] Throughout the twentieth century, according to Morris Janowitz, military leadership changed from a direct style of authoritarian "domination" to an indirect style of collaborative "manipulation."[86] Officers increasingly focused more on the technocratic coordination of complex weaponry and operations rather than heroically leading masses of men in battle. Increasing bureaucratization and greater interdependence across specialized units tended to encourage a more collective management style.[87] The advent of nuclear weapons put an additional premium on civilian skill sets for developing strategy, engineering systems for communications and safety, gathering and analyzing intelligence, and planning for the employment of long-range precision munitions. American politicians also became more casualty averse after the trauma of conscription during the Vietnam War. The rise of the all-volunteer force then encouraged even more military professionalization and specialization.[88] The importance of joint, interagency, and coalition coordination work has further increased in the modern era of counterinsurgency, counterterrorism, and cybersecurity.

The historical roots of the informational turn in military practice are deep and wide. While the nature of war has not changed, its conduct has gradually changed over time to emphasize intellectual over physical labor. Military outcomes have become more sensitive than ever before to informational factors like intelligence and targeting, but critical information has to be processed by organizations that are more complex than ever before. The increasing technological mediation of the human experience of war has not translated into lasting military performance advantages. The task ahead is to synthesize these various perspectives on information technology and military power into a simple framework for understanding information practice.

# A Framework for Understanding Information Practice

> We find certain things about seeing puzzling, because we do not find the whole business of seeing puzzling enough.
>
> Ludwig Wittgenstein, *Philosophical Investigations*

Figure 2.1 provides an overview of this book's theoretical framework with key explanatory factors highlighted in bold. In the language of social science, the "dependent variable" is *information practice*, which coordinates an organization's representations with its world. Practice itself is an intervening process that shapes military innovation and battlefield performance. The "independent variables" are the external problem posed by the operational environment and the internal solution adopted by the military organization. The dependence or independence of these variables is ultimately an analytical convenience. Endogeneity (reciprocal causation between dependent and independent variables) is inevitable because the members of an organization try to modify the same technologies that strategic competitors try to counter. Yet the different timescales of all of these processes make it possible to get some explanatory traction in any given case by focusing on specific problems and solutions that appear in a particular, albeit temporary, configuration. My basic argument is that the interaction between operational problems and organizational solutions gives rise to four different patterns of information practice, two of which improve and two of which undermine performance.

Note that there is no direct pathway from technology to innovation or performance. Scientific concepts, physical devices, and commercial products simply create technological possibilities. External and internal actors have to exploit possibilities in accordance with their interests and abilities. The problems and solutions that they thereby create are thoroughly *sociotechnical*—they have both institutional and infrastructural aspects. An *information system*, accordingly, should be understood as not just the

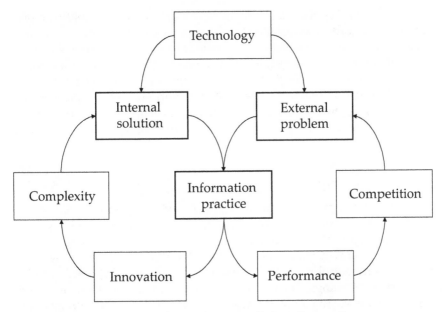

Figure 2.1. The organizational and strategic context of information practice

computing devices and software applications in an organization, but also the people and processes that generate, transform, and communicate information. I also fold technological and cultural considerations together into the notions of the external problem and the internal solution. All of the manifold factors that shape innovation and performance are ultimately expressed in terms of operational problems and organizational solutions.

This chapter presents an interdisciplinary synthesis. No single intellectual field—not philosophy, computer science, economics, political science, or biology—has managed to get its arms fully around the problem of information. Our scientific understanding of information is itself full of fog and friction. Readers from particular disciplines may be concerned that I ride roughshod over some important nuances, but to address them all would result in a much longer and less accessible book. I believe that a small number of simple, time-tested distinctions can still be useful, to a first order of approximation, for explaining complex phenomena. My primary goal here is to open up a space where it is possible to see vital patterns of human-computer interaction in military operations that have been hiding in plain sight.

## The Nature of Information Practice

Information practice is the sociotechnical pattern of organizational behavior that coordinates the relationships between internal representations and the external world. People and machines gather up data into simplified representations (records, reports, files, forms, diagrams, displays, models, charts, images) that provide viewers with a sense of mastery over things that are too difficult (too far, too fast, too many, too big, too small, too dangerous, too abstract) to perceive directly. Yet if representations are poorly constructed or otherwise uncorrelated with the world that they represent, any sense of mastery will be an illusion. The decisions that organizations make based on their representations, therefore, can produce either control or confusion.[1]

## WORKING FRICTION

Rodger Winn, an officer in the Royal Navy's Operational Intelligence Center (OIC) during World War II, called his submarine tracking plots a "working fiction." The room received radio direction finding reports, decrypts of German communications from Bletchley Park, U-boat sightings from friendly patrol boats and airplanes, and reports of attacks on Allied convoys and sinkings. The OIC represented every known German U-boat with a pin displaying a unique two-letter code (AA, BB, etc.). Winn's team transformed incoming data, constrained by reported geocoordinates or numbers, into pencil marks and pushpins on the corresponding region of the chart. The combination of these marks, the configuration of the pins, and the cognitive labor of experienced analysts resulted in estimated positions of ships that personnel would track by moving the pins. The "working fiction" thus constructed became the basis for routing convoys away from German wolf packs and intercepting them with Allied destroyers and aircraft. As Winn recalled, "What could only be an estimate and a guess was to be taken as a fact and acted upon." The deliberate construction of internal representations that reliably correlated with relevant features of the external situation provided a real advantage. Commanders could use the sea plot to place scarce assets where they could make the most difference and avoid the places where they would be the most vulnerable. This graphical fusion of intelligence from multiple sources thus enhanced both efficiency and effectiveness in the Battle of the Atlantic.[2]

Yet working fictions can produce tragic facts. On July 2, 1988, the USS *Vincennes* shot down Iran Air Flight 655, killing nearly three hundred civilians. The Ticonderoga cruiser, built around the state-of-the-art Aegis fire control system, was engaged in a firefight with Iranian gunboats. When its Combat Information Center detected an approaching aircraft, Aegis operators misinterpreted the ascending Airbus for a descending Iranian F-14 fighter on an attack profile. They also failed to notice Flight 655 when they checked the published listing of commercial flights. Although the Aegis radar worked perfectly, the entire man-machine system failed. It was primed to discriminate friendly from enemy aircraft and to facilitate rapid decision-making, not to calmly identify an innocent civilian airliner approaching from a country that had just ordered its gunboats to attack.[3]

Representational pathologies are not new to the digital age. Soldiers and sailors have long struggled to make sense of maps, reports, and orders. On May 31, 1916, the director of the British Admiralty Operations Division asked his cryptologic center in Room 40 where direction-finding stations placed the German radio call sign "DK." The cryptologic officer told him that it was in port at Wilhelmshaven. The analysts in Room 40 knew very well that the German commander, Admiral Scheer, transferred DK ashore and assumed a different call sign when at sea; yet the director had asked explicitly about DK, not Scheer's whereabouts. The director then signaled to the British commander, Admiral Jellicoe, that the German fleet had not yet left port. Jellicoe was somewhat surprised to find himself confronting Scheer's entire High Sea Fleet four hours later in the Battle of Jutland, the first major naval engagement in history conducted entirely beyond visual range. Jellicoe tried to pursue the battered Germans, but when Room 40 passed him an intercepted German position report, it turned out that the ship's faulty navigation system had reported its own position incorrectly. These two errors shattered Jellicoe's confidence in Room 40, so he ignored later intercepts that indicated the correct position and heading of the enemy fleet. Jellicoe sailed off in the opposite direction, and Scheer escaped. As the relationships that link inscriptions to situations changed, true information from Room 40 became false, then false information was mistakenly validated, and, finally, true information could not be perceived.[4]

Clausewitz famously likens uncertainty in war to "a fog."[5] Technology was already a source of fog in his day insofar as massed musketry created thick smoke that obscured the battlefield. Clausewitz also likens organizational breakdowns to mechanical friction: "When much is at stake, things no longer run like a well-oiled machine. The machine itself begins to resist."[6] The sources of friction are legion. According to Clausewitz,

> Everything in war is very simple, but the simplest thing is difficult. The difficulties accumulate and end by producing a kind of friction that is inconceivable unless one has experienced war.... Countless minor incidents—the kind you can never really foresee—combine to lower the general level of performance so that one always falls short of the intended goal.... Friction is the only concept that more or less corresponds to the factors that distinguish real war from war on paper.[7]

Clausewitz's *On War* describes sources of friction including the encounter with danger, physical exertion, unreliable intelligence, and equipment breakdown. All of these were major problems in Napoleonic warfare, but today technology ameliorates some of them. Long-range precision weapons, motorized transportation, and armor plating reduce exposure to danger and physical exertion for many personnel. Satellites, sensor networks, and the internet provide a flood of detailed intelligence that was inconceivable in Clausewitz's day. Yet all of these improvements come at the price of greater dependency on ponderous bureaucracies and complicated data systems. Military personnel still work long, exhausting hours in front of computers and in staff meetings. Networked units still fail to coordinate their plans or get caught in dust storms. Unreliable databases, incompatible interfaces, and malicious hacking can still produce "a kind of friction that is inconceivable unless one has experienced war."

The mechanical metaphor of friction remains apt in the digital era. Friction wears down surfaces, generates heat, and impedes movement. Tropes about the virtual or weightless nature of cyberspace are profoundly misleading. Even as the same data can be represented in many forms, data are always embodied as marks on a page, magnetic charges, synaptic connections, pulses of light, or sound waves. Determining the sameness of any data, moreover, requires some actor or physical process to make contact with them for interpretation and comparison. Data formats are always the result of particular decisions made in previous circumstances, and those circumstances may be hard to discern in the present. Paul Edwards thus introduces the concept of "data friction" to describe the resistance that people encounter when "dissimilar data surfaces make contact. Some of those surfaces are human; they make mistakes, argue, and negotiate. Others require technical negotiation across protocols and runtime contexts. All that friction generates error and noise."[8] Oliver Williamson, similarly, likens economic inefficiency to friction: "The economic counterpart of friction is transaction cost: do the parties to the exchange operate harmoniously, or are there frequent misunderstandings and conflicts that lead to delays, breakdowns, and other malfunctions?"[9] Edwards and Williamson point to political and organizational sources of friction other than the stress, exhaustion, uncertainty, and malfunctions cited by Clausewitz. Yet according to Peter Paret, even Clausewitz's first use of the term was "during the campaign of 1806 to describe the difficulties Scharnhorst encountered in persuading the high command to reach decisions, and the further difficulties of having the decision implemented."[10]

Information friction, therefore, can be caused by external factors such as enemy action or bad weather or by internal factors such as miscommunication or equipment failure (table 2.1). It can also result from either unintended breakdown or deliberate disruption, manipulation, interference, politicization, obfuscation, or lying. Organizational politics is a major source of information friction simply because networked organizations connect many different actors.

Actors often mobilize representations rhetorically to advance parochial interests or deflect criticism. Bruno Latour points out that networks of inscription can "force dissenters into believing new facts and behaving in new ways," especially if they are unwilling or unable to do the work to deconstruct those networks.[11] The Navy Special Projects Office, for example, created the Program Evaluation and Review Technique (PERT) during the development of the Polaris ballistic missile submarine. PERT became famous in later years as a model for scientific management, but its real utility for the Special Projects Office was something else entirely. In a multimedia management center showcasing PERT, flashy computerized graphics created an aura of efficiency and diverted critical attention from unconventional management practices. One contemporary recalled that PERT "had lots of pizzazz and that's valuable in selling a program. . . . The real thing to be done was to build a fence to keep the rest of the Navy off us. We discovered that PERT charts and the rest of the gibberish could do this. It showed them we were top managers."[12]

The political power of a state or organization is a product, in part, of the representational system that produces actionable knowledge about the world.[13] Latour highlights the circulation of inscriptions that enable the exercise of power:

> A man is never much more powerful than any other—even from a throne; but a man whose eye dominates records through which some sort of connections are established with millions of others may be said to *dominate*. This domination, however, is not a given but a slow construction and it can be corroded, interrupted, or destroyed if the records, files, and figures are immobilized, made more mutable, less readable, less combinable, or unclear when displayed. In other words, the scale of an actor is not an absolute term but a relative one that varies with the ability to produce, capture, sum up, and interpret information about other places and times.[14]

Likewise, in military operations, if a commander hopes to dominate the battlefield, data from intelligence sensors, field reports, and supporting agencies must be gathered up into simplified representations of situations, targets, and plans. Information practice coordinates the "slow construction" of representations, but this process "can be corroded, interrupted, or destroyed" by information friction (table 2.1). Friction for whatever reason can undermine performance. At the same time, nevertheless, friction also provides traction for developing new understandings.[15]

Table 2.1  Sources of information friction

|  | *Internal* | *External* |
|---|---|---|
| Unintended | Malfunction and misunderstanding | Rough environment and bad intelligence |
| Deliberate | Politicization and manipulation | Enemy deception and disruption |

### BREAKDOWN OF DISTRIBUTED COGNITION

Breakdowns alert practitioners to problems in their information system and provide opportunities for repair. As an everyday example, consider a mobile application like Google Maps that you might use to navigate traffic. As you drive to work, you probably do not consciously think about all of the cellular networks, internet routers, data warehouses, satellite constellations, software architectures, and cartographic labor that make your trip possible. Only if you suddenly find yourself in heavy traffic on a road that is colored green rather than red do you begin to ask questions. Is the cellular network overloaded with connections? Are other drivers not sharing locational data? Have network operators gone on strike? Did programmers make an error? Has the Global Positioning System failed? Are hackers injecting bogus data? In asking questions like these, your attention shifts from your preoccupation with driving and making a living to concerns about the infrastructure that makes this possible. Your attention shifts from "what information means" to "how information works." Breakdown reveals some of the constitutive assumptions that enable technology to help you understand the world, even as much remains hidden or only dimly perceived.[16]

*Mediated Experience.* Just as spectacles refocus the world for blurry eyes, information technologies represent ideas and perform calculations that are too complicated for the unaided mind. Information technologies act like epistemic prosthetics that enhance our mental capacity for computation and communication. In a pioneering study of "distributed cognition" aboard a U.S. Navy vessel, Edwin Hutchins describes the computational work involved in open ocean navigation.[17] The crew shoots bearings, plots fixes, and charts courses by mapping the physical constraints of the maritime environment onto the physical configuration of measurement instruments (alidade, sextant, etc.) and then combining these constraints as markings on various representations (chart, log, etc.). Navigation is thus enabled by the cognitive scaffolding of the environment. More generally, we use databases, forms, and maps to augment human memory and precompute tasks that ease the burden on biological cognition.[18] Kim Sterelny argues that human cognitive evolution is distinguished by "niche construction" and "epistemic engineering" in which "agents often manipulate their environment to turn [difficult] memory problems into [easier] perceptual problems."[19] Charts, diagrams, and arrangements of icons or tokens, similarly, transform computationally intensive cognitive tasks into simpler visual tasks. As David Kirsh and Paul Maglio explain, the physical rearrangement of symbols ("epistemic action") can make it easier to see what must be done to solve a problem ("pragmatic action").[20] In general, engineered devices have "affordances" that suggest particular physical interactions—handles afford grasping, triggers afford pulling—by virtue of the know-how that gets socialized into human beings through communities of practice.[21] Scientific instruments thus embody sophisticated theoretical knowledge not captured in textbooks and provide insight into previously invisible phenomena.[22] Reliance on epistemic prosthetics and cognitive scaffolding is so crucial for human psychology that many cognitive scientists argue that the mind extends quite literally beyond the brain.[23]

To build up an intuition for the technological mediation of military experience, consider the stylized problem of tracking enemy aircraft. Figure 2.2 depicts an observer looking directly at an incoming aircraft with the naked eye. A column of electromagnetic energy physically connects the airplane to the observer's retina, and neural circuitry filters and amplifies the relevant information. The observer can track the movement of the airplane through the sky through a biomechanical servo mechanism that maintains a visual connection.

Now imagine an observer inside an air operations center that receives data from outlying radar stations, for instance in Fighter Command headquarters during the Battle of Britain, the subject of chapter 3. In figure 2.3 the operator looks at a wooden track marker that represents a group of enemy aircraft.[24] The observer and representation are depicted in brackets because they form a sort of compound subject with its attention squarely on the remote object. The observer's eyes look at the track marker, but her mind is preoccupied with the raid. By virtue of its physical position on a map table, the marker represents the distance and bearing to the contacts and their heading, along with some additional information about aircraft affiliation (hostile), number (twelve), and altitude (twenty thousand feet).

Figure 2.2. Direct perception

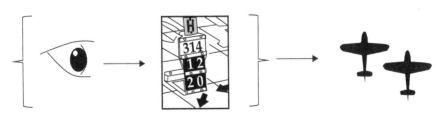

Figure 2.3. Indirect perception via representation

The operator thereby experiences more entities, and more abstract judgments about them, than she could possibly perceive by just looking at the sky. Yet this very disconnection provides her with the freedom to move around the map, confer with colleagues, and update or reposition the track marker without actually making physical contact with the target. She exploits the flexibility of the representational apparatus and her direct access to it to establish an indirect, referential connection to the target.[25] By maintaining a correlation between the structure of the representation and the structure of the world, the Ops Room can scramble interceptors to effect a more direct connection in combat.

*Reflexive Experience.* Figure 2.4 depicts a shift in the observer's attention from what information means to how information works. The shift to reflexive experience can be effected through surprise (interceptors scramble but find no incoming raid) or deliberate effort (scientists examine data processes to improve interception rates). Through the process of breakdown, the compound subject in figure 2.3 comes apart. The subject in figure 2.4 now looks *at* rather than *through* the raid marker to consider the enabling relationships that link representation and reality. She might think about whether the track is spurious, whether she trusts the radar operators who provided the data, whether the data are too old to be useful, or whether she is too exhausted to think coherently about the plot. In the situation depicted in figure 2.4, breakdown might enable her to discover that the referential integrity of the token is unreliable, since it depicts twelve aircraft when there are really only two.

Periodic alternation between mediated experience (what information means) and reflexive experience (how information works) is a fundamental feature of information practice. The breakdowns that prompt this shift can result from either unexpected friction or deliberate efforts to imagine, investigate, or inspect system behavior. Without some sort of breakdown, system design requirements cannot be articulated and systems cannot be built, modified, or repaired. The attentional shift discloses what is at stake pragmatically in the structure and process of representation and enables practitioners to debug, fix, and improve it. Moreover, as practitioners

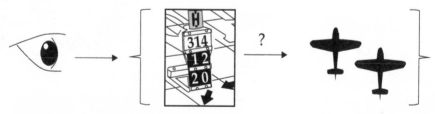

Figure 2.4. Reflexive perception of representation

consider the assumptions, tools, and interactions that produce their working knowledge, there is always more to be uncovered. Like the submerged part of an iceberg, there are always more assumptions in the background of experience that can be revealed through further breakdown.

*Sociotechnical Institutions.* Breakdowns in complex information systems reveal not only computational infrastructure but also coordinating institutions. If cognition is distributed, it is also social. Many different actors are involved in building the technologies and maintaining the relationships that enable mediated experience. Successful distributed cognition relies on the heedful effort of culturally situated practitioners who anticipate, react to, and repair breakdowns. Karl Weick and Karlene Roberts show how sailors on an aircraft carrier employ a variety of formal and informal measures, including telling each other sea stories about accidents and mishaps, to enable crew members to proactively identify and correct errors before they turn into tragedies.[26] The social nature of the control cycle is manifested even in very simple information systems. David Mindell thus describes a tactical artillery team in World War II:

> Fire control systems sought to solve the basic task of aiming a gun by breaking it into three acts: perception (looking through the sight), integration (leading the target, estimating the trajectory), and articulation (pulling the trigger). . . . Throughout, people were the glue that held the systems together. By converting, interpreting, and communicating the data, they worked to maintain the veracity of the abstractions and their faithfulness to the outside world.[27]

Mindell's "three acts" of perception, integration, and articulation describe the basic movements of a cybernetic feedback loop. Data come in through an apparatus of sensors (intelligence collection). They are combined with data stored in the organization and compared with the organization's goals (internal memory). Data go back out to influence the behavior of agents in the organization (operational communications). More incoming data provide feedback about the organization's performance. Feedback enables people in the system to make further interventions that narrow the gap between their goal state and the observed state of the world (or avoid opening up a gap in the first place). The military OODA loop—observe, orient, decide, act—describes these same functions, albeit with an extra phase that breaks "integration" into "orient and decide." This distinction usefully highlights the difference between data management and organizational goals. Integrating data is not just a mechanical process of assimilation but an active interpretive effort to make sense of situations and figure out what is important. The OODA loop can either alter or preserve the status quo depending on the commander's intent that guides it.

In this book I use the terms *measurement, coordination,* and *enforcement* to describe the three basic cybernetic functions that Mindell describes, and the term *intentionality* to describe the preferences that guide all of them (figure 2.5). I use this institutional nomenclature because I want to underscore the fact that collective action is always at stake whenever, as Mindell puts it, "people [are] the glue that [holds] the systems together." Institutions, according to the Nobel laureate Douglass North, are the human-devised "rules of the game" that structure human interaction.[28] Material infrastructures and devices, by extension, can be likened to the playing field and equipment. The same teams will play differently with the same rules if they play on artificial turf or grass, in rainy or sunny conditions. Athletes who observe the rules as written might still modify their equipment or uniforms for additional advantage, unless the rules explicitly prohibit such modifications, or the referees fail to enforce the rules. *Sociotechnical institutions,* accordingly, are the human-built "rules and tools" that constrain and enable human and machine interaction. When Napoleon III wanted to prevent revolutionaries from building barricades in Paris, he did not just outlaw them; he bulldozed wide boulevards through restive neighborhoods to improve the relative power of security services. When civil architects in New York built low bridges on Long Island parkways to exclude public buses, they also reinforced patterns of racial segregation.[29] Social institutions have material heft.[30] John Law thus describes the efforts of entrepreneurs to construct both material architectures and organizational and political support for them as "heterogeneous engineering."[31] Likewise, many breakdowns that create information friction are political or organizational rather than simply technological. Debugging an information system

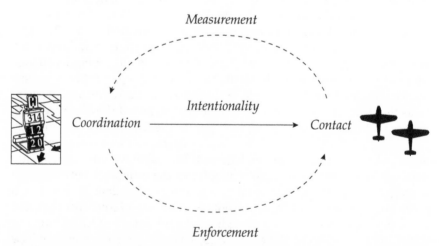

Figure 2.5. Collective action problems in information practice

thus entails paying as much attention to authority relationships and bureaucratic processes as to hardware and software architectures.

## THE CONTROL CYCLE

The cybernetic paradigm for information and control grew out of the problems of fire control, navigation, and cryptography that emerged before and during World War II. The postwar cybernetics movement went on to influence the development of computer science, automation engineering, game theory, and strategic command and control.[32] A thermostat is an engineered control device that turns a heater on or off when the temperature falls below or above a given setting. Living organisms rely on sensorimotor loops to regulate breathing, stabilize visual objects, walk or swim about, and do just about anything else. Bureaucracies are cybernetic systems that rely on standard operating procedures and cultural habits to produce reliable, routinized behavior.[33] Control cycles direct the movements of an infantry squad as much as they do the policies of a nation. Cybernetic concepts have also been applied to the self-organization of entire economies and societies.[34]

While the cybernetics movement tried to mechanize politics, information practice reveals the politics in machines. Cybernetic models like the OODA loop or rationalized command and control diagrams elide the confusion and workplace disputes that occur in actual military organizations.[35] There are two major flaws in the cybernetic legacy of control. The first is insufficient attention to the problem of intentionality, or the goals, values, interests, and meanings that guide all human behavior and thus sociotechnical control. Controversies and misinterpretations about command intentions can be major sources of friction. The second is insufficient attention to the collective action problems in the sociotechnical implementation of control. Reliable computation in a distributed organization relies on common protocols and common understandings, but anyone who has spent any time at all in a large bureaucracy knows that uncommon processes are common and disagreements across work centers are inevitable.

The sociotechnical institutions of measurement, coordination, and enforcement link organizational intentionality to physical contact with the environment (figure 2.5 and table 2.2). They constitute different aspects of a holistic system. Intentionality and measurement are more focused on creating knowledge for the organization. Coordination and enforcement are more focused on influencing the behavior of people and machines, inside or outside the organization. Measurement and enforcement are externally focused on detecting and shaping the state of the world. Intentionality and coordination are internally focused on figuring out how the organization should make sense of the world. All four of these functions pose collective action problems that must be solved to enhance the efficiency or effectiveness of the control cycle.

**Table 2.2 Control functions in information practice**

|  | *Interaction within the organization* | *Interaction with the environment* |
|---|---|---|
| Shaping practitioner understanding | *Intentionality*<br>Negotiation and socialization of the goals and values of the organization that guide representational processes | *Measurement*<br>Translation of physical contact with the outside world into representations of entities, events, and relationships in it |
| Shaping sociotechnical systems | *Coordination*<br>Cooperation to implement the technical systems and organizational processes used for representation | *Enforcement*<br>Translation of representations of potential action into physical contact in order to change the world and/or measure it again |

*Intentionality.* Intentionality directs an organization toward whatever it wants to do or considers important. In a military context, intentionality refers to commander's intent, concepts of operation, rules of engagement, as well as the military organization's goals, beliefs, values, preferences, norms, and identities. The origin, durability, and socialization of intentionality is a matter of organizational culture and politics, and these things must be explained with other theories. Whatever its source, intentionality guides the enactment of measurement, coordination, and enforcement, which together enable the organization to close a gap (or avoid one from appearing) between what it wants and what actually exists. If it wants nothing, control is meaningless. Intentionality provides the *why* of control; the other functions are the *how*. The pragmatic problem for sociotechnical control is to manage the gap between what ought to be and what is.

Philosophers use the concept of intentionality more generally to describe the directed nature of all referential phenomena such as beliefs and perceptions. For instance, we always believe or perceive *something*, and this something is usually *other* than the belief or perception itself. Franz Brentano argues that mental ideas have an "intentional inexistence" because they can be about or directed toward things that have never existed (mermaids), no longer exist (dinosaurs), or do not yet exist (death of the sun).[36] There is a similar "intentional inexistence" in the meaning of any technologically embodied representation because it directs our attention to something other than the representation itself (e.g., friendly and enemy formations, political events, weather patterns, desirable states of the future, etc.). I have been casually describing the referent of representations as "reality," but I do not mean to imply that representations are not real. On the contrary, the materiality of data is a basic theme that I stress throughout this book. Representational content is "irreal" in Brentano's sense only insofar as an internal representation refers to some other external reality that is physically out of reach.[37]

*Measurement.* The battlefield is what it is, full of warriors and weapons bumping and shoving each other. It is as inaccessible in itself as Kant's noumenal world. Reality, as William James notes, is a "great blooming, buzzing confusion" prior to cognitive interpretation.[38] Measurement, analogously, is the distributed cognition that parses battlefield confusion into discrete symbols that can be endowed with meaning. Every physical connection of a sensor, via direct contact or through sound waves or electromagnetic energy, to some particular entity in some situation is a localized, material event. Measurement is a transduction process that translates this physical contact with the environment into movable symbolic records. The act of inscription enables material traces of this contact to persist beyond the moment of physical transduction. Further transformations of this trace can produce ever more refined inscriptions that enable people (and machines) that are situated more remotely to filter, compare, combine, and analyze these records. Latour evokes images such as a "cascade of inscriptions" or "circulating reference" to describe the propagation from measurement devices to remote observers.[39] Measurement processes separate the relevant signals from the ambient noise, where relevance is defined by the intentionality of participants in the process, by amplifying traces of physical contact into more abstract, movable, and discriminable records.

*Coordination.* Latour describes the movable data produced through measurement as "immutable mobiles" and describes their confluence in "centers of calculation" where scientists can make sense of them.[40] When people can gather around a miniature model of the outside world—a sand table of battlefield terrain, a plot of ships at sea, a depiction of satellites orbiting the earth—they gain a sense of mastery over it. Practitioners before a synoptic representation can imagine how their interventions in the world will play out. Coordination, in the context of the control cycle, is the process of assembling charts, diagrams, imagery, and other authoritative depictions of the world, to include the preparatory work of arranging the devices and work processes that make this possible. This work enables an organization to construct ever more refined and abstract representations of the outside world by comparing and combining incoming data with stored data. Coordination structures work in both space and time by gathering together many different records, that were produced previously in other places, in one locale in the present from which personnel can then initiate action that will fan out into the future. A lot of the sociotechnical work of coordination thus focuses on precomputing representations in advance to ease the runtime burden of distributed cognition. For example, cartographers create maps, and soldiers annotate them, before a mission begins. Precomputation of parts of the representational solution saves the soldiers a great deal of effort in determining their location in the field. Coordination encompasses all the work that produces, in advance, internal representations that will provide practitioners with leverage over external situations, at some later time.

*Enforcement.* Enforcement is the business end of the control cycle. It attempts to bring the state of the world into line with the intentional goal state (or keep them from diverging). Enforcement, as I use the term here, works like measurement in reverse. The cascade of inscriptions moves from internal representation toward external contact, progressively transforming symbolic abstractions into increasingly particular local situations, culminating in contact with the battlefield. The notion of contact here describes the physical connection of sensors, weapons, troop movements, humanitarian airdrops, or communications to some actual entities in the external environment. Contact for the sake of intelligence aims to make adjustments to representations (such as reports and assessments) in order to make them correspond better to reality. Contact for the sake of operations aims to make adjustments to reality in order to make it correspond more with representations (such as orders and regulations).

Feedback via contact—enforcement that enables measurement—is what enables an organization to adjust its future behavior based on the results of past behavior. Combat enforcement to disrupt enemy activity also creates opportunities for gathering intelligence about enemy behavior and disposition. Intelligence measurement, conversely, relies on some sort of operational intervention (enforcement) to position sensors and monitor sources. Information practice relies on both negative (balancing) and positive (amplifying) feedback to correct undesired errors or highlight desired features. Feedback enables an organization to improve the correlation of the structure of its representations with the structure of the world. Tighter correlation, in turn, makes it easier to achieve intentional results. The progressive reduction of uncertainty through iterated cycles of connection (via enforcement) and disconnection (via measurement) enables an information system to progressively triangulate on the intentionally relevant characteristics of its operating environment. The ongoing correlation of representations and reality through iterated cycles of feedback is what makes targets actionable.

## REFERENTIAL INTEGRITY

A military organization may attempt to move forces into place, rendezvous with allies, monitor enemy activity, or attack targets. It will also want to avoid unpleasant surprises like running into an ambush, becoming isolated, or suffering a catastrophic breakdown. To accomplish all these goals, a unit must rely on a lot of circulating reference that it cannot directly or immediately verify. These data are valid only to the extent that the unit, and other organizations supporting it, have expended effort to combine, corre-late, transform, or communicate representations. Continuous feedback from multiple operations and multiple sources of information can enable practitioners to triangulate on entities of concern, which reduces the likeli-hood of being surprised when they move to contact. Military knowledge is profoundly pragmatic in this sense: the truth value of information is rela-tive to the intentional behavior it enables and the unintentional breakdowns it avoids.[41] More simply, information is relational.

Information is more than just bits and bytes in a physical medium. The reason that we care about any given pattern of bits is because it represents some intentional content that makes it meaningful to someone about some-thing in some pragmatic context. In Gregory Bateson's compact formula, information is "a difference which makes a difference."[42] Pragmatic meaning emerges from a complex set of relationships between systems of representation and relevant entities in the world, between people who design and maintain representational systems and people who use them for their projects, and between measurement, coordination, and enforce-ment processes. Digital computing systems, and the engineers who design them, are preoccupied with maintaining the referential integrity of a com-plicated system of pointers, identifiers, models, records, and representa-tional states that are at constant risk of falling out of sync as runtime circumstances change.[43] Every computer user has experienced the frustra-tion of trying to reset a forgotten password on a web page, which then sends them to another web page to verify their identity, where they must reset another password with a text to a mobile device, all before the first page times out, and so on. The proximate cause of almost all information friction, whatever its ultimate source in the organization or environment, is the breakdown of referential integrity.

Given the relational nature of information, the truth value of any given information product is always a promise deferred. For example, you might use a train schedule to rendezvous with a train at the station tomorrow, or browse a book catalog to discover a new book by a favorite author. Yet you can only verify the truth or falsity of these representations by going to the station or buying the book, or accepting on faith the reports of another's experience in a similar situation. William James points out that it is not practical to verify or even inventory all of the knowledge that we use on a daily basis. Thus we have to take for granted that someone else has ensured the referential integrity of most of the data on which we rely to make impor-tant judgments. As James observes,

Truth lives, in fact, for the most part on a credit system. Our thoughts and beliefs "pass," so long as nothing challenges them, just as bank-notes pass so long as nobody refuses them. But this all points to direct face-to-face verifications somewhere, without which the fabric of truth collapses like a financial system with no cash-basis whatever. You accept my verification of one thing, I yours of another. We trade on each other's truth. But beliefs verified concretely by somebody are the posts of the whole superstructure.[44]

James's financial metaphor suggests that a political economic perspective might be useful for analyzing information practice. Indeed, the sociotechnical institutions of control are the conditions for the possibility of referential integrity. Measurement, coordination, and enforcement pose challenging collective action problems in the production and maintenance of organizational knowledge.

To sum up, information is a complex pragmatic relationship between representations and reality, broadly construed. This relationship is enacted by sociotechnical institutions that implement control cycles, which attempt to narrow the gap between intentional goals and operational reality. Information friction—the compromise of referential integrity resulting from collective action problems in the sociotechnical institutions of control—can afflict any part of the control cycle. Practitioners in the control cycle are active agents who periodically shift their attention between mediated experience (what information means) and reflexive experience (how information works). This reflexivity enables them to respond to and repair friction. Breakdowns in practice, whether they are the result of unexpected events or deliberate investigations, enable practitioners to consider the validity of their representations and to make adjustments to the rules and tools that generate them. Control can never fully be taken for granted because internal or external actors can always disrupt or distort the sociotechnical institutions that make control possible.

## The Determinants of Information Practice

So far this discussion has focused on descriptive concepts that reveal how information practice works. Now I will shift to explaining the conditions that make knowledge and control more or less reliable. If operational information is a relationship between a military organization and the battlefield environment, then it follows that factors on either side of the relationship can affect referential integrity. Information practice is like a bridge spanning a chasm. If both ends are securely anchored to the canyon rims, then traffic can journey across safely. The bridge both constrains the pathway of travel and enables a successful transit. If one side of the chasm buckles and shakes, however, then the other side must adjust to compensate for it. The journey can still be made successfully, but it will require a bit more effort along the way. The real difficulties emerge when there is a structural mismatch that prevents movement back and forth. Enough shear between the two sides of the canyon may even destroy the bridge, leaving traffic piled up on either side or plummeting into the abyss. Like all metaphors, this one is imperfect. Information practice transforms, combines, and manipulates representations as they move through an organization, which doesn't usually happen to vehicles on a bridge. The bridge metaphor is useful, however, for underscoring (1) the fundamentally relational aspect of informational phenomena, and (2) the physical movement of inscriptions across organizational boundaries.

Internal and external structures both constrain and enable the flow of inscriptions that underwrite the referential integrity of military representations. Unfortunately, as Clausewitz points out, "war is a flimsy structure that can easily collapse and bury us in its ruins." It is perhaps no surprise that Clausewitz was skeptical about the value of intelligence: as he famously puts it, "Many intelligence reports in war are contradictory; even more are false, and most are uncertain."[45] How, then, is knowledge in war ever possible? The answer, in short, is that knowing a target that does not want to be known is an organizational achievement. Reliable operational knowledge is possible *if and only if* an information system can establish and maintain a dynamic correlation between internal and external structures. This feat requires the organization to confront two different control challenges. The first is the external challenge of understanding and overcoming a willful and reactive opponent in a particular geographical situation. The second is the internal challenge of coordinating all the intellectual and material resources that make it possible to solve the first challenge. To control the enemy, therefore, a military must also control itself.

THE EXTERNAL PROBLEM

Control cycles attempt to correlate the constraints of the internal representation with the constraints of the external situation. A hostile competitor, however, usually does not want to be constrained. The enemy seeks freedom of action to hide, plan, attack, and defend. Thus many of the most important referents of a military organization's internal representations will actively avoid being constrained by employing maneuver, camouflage, and deception. Nevertheless, the enemy cannot eliminate all constraints. Indeed, the enemy organization *must* constrain itself if it wants its own information system to work at all. Moreover, the enemy is constrained by the physical topography and sociotechnical structure of the world. Even total anarchy is not free of constraint—far from it. Geography, physical processes, material infrastructure, and organizational behavior take time and effort to alter. Terrain canalizes movement. Physical barriers and defensive formations reinforce geographical constraints. Resource streams limit the speed, maneuverability, and firepower of combatants. Doctrinal routines create predictable patterns of force deployment and employment. Combatants can impose additional constraints on enemy behavior through their own force deployments, defensive arrangements, and reconnaissance patrols.

War is a contest of control in which combatants compete in the construction and destruction of the systems that enable them to know and influence each other. Figure 2.6 depicts two competing control cycles. Each combatant is part of the environment for the other, and each is the other's primary object of concern. Both try to avoid, disrupt, or redirect the other's efforts to exert control. Organizations at war struggle to build better information systems while their adversaries struggle to break them. In order to accomplish these positive and negative intentions (i.e., asserting and denying battle-field control), each combatant must be able to sense and respond to enemy actions, and each must persist as a coherent fighting force. Each tries to improve and protect the data systems on which they rely to measure the world, coordinate work processes, and enforce their intentions. They can further try to deceive, disrupt, or avoid the adversary's information system.

The primary determinant of the external information problem is the sociotechnical structure of the control contest. A thorough analysis of strategies to influence or protect control—including second-order forms of information practice to protect or exploit information practice (counterintelligence and cybersecurity)—is beyond the scope of this book.[46] The empirical chapters focus mainly on the world as seen from one side of a control contest in order to illuminate the difficult organizational trade-offs that military actors have to manage. In each case study, I will describe the strategic interactions that constitute the information problem, even as I will not go into the specifics of enemy organizations and strategies in the same level of detail. The key point here is that strategic interaction with other

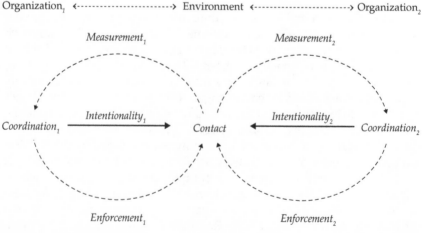

Figure 2.6. The control contest

actors in the environment, and their freedom of action within it, defines the problem that a military information system has to solve.

Any source of nonrandomness in the control contest is a potential source of traction for military measurement and enforcement. Table 2.3 summarizes some of the external factors that constrain the information problem. The most important is the ontological definition of the operational domain. W. V. O. Quine defines an ontology as an abstract account of what sort of things and attributes of things exist in any given domain.[47] In computer science an ontology is a formal description of entities and relationships that enables data exchange across applications and databases.[48] An ontology as I use the concept here is the set of entity types and properties that matter for a given pragmatic project. Parsimonious models that correspond to durable features of the world make it more likely that representational assumptions will be validated through enforcement (i.e., movement to contact). If, however, entities and properties that matter in combat are subject to change, or if the relationships between them are multivalent, then it is harder to reliably represent the world. In general, if an ontology has to account for many types of entities, many properties of types, or many types of relationships among entities, then representations are more likely to fall out of sync with the world.

Representation is also more difficult if ontologies are fluid. When friendly and enemy forces are sorted into a standardized order of battle, then a finite set of symbols can represent relevant entities on the battlefield. When new types of things or new groupings of things emerge in the midst of conflict, however, categorization and targeting processes become equivocal. The enemy or friendly affiliation of battlefield entities is another source of instability, especially as insurgents or covert operatives forge identity documents, switch or abandon uniforms, disguise allegiances in order to penetrate enemy lines, flip to become double agents, or use commonly accepted symbols of neutrality like the Red Cross to cover hostile activities.

Different geographies can be more or less constrained. The earth's atmosphere, oceans, and outer space are emptier and relatively simpler spaces than terrestrial geography. The "human terrain" is full of people, complex landforms, and cultural features. The air, sea, and space domains all require specialized engineered technology to access, so objects within them tend to

**Table 2.3 External problem**

| Constrained information problem | Unconstrained information problem |
| --- | --- |
| Simple, defined, fixed types of entities | Complex, ambiguous, dynamic types |
| Few numbers of relevant entities | Many potentially relevant entities |
| Cooperative or predictable actors | Competitive or unpredictable actors |
| Tightly coupled dependencies among entities | Loose coupling and ambiguous relationships |
| Scientific feasibility of physical contact | Physical inaccessibility of relevant entities |

be more standardized by design. While the high speeds of weapons moving through air and space complicate representation somewhat, their trajectories are easier to model mathematically, which simplifies representation considerably. By contrast, terrestrial societies full of heterogeneous institutions and willful actors are more difficult to formalize. Combat interactions also have different perturbing propensities in different domains. Projectiles that overshoot targets at sea splash harmlessly, and the ocean returns to its prior homogenous state. Ground fires create more risk of inadvertently killing civilians, which can change the political makeup of the conflict. Similar differences across domains can also manifest within domains. Urban or rugged territory is harder to model and control than unpopulated open desert—an ocean of sand. Some geographic processes occur slowly enough that cartographic maps tend to remain stable on tactical timelines. Yet rivers dry out, vegetation patterns change, and important terrain variation may be hidden by the scale of a map. Some microfeatures, like furrowed washes in a badland that enable enemy troops to infiltrate or escape, may be discovered only as patrols come into contact with enemies hidden in them.

Cultural features can also be more or less durable or accessible. Durable infrastructure imposes constraints on the movement of even irregular groups on the ground. Engines generate infrared signatures and sharp corners reflect radar energy, and these facts are hard to get around without substantial engineering effort. Constructed road and rail networks, cellular communications, and power grids are more amenable to representation precisely because they are built out of standardized components and procedures. Transactional data such as email and telephone communications, therefore, can facilitate systematic analysis of traffic patterns. Credit card purchases, social media updates, locational metadata, and so on create "data exhaust" that can be leveraged for targeting. By contrast, maps and diagrams of messy social networks with nonstandard entities, roles, and relationships will tend to be less stable. Families, clans, tribes, or organizational cultures may not change overnight, moreover, but their norms and behaviors are often opaque to outside observers, or even indigenous informants. More robust and stable institutions, such as a professional military bureaucracy, will generally be easier to model than a fissiparous community on a lawless frontier. An enemy that adheres to doctrinal concepts or follows regular habits also enhances external definition.

Measurement and enforcement rely on physical contact with entities of concern, but some relevant entities may simply remain out of reach given the scientific state of the art. For most of human history, records circulated no faster than animal muscles or the wind could carry them. The Battle of New Orleans was fought two weeks after the Treaty of Ghent ended the War of 1812 because the opposing commanders had not yet received the news from across the Atlantic. Technological progress in electronic and satellite communication improves the ability, in principle, to establish connections farther and faster. New modes of communication and computation make it possible, in principle, to correlate the objective state of the world with the state of a more complex representation. Technology supports a degree of optimism about intelligence that Clausewitz could not possibly have entertained.[49] Advances in the scientific state of the art make it possible to illuminate the dead space where the enemy hides from measurement and enforcement. It has generally become harder to hide big pieces of metal on the earth, let alone large troop formations. Yet surveillance improvements also create incentives for competitors to seek out more constricted forms of dead space, for example by hiding insurgent combatants within the civilian population, moving sensitive facilities deep underground, or inserting malicious code into commercial infrastructure. The ability to see more things also increases the number of places one must look to find anything. It must be emphasized that the scientific potential for improving the precision, range, speed, bandwidth, and fidelity of measurement or enforcement is only one aspect of the information problem. The technological potential for contact with remote entities can be actualized only through a complementary information solution.

## THE INTERNAL SOLUTION

The external pole of the informational relationship is often poorly constrained in war as a result of enemy action, bad weather, and chaotic battlefields. Military organizations can attempt to compensate by increasing the representational capacity of the internal pole. In effect, military organizations import their own structure into an unstructured environment. Unfortunately, doing so can exacerbate the collective action problems inherent in the sociotechnical institutions of control. It is helpful to adopt a political economy perspective to understand the trade-offs here.

Two basic alternatives have been used to analyze the organization of nearly every type of social system, from families to the global economy. Governments and markets, or hierarchical and self-organized systems, each have characteristic strengths and weaknesses.[50] Self-organized markets enable a decentralized search for new ideas and innovations. Yet they create problems when there are negative externalities (pollution), information asymmetries (imperfect knowledge), or unfair competition (monopoly and cartelism); in the worst cases these market failures combine to create panics and crashes.[51] Governance institutions can mitigate these problems through deliberate planning and redistributive policies (mobilizing large-scale resources), development of common standards and protocols (harmonizing actor expectations), and enforcement of norms and contracts (improving the rule of law).[52] Yet regulatory policy can be captured by rent seekers who pursue private ends with public means, and the administrative burden of management can impede adaptation to new circumstances.[53]

Policymakers may try to harness the strengths of each mode of organization while avoiding characteristic weaknesses.

Even though military organizations appear to be hierarchical and bureaucratic to the extreme, modern militaries are also highly distributed and diverse communities that depend on a lot of informal coordination. The national security apparatus of most states is actually a society of many different agencies, services, branches, and units that have heterogeneous interests and worldviews.[54] Doctrines such as "combined arms warfare" in the army and "centralized control/decentralized execution" in the air force emphasize the importance of seizing the initiative and adapting to circumstances, much like decentralized firms in a market, even as they stress the importance of unity of command and common tactics, techniques, and procedures, much like a centralized regulator. The venerable distinction between governments and markets is thus a reasonable place to start looking for insight into the trade-offs between top-down and bottom-up coordination in security institutions.[55]

Just as the external problem can be more or less constrained, the internal solution can be more or less constrained. I describe organizations in which the behavior of subordinate components is highly constrained as *institutionalized* to emphasize the human-built nature of sociotechnical constraints. Organizations that provide subunits with more latitude for self-organization and exploratory adaptation I describe as *organic*. The previous chapter highlighted both the top-down bias in the literature on military innovation and the pervasiveness of bottom-up adaptation in modern war. One reason that I clustered the literature into four categories is that they implicitly reflect the advantages and disadvantages of institutionalized governance and organic markets, respectively. The trade-offs summarized in table 2.4 are enduring and recurring. The next section will describe how different patterns of

**Table 2.4  Trade-offs in political economy with corresponding military innovation literature and types of information practice**

|  | Advantages | Disadvantages |
|---|---|---|
| Government trade-offs | Economies of scale | Policy lock-in |
|  | Common standards and protocols | Rent seeking and policy capture |
|  | Accountability | Administrative overhead |
| Category of literature | Contingent doctrine | Organizational culture |
| Type of practice | Managed practice | Insulated practice |
|  |  |  |
| Market trade-offs | Innovation and adaptation | Negative externalities |
|  | Efficient search | Unfair competition |
|  | Knowledge diffusion | Information asymmetries |
| Category of literature | User innovation | Fog of war |
| Type of practice | Adaptive practice | Problematic practice |

information practice map onto these same four categories. First, however, it is important to understand what makes a military information system more or less institutionalized.

Table 2.5 summarizes factors that constrain the internal solution. An institutionalized system is distinguished by consensus on goals and purposes, consistent definition of bureaucratic interfaces, resources targeted to the maintenance of representational protocols, and stakeholders who contribute to public goods without excessive mutual interference. Institutionalization enables organizations to rationally partition work across various centers of calculation with a minimum of informal negotiation. Standardized protocols enable personnel to collaborate to construct reliable representations of the world. Abstraction, which is essential for measurement and coordination, breaks down hard problems into more computationally feasible subproblems. Comparable abstractions make the world more countable and controllable. Representations that are more communicable, commensurable, comprehensive, combinable, impersonal, and quantified enable more far-flung networks of representation to be gathered together. Conversely, organic protocols that are incommunicable, incomparable, fragmented, confusing, and qualitative make it harder to coordinate reliably with distant centers of calculation. If different parts of an organization have divergent goals or understandings, and if the division of labor is ill defined or contested, then the adoption of technical standards and protocols is more likely to become a source of friction.[56]

Larger and more diverse organizations have more incentives to institutionalize information systems to get members on the same page, but by the same token they face more difficulties in actually doing so. Some components will be more likely to adopt different interface and communication protocols. They might have different ideas about how to operate or which missions they ought to pursue. Units will be concerned about differential implications of any given mission for their prestige, budgets, and

### Table 2.5 Internal solution

| Institutionalized information solution | Organic information solution |
| --- | --- |
| Centralized authority | Decentralized authority |
| Fewer members in fewer organizational units | Many members, units, or different agencies |
| Complementary specialization across subunits | Competing or interfering specialization |
| Common identities, goals, values, or understandings | Heterogeneous intentionality across subunits |
| Standardized formal protocols and interfaces that connect subunits | Informal or ad hoc data processes or idiosyncratic implementations |
| Reliable availability of resources | Resource scarcity |

authorities. Within units, computer users will devise informal workarounds when the official systems they are supposed to use get in the way of the mission as they understand it. Over time, military data networks thus tend to become a tangle of user-developed databases, scripts, websites, and applications running right alongside officially sanctioned programs procured from defense contractors. The centralized, institutionalized administration of systems competes with the decentralized, organic adaptation of components.

More resources can aid institutionalization. It takes a lot of effort to implement, maintain, and enforce common procedures given the tendency of subordinates to do their own thing. A steady flow of resources empowers managers to keep personnel in line. The availability of resources also reduces the incentives for subordinates to do things differently. Why innovate and potentially fail if habit will be rewarded? Scarcity, by contrast, encourages units to make do with whatever they can beg, borrow, or steal. Competition over resources can also energize self-organization. Indeed, interservice competition was an engine of innovation during the Cold War. The institution of a stronger Joint Chiefs of Staff after the Goldwater-Nichols reforms curtailed competition as service logrolling assured that all services received funding for cherished projects.[57]

## The Quality of Information Practice

The interaction of external problems and internal solutions produces four different patterns of information practice. This is the primary proposition of the theory, as summarized in table 2.6. Two of these patterns enhance military performance (managed and adaptive), and two undermine it (insulated and problematic). They correspond to the strengths of governments and markets and their weaknesses, respectively, as depicted in table 2.4. The secondary proposition of the theory, which I will discuss further in the next section, is that organizations tend to cycle through all four patterns as external and internal actors engage in disruptive exploitation and stabilizing reform. The challenge for leaders at all levels is thus to match the structure of their internal solution to the structure of the external problem they face. As Lex Donaldson points out, "There is no single organizational structure that is highly effective for all organizations. . . . Thus, in order to be effective, the organization needs to fit its structure . . . to the environment."[58]

**Table 2.6 Theory of information practice**

| | Constrained problem | | Unconstrained problem |
|---|---|---|---|
| Institutionalized solution | Managed | | Insulated |
| Organic solution | Problematic | ◀ Exploit and reform ▶ | Adaptive |

This framework is broadly consistent with Charles Perrow's theory of normal accidents, but I combine the concepts of coupling and complexity, whereas Perrow distinguishes between them.[59] Perrow argues that centralized management is appropriate for tightly coupled yet linearly decomposable systems; managers can control the risks of tight coupling with clearly codified procedures. Conversely, decentralized operations are more appropriate for loosely coupled but complex nonlinear systems; personnel can exploit nuanced local knowledge without inflicting harm on the rest of the organization. Either arrangement can work well in loose and linear systems, but both run into problems in tight and complex systems. Coupled complexity is the most difficult case because centralized management is needed to control catastrophic risks but decentralized operations are needed to make sense of dynamic situations. Networked militaries are particularly complex and tightly coupled systems, which creates difficult trade-offs between centralization and decentralization in war.[60] I collapse Perrow's variables of coupling and complexity into the single concept of the external problem. The external problem, in this framework, can thus range from coupled and simple at one extreme (constrained problems) to loose and complex at the other (unconstrained problems). This range highlights the regions where organizations have the best hope of achieving a good fit.

Any theory about complex systems must make simplifications. My combination of Perrow's two dimensions into a single one, along with a few other concepts in table 2.3 that Perrow does not stress, does sacrifice some explanatory power. My concept of the internal solution, similarly, combines Perrow's emphasis on the centralization of authority with several other administrative features that Perrow does not stress (see table 2.5). The constitutive factors of problems and solutions do not necessarily co-vary in every case. Explanatory parsimony usually does some violence to complex phenomena. Nonetheless, I believe that these simplifications are worth making, as a first-order approximation, because they highlight the similar and recurring political economic trade-offs that can emerge in any social organization for many different reasons. Dialectical tension between ideal types, moreover, can be revealing. The active agency of practitioners to reshape solutions and problems along these dimensions becomes an engine for generating the very complexity that a two-dimensional framework omits. Simple mechanisms can generate complex systems. I will consider the exceptions, nuances, and interesting mixed cases in each of the empirical case studies. First it is important to understand the ideal types of information practice.

## MANAGED PRACTICE

The best situation occurs when the warfighting problem is well structured and organizational systems are well adapted to it, hand in glove.

The military organization fights as it trains, and its training is presciently suited to the nature of the war that it ends up fighting. Managed practice is distinguished by a strong distinction between the design and use of information systems. Systems of doctrine and technology are designed in peacetime to be used in wartime. "Breakdown" still occurs in managed practice, but it is a deliberately planned process of identifying requirements and linking them to capabilities, precisely to avoid experiencing unpleasant, unanticipated breakdowns during wartime operations. System architects thus worry about how information works (reflexive experience) in peacetime so that system users can focus just on what information means (mediated experience) in wartime. Because warfighting requirements tend to be stable, furthermore, it is easier to clearly articulate them in advance. The military organization thus knows who, where, and how it will fight. Managers and engineers can then undertake systematic research, development, test, and evaluation (RDT&E) programs. Senior officers can hold periodic performance reviews to monitor progress and make top-down adjustments to information systems.

When the fighting starts, thanks to all this careful preparation, information systems tend to behave as expected. Breakdowns are minimized, so practitioners have little reason to abruptly shift their attention from mediated to reflexive experience. The ontology of representation generally corresponds to all of the relevant types of entities on the battlefield. The organization is able to quickly and accurately identify the locations, affiliations, and intentions of friendly, enemy, and neutral entities. It can efficiently share this information with relevant and authorized people throughout the organization, who can then mobilize resources quickly and effectively to accomplish a common mission. Control loops efficiently and effectively make contact with their intended targets. Managed practice is thus able to substitute information for mass, that elusive dream of technology theory. This is possible only because sociotechnical control protocols are congruent with the nature of the military task. Standardized representations of a well-structured environment enhance bureaucratic control of both the organization and battlefield events.[61]

Bureaucratic management improves control by mobilizing economies of scale, mandating common doctrine and standards, creating synchronized plans based on detailed intelligence, rationalizing logistics, managing personnel, and ensuring accountability up and down the chain of command. Modern command staffs have extremely specialized functions with dense interdependencies across them. Staff sections are appearing at even the lowest echelons of military formations, with intelligence and operations analysts working at the company level or below. A vast administrative infrastructure enables the U.S. military to coordinate intelligence agencies and units from all of its services, carry out complicated campaign planning, integrate logistic supply chains, and interface with other government agencies and coalition allies.

In large-scale engineering projects, a designated lead system integrator usually helms multistakeholder projects.[62] Frequent meetings and detailed policy coordinate the stakeholders and subprojects over a long period of time. Large-scale data systems are particularly dependent on shared data models to facilitate semantic interoperability across distributed units and referential integrity across system boundaries.[63] For example, four different defense contractors used information technology to coordinate construction of the B-2 stealth bomber; common adherence to rigid standards and a shared technical grammar enabled them to control transaction and agency costs despite major scientific uncertainties.[64] Importantly, intervention from the air force program office to set process standards was critical. Even open industry standards need to be interpreted and implemented by some governing body for specific projects. Engineering abstractions with a well-defined and centrally stabilized ontology can enable efficient coordination.

Formal management also enhances accountability by depersonalizing systems and improving the transparency of their construction. Surveillance techniques provide the means to monitor an organization's members and enforce compliance with organizational policy. "Two person integrity" requirements for the handling of top-secret data or "two person control" for launching a nuclear weapon help to ensure that no single person can abuse access to sensitive data or controls. Target review boards audit the identification of bombing targets and coordinate mensuration. Gun camera video provides forensic evidence to exonerate or prosecute operators who shoot civilians. Security access protocols—passwords, ID cards, biometrics, and so forth—enable counterintelligence units to track the comings and goings of personnel under investigation. Accountability measures thus harden measurement and enforcement processes against enemy disruption, ensuring that representations in "centers of calculation" are more trustworthy.

Unfortunately, personnel in more legible systems also tend to become more risk averse. Mistakes become more visible, and the punishment of deviations becomes more likely. Management begins to court insulation when its systems discourage personnel from taking the initiative in changing circumstances.

INSULATED PRACTICE

Military performance suffers when there is a poor fit between the military task and the military organization. Mismatches become more likely when the environment is dynamic or ambiguous yet the organizational culture is rigid. An unconstrained environment gives the organization the latitude to double down on its preferred forms of intentionality and coordination. Gaps between the state of the world and the goal state go unperceived or ignored, or awareness of them is suppressed, by organizational leadership. Practitioners fail to adopt a critical or reflexive attitude toward their own information system. Their attention thus remains directed toward entities and events in the world, as in managed practice, but the referential integrity of the representations becomes less reliable. Practitioners remain confident in their models, but their confidence is misplaced. Representations become reflections of the organization's own biases rather than of objective features of the world.

J. F. C. Fuller disparaged British General Headquarters in World War I as a "lamaistic system" where General Douglas Haig worked "like a mechanical monk" with "little or no contact with reality." As a result, he said, "the cutting edge of the army . . . time and again . . . was blunted, jagged or even broken."[65] One study of the differences between line and staff information processing in the Battle of Passchendaele found "an almost complete disassociation of strategic and tactical thought. . . . Neither was in a position to guide the other. It was rather a matter of outright dominance, and the framework of organization gave staff the upper hand."[66] Visiting the front after the battle, the British chief of staff suffered a nervous breakdown upon realizing that he had ordered men to advance through a sea of mud that devoured tanks and infantry units without a trace. From the perspective of the insulated information system at the headquarters, the British scheme of maneuver had seemed reasonable enough.[67]

A modern air-conditioned joint operations center can become a digital lamasery. Panoptic representations become a "closed world" as institutionalized abstractions fall out of sync with a more ambiguous reality.[68] Insulated control systems prioritize their internal solution over the external problem. For example, as the First Marine Division's after-action report from 2003 observes, "data on the [common operational picture] was often untrustworthy" and "other track management systems did not appear to function at all," in part because "the enemy did not conform to our expectation of a conventional line and block organization for combat."[69] Insulated militaries tend to conduct the same types of operations repeatedly without being able to perceive the counterproductive side effects they generate. Units in the field report what they believe headquarters wants to know.[70] Headquarters aggregates performance metrics that demonstrate progress along preferred performance vectors, whether or not they meaningfully relate to the strategic situation. The organization's cascades of inscription tend to reproduce and amplify organizational preferences, filter out discrepant signals, suppress corrective feedback, and lock in behaviors that become counterproductive. The insulated organization fails to recognize the mismatch between the state of its knowledge about the world and the state of the external world itself.

Political interest groups or lobbies often seek private gains at public expense. Similarly, units may develop data feeds and displays that support their most favored tactical options, thereby revealing and advertising more opportunities to exercise them. Quantified statistics of conflict casualties, refugee flows, drug trafficking, and militant force levels are particularly vulnerable to politicization; they provide an air of scientific authority for estimates that are actually quite unreliable because of the self-hiding nature of illicit phenomena.[71] Intelligence agencies are particularly attractive targets for politicization because they enjoy a monopoly position as the government's keeper of secret knowledge, which enhances the rhetorical value of intelligence in making the public case for a preferred policy, or silencing critics.[72] The monopsony, secrecy, uncertainty, and huge financial sums involved in defense acquisition tempt firms and lawmakers to prioritize lucrative contracts over mission performance.[73] The great expense of modern weapons makes great temptation for rent seeking.

Even in the absence of corruption or political agendas, complicated acquisition regulations and engineering processes make it difficult for procurement offices to understand emerging warfighting requirements. One study by the Defense Science Board found that "software is rapidly becoming a significant, if not the most significant, portion of DOD [Department of Defense] acquisitions," and yet "programs lacked well thought-out, disciplined program management and/or software development processes."[74] Software architecture is inherently difficult to understand because nonlinear interactions across abstract and arbitrary logical functions are hard to audit. Assuring the quality of operational code is an excruciatingly slow and experimental process. As Frederick Brooks argues in a classic book on software engineering, "Adding manpower to a late software project makes it later."[75] The integration of hundreds of engineers and managers, distributed across procurement offices, subcontractors, and geographical locations, results in a perpetual churn of PowerPoint presentations, meetings, emails, Gantt charts, and configuration control documents. It thus becomes challenging to incorporate new technological opportunities that cut across the grain of modular architectures; legacy code becomes locked in through interdependence with other systems and components of the same system.[76]

Computerization in the name of administrative efficiency can also become counterproductive for military effectiveness. Administration by computer system can create the unintended equivalent of a work-to-rule strike, in which employees do only and exactly what is explicitly described in their employment contracts.[77] The striking workers take all of their allotted breaks, follow every procedure to the letter, employ no creative workarounds, and make no extra effort. Punctilious obedience to corporate policy holds productivity hostage without the defiant provocation of picketing or other work stoppage. The strike thereby highlights the inherent value of creative, motivated, interpretive labor. Computerization may unintentionally implement a work-to-rule strike simply because computers are deterministic agents that perform their routines precisely as coded. Personnel become accountable to different databases tracking training and readiness, security clearances, equipment provision, travel, fitness, medical readiness, financial health, family situation, and career progression. These databases interoperate imperfectly, if at all. All of these different systems are subject to technical configuration and security regulation by network administrators who control accounts, network access, and application certification. Managers can shut down services at the first sign of cyber intrusion or classified information spillage. As Weber's iron cage becomes gilded in silicon, the interests of the human mission are held hostage by automated functionaries.

One study found that "administrative overhead, far from being curtailed by the introduction of office automation and subsequent information technologies, has increased steadily across a broad range of industries."[78] Gene Rochlin argues that the substitution of automation for human labor deprives organizations of the slack resources people need to compensate for unexpected breakdowns: "What is lost in many cases is not just variety, and specific human skills, but the capacity to nurture, enhance, and expand them through the messy process of direct, trial and error learning."[79] Enterprise software systems thus increase the "brittleness" of bureaucracy by forcing practice into well-worn formal pathways.[80] Clever algorithms can enable artificial intelligence to fake interpretation in defined situations for which training data are available, but common sense and pragmatic judgment continue to elude formalization.[81] When people or machines heedlessly execute routines without adjusting them to the context of performance, taking into account intentional goals and values, then efficiency decreases and the potential for mishaps increases.[82]

ADAPTIVE PRACTICE

One reason that bureaucracies do not seize up completely is that their members often defect from the orthodox system to come up with creative workarounds. Organizational performance can improve when organic solutions emerge to address more ambiguous problems. Military personnel can take advantage of the autonomy of their wartime deployment, where administrative controls are looser, to exploit discretionary resources for exigent needs. They might download software and purchase commercial devices, or have family members send them by mail, to address idiosyncratic computational problems that are not being addressed by the official technology procurement office. Local degrees of freedom in the work situation and technical architecture provide space for personnel to experiment. They can test out new ideas and techniques without triggering managerial suppression of variation.

Adaptive practice is a breakdown condition in which the design of the information system moves from the background to the foreground of practitioner awareness (reflexive experience). Unlike managed practice, where there is a clear distinction between the design and use of systems, in this case users engage in design activities themselves. Whereas software engineers distinguish between "design time" programming and "runtime" execution, adaptive practice might be described as "runtime design." Some local breakdowns are not necessarily harmful to the rest of the organization if the unconstrained or loosely coupled nature of the problem provides a buffer for experimentation. Practitioners can take advantage of slack control relationships to make local modifications that do not necessarily cause problems elsewhere. Adaptive practice is characterized by a ferment of informal activity that produces jury-rigged solutions, field expedients, working prototypes, and custom-tailored systems. These interventions are most likely to be performance enhancing when practitioners take an active, mindful role in designing, redesigning, hacking, and modifying their technologies and processes.

Friedrich Hayek highlights the epistemic advantages of self-organization: "Knowledge of the circumstances of which we must make use never exists in concentrated or integrated form, but solely as the dispersed bits of incomplete and frequently contradictory knowledge which all the separate individuals possess."[83] Hayek argues that decentralized markets make the most efficient use of local knowledge. Champions of "open source software" or "peer-to-peer production" argue similarly that decentralized knowledge production is the most efficient way to allocate diverse intellectual resources to create novel ideas, or repair mistakes.[84] Eric Raymond's battle cry for open source software—"Given enough eyeballs, all bugs are shallow"—emphasizes the efficiency of decentralized search.[85] Indeed, open source communities have created some of the most reliable components of the internet ecosystem, such as the Linux operating system and the Apache web server.

According to Eric von Hippel, user innovation is most likely when the costs of inventing something that users will personally benefit from using are less than the transaction costs of finding a manufacturer to make it.[86] Whenever user needs are hard to articulate or highly changeable (i.e., information is "sticky"), and when future demand for the manufacturer's effort is ill defined, then it is difficult for firms to connect their engineering expertise to actionable user requirements. An alternative to contracting out a solution is to look for "lead users" who will use their local knowledge and materials to create low-cost prototypes. Such individuals tend to enthusiastically share inventions with other users who have similar interests. Peer inventors are willing to further improve and share the designs, thereby generating increasing returns in a self-reinforcing innovation community. User innovation has improved product quality and social welfare in a variety of industries, notably in software development, extreme sports equipment, and scientific instrumentation. User innovation can also play a search and prototyping function for manufacturers. Once users have clarified their own needs by using their own prototypes, firms can apply more capital and administrative expertise to institutionalize production.

Military personnel, similarly, often adapt data management procedures, reconfigure commercial technologies, or even code up software prototypes on their personal computers. The first digital aid for naval antiair warfare, the Naval Tactical Data System, was developed by a savvy user group with support from senior officers.[87] Even after the navy established a formal electronic system procurement office to regulate the system, members of the USS *Eisenhower* carrier battle group cobbled together an alternative tactical decision-aid during a large exercise in 1981. Embarked sailors even wrote some of the software. Their prototype—initially known as the "Jerry O. Tuttle System" after the battle group commander and later redubbed the "Joint Operational Tactical System"—replaced a costly computer system that had been under contract for a decade. Admiral Tuttle later commanded the Naval Space and Electronic Warfare Command, completing the transition of his system from an informal experiment to a regulated program of record.[88] In practice, user innovation and bureaucratic management can be complementary. Chapter 4 explores this fraught relationship in depth.

While user innovation can and does occur in military organizations, it also has to overcome much more bureaucratic resistance than civilian user innovators ever experience. Military users live in a controlled world of classified information and austere system configuration policy. Their short tours of duty and warfighting responsibilities limit their ability to provide technical support for their inventions. Warfighter prototypes lack the legitimacy of official acquisition and certification regimes. Cybersecurity is a growing concern, furthermore, and for good reason. Amateur software can introduce vulnerabilities that an experienced software engineer would know to avoid (e.g., buffer overflow or unencrypted credentials). The systems management bureaucracy, preoccupied with configuration management and information assurance, has the power to punish users for illegitimate deviations from standard operating procedures. Procurement offices and network managers rightly worry about the quality, scalability, and reliability of amateur code. Rube Goldberg contraptions and expedient fixes can inadvertently jeopardize the reliability, security, compatibility, efficiency, or maintainability of enterprise information systems.

Organic exploration that seems helpful in the short run can thus be destructive in the long term.[89]

## PROBLEMATIC PRACTICE

Military performance suffers when the relevant entities and relationships are so tightly coupled, or structural constraints are so rigid, that uncoordinated events in one part of the system create deleterious effects elsewhere. Friction creates shear between the internal and external poles of informational relationships. Insulated practice and problematic practice both heighten the potential for undesirable performance outcomes; the difference between them is that practitioners remain unaware of or indifferent to the problems in the former but are alerted to and concerned about them in the latter. Problematic breakdowns manifest as information overload, interoperability failure, security compromise, accidental collision, and targeting error. The characteristic experience of problematic practice is sudden surprise or horror. Fratricide (friendly fire), equipment casualties, missed rendezvous, unexpected terrain conditions, betrayals by allies, and enemy ambushes are all examples of sudden breakdowns that are both unexpected and immediately recognized as problems. As practitioners realize that something is wrong, their attention shifts from mediated to reflexive experience. Practitioners may also shift their attention when they receive warning or indication that there may be such a problem in their control system. That is, practitioners may be able to imagine or simulate an unwanted breakdown. In such situations, practitioners consider what it would be like to experience their information systems as broken or dangerous. The severity of problems can range from minor annoyance to catastrophe depending on the scale of the system under consideration and their functional purpose.

Desirable information system qualities like security, reliability, interoperability, organized files, and common database ontologies can be considered to be public goods for a given community of computer users. Mancur Olson argues that in the absence of some public-minded coercive authority, such as a regulatory agency with effective enforcement mechanisms, public goods will be underprovided and concentrated interests will be overrepresented.[90] Many of the frustrating problems that users experience in a shared computing environment result from failure of the user community to provide these public goods.[91] Email inboxes fill up with low-priority traffic. The organization of the shared file server becomes inscrutable. The meaning of database fields becomes ambiguous. Personnel struggle to connect incompatible systems. The representations they create become illegible or unreliable. Units opportunistically appropriate technologies for their own needs while ignoring the effects on the quality of shared databases.

Incompatible systems and conflicting data formats are a recurring nightmare in the history of command and control. Different services or agencies procure information systems for their particular missions, but their "silos" and "stovepipes" fail to interoperate when they need to work together. Development, maintenance, training, and quality control are often the responsibility of the subunits that procure these systems, but referential integrity is a public good that spans them. As a result their radios may be unable to connect. Databases use inconsistent ontologies and idiosyncratic formats of dates, times, and geocoordinates. Proprietary licensing restrictions further preclude data sharing. Command and control incompatibilities have contributed to many high-profile military disasters, including the Bay of Pigs fiasco, the Mayaguez incident, and the botched Iranian hostage rescue attempt. Targeting errors resulting in civilian deaths and fratricide are, more often than not, a consequence of breakdowns in decentralized sociotechnical practice. Even simple errors can have tragic consequences. For instance, if different geodetic datums compute the "same" coordinates at physical locations hundreds of yards or even miles apart, then soldiers who neglect to check the map datum might inadvertently call for fire on their own position.[92]

Information systems that solve coordination problems at one level of abstraction may still generate discoordination at another. Digital networks that everyone can access because they use the same applications and communication protocols can become echo chambers for unreliable data. Because military personnel are public servants working with public data, norms about preserving authorial provenance (i.e., avoiding plagiarism) tend to be weaker in the military. Personnel cut and paste bullet points and catchy diagrams from one document or PowerPoint slide into another, rearranging fonts and layouts but deleting provenance markers. Sourcing data often does not survive the cut-and-paste operation, which complicates the discrimination of relevant data from the flood of noise. As the metadata that protect the provenance of the chain of custody is clipped away or ignored, questionable reporting becomes incorporated into other products downstream. Unreliable intelligence claims or bad target nominations reappear even after one part of the organization has raised concerns. When buyers cannot determine the difference between high-quality and inferior products, then quality sellers will be reluctant to bring their wares to market and skeptical buyers will be unwilling to pay the higher prices they demand.[93] Shared computer networks, similarly, can become digital lemons markets where bad representations drive out the good.

Competently sourced and evaluated intelligence is a high-quality good that is too often undervalued in a decentralized market for knowledge. Prior to the financial crash of 2008, for example, some financial derivatives were difficult to accurately value because they incorporated clusters of asset-backed securities and collateralized debt obligations that were rated by different agencies using inconsistent criteria; complex tranches included extremely risky housing loans yet they received AAA ratings, which in turn suckered risk-averse investors who normally would have stabilized derivative markets.[94] Friction-laden analytical assessments act something like inscrutable bond ratings, compartmented behind security walls that protect prior assessments from inspection. Assessments become detached from any correspondence with reality, yet they continue to trade with authority. The intelligence judgments surrounding the assessment of Iraq's weapons of mass destruction appear to fit this pattern: a long series of reasonable judgments in their local context, an accumulating mass of glittering analytical products describing the overall chemical and biological programs, failure to properly vet dubious sources like the human- intelligence asset "Curveball," and political pressure to produce a positive intelligence judgment. All contributed to substantial willingness among analysts and policymakers to invest in the weapons-of-mass-destruction hypothesis in late 2002.[95] As before the financial crash of 2008, there were Cassandras aplenty, but the bubble of toxic assumptions proved too tempting an investment for many. Without reliable verification somewhere, in William James's analogy, "the fabric of truth collapses like a financial system with no cash-basis whatever."

## The Complexity of Information Practice

Contrary to the assumptions of technology theory, war has neither changed its nature nor become any more decisive in the information age. Fog and friction are still serious problems, created in part by the very systems that are supposed to reduce uncertainty and improve control. What is historically novel, however, is the complexity of the information systems that mediate the human experience of war. Indeed, nested and overlapping complexity is a fact of life in any social system. Almost all organizations are made up of other organizations. Organizations operate in an environment populated with still other organizations. Employees work in informal social networks, embedded in managed firms, competing in decentralized markets, regulated by federal governments, negotiating in political anarchy, altering the climate of a common planet, and so on. The dilemmas of governments and markets, or hierarchy and anarchy, or institutional and organic organization, reappear at many different scales. Positive and negative feedbacks across and within organizations can drive systems in unexpected and nonlinear directions.[96]

Any empirical analysis of information practice, therefore, must take care to define the scope of the relevant problems and solutions. Managed practice that improves control capacity within one component of an organization will seem like opportunistic adaptation from the perspective of the larger organizational system. Personnel who adapt commercial software packages to create new databases and visualizations may be engaging in adaptive practice from the point of view of bureaucratic program offices. Yet from the perspective of the local work center that wants to share its data and visualizations, its personnel are trying to coordinate common standards to enable managed practice for the center. The nesting of different patterns can even be applied to conflict between noncooperating belligerents. Beneficial adaptation for one side creates harmful problems for the other. An IED attack and a radio incompatibility are both examples of uncoordinated adaptations; the difference is that harm is explicitly intended in the former but not the latter. Conversely, a system of cooperating allies that forges an agreement on common contracts and protocols may implement something like managed practice within its coalition.

War features extreme competition between opposed organizations of cooperators. People in a military organization try to cooperate to avoid control failures, while their external competitors seek to cause them. More benign competition can also occur across the components of each organization, even as components are cooperating in their fight against the enemy. Adaptive practice at any scale exploits, defects from, or works around more insulated practices. User innovation exploits discretionary resources to meet local requirements. Since anarchy is defined as the lack of overarching institutions, moreover, systems of belligerents can be considered to be extremely organic organizations. In combined arms warfare, exploitation is the attack that follows a breakthrough of enemy defenses. In cyber operations, exploitation is the covert subversion of a target's computer networks. Exploitation is thus an opportunity or a threat depending on whose intentions are being realized. The converse movement is managerial intervention to repair, replace, or reform problematic practice. The upgrading of computer network defenses by patching software and enhancing network surveillance is a reform that protects against problems caused by certain forms of cyber exploitation. Organizations may also attempt to identify, improve, and institutionalize grassroots initiatives that seem to be paying off. Something like this happened with the development of counterinsurgency doctrine during the Iraq War, as discussed in chapter 5. Counterinsurgency is a very ambitious reform strategy that aims to transform not only military doctrine but also an entire host society, replacing problematic anarchy with new governing institutions. Yet if the new mode of managed practice improves combat performance, then enemies also have new incentives to adapt and change the problem again.

*Exploitation*, as I use the term here, is the movement from insulation to adaptation.[97] *Reform* is the movement from problems back to management. Problematic practice is the failure mode of adaptation, and insulated practice is the failure mode of management. Pursuit of the advantages of management invites the disadvantages of insulation. Pursuit of the advantages of adaptation invites the disadvantages of coordination failure. Some parts of the system will exploit the insulation of the whole. Others will try to negotiate reforms to recoordinate the whole. Bottom-up and top-down initiatives may both try to improve efficiency and effectiveness, but they may have different understandings of what that means.

Exploitation and reform act together like a piston that drives up system complexity (figure 2.7). The collective behavior of actors in a distributed system thus tends to cycle through managed, insulated, adaptive, and problematic practice. Like a Bach fugue, this canonical cycle repeats indefinitely with variations. Over time, cycles of exploitation and reform build in new features and controls, add more devices, connect more outside organizations, impose more policy constraints, and rely on more sophisticated data abstractions. More complexity enables organizations to represent more numbers and types of relevant entities, properties, relationships, and interactions, yet at the price of more difficult coordination problems.

This complexity ratchet is analogous to a mechanism proposed by Scott Snook to explain military accidents.[98] In Snook's theory of "practical drift," practitioners tend to relax rule-based logics of action (management) and adopt more task-based logics (adaptation) in loosely coupled environments. If, however, the system becomes tightly coupled for whatever reason, then unexpected collisions (problems) become more likely. A tragedy usually prompts investigations and reforms to enforce new rules (more management). Snook's book examines a friendly fire incident from northern Iraq in 1994, where an E-3 controller cleared a section of F-15

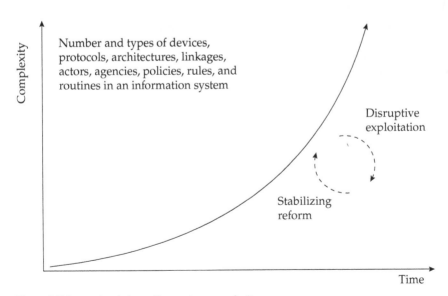

Figure 2.7. Increasing information system complexity

fighters to engage two Black Hawk helicopters. Each had relaxed formal procedures and adopted idiosyncratic practices for convenience. Yet they each assumed that all of the others were still adhering to standardized practices. The Black Hawks deviated from their posted schedule; the E-3 had inexperienced crew members and relaxed its data-checking protocols; the F-15 pilots were primed to see Iraqi aircraft violating the no-fly zone. The airborne controller thus cleared the fighters to engage the Black Hawks after the fighters mistakenly reported them as hostile Iraqi aircraft. The practical drift of local sensemaking in a decoupled environment resulted in fratricide when the actors were recoupled together. My general mechanism is analogous to Snook's, but I consider a wider range of inadvertent and deliberate breakdowns, including adversaries who create malicious breakdowns.

Cycles of exploitation and reform proceed by two steps forward and one step back. As control problems are solved, the solutions create new problems. Then new solutions are piled on top. Exploitation destabilizes what reform stabilizes. The resulting development, like the evolution of any other type of institution, is path dependent.[99] Increasing returns to legacy practices and infrastructure, together with the friction generated by adaptation, create an evolutionary bias for complexification.[100] Spiraling complexity is not inevitable, as some reforms may intentionally aim for simplification. Depending on political fortunes, scientific progress, and resource budgets, complexification may occur haltingly, get stuck in one state for a while, or even backslide. Yet it is precisely the contingent interactions between actors and artifacts at the microlevel that power the complexity ratchet at the macrolevel.

Supplementary theory is needed in any particular case to explain why one group rather than another decides to innovate around insulation, why innovators gain advantages over incumbents (or vice versa), why conflicts between them persist or resolve, or why one group manages to reform its preferred solution over others. The following chapters employ this simple conceptual framework to make sense of complex information systems in a series of detailed case studies.

# Strategic and Organizational Conditions for Success

## The Battle of Britain

The official history of the Battle of Britain describes "the most efficient scheme of air defence in the world at the time."[1] One scholar calls it "the classic modern C2 [command and control] success story, the example usually cited when the topic is the force multiplier effect of capable C2."[2] Any theory of information technology and military power must be able to explain this intrinsically important case. While the details of the battle are well known, the importance of structural factors in explaining its outcome has been less appreciated. In this chapter I will show how a constrained problem and an institutionalized solution enabled the Royal Air Force (RAF) to successfully manage the air battle. This case also has a special relevance for the American cases in the next chapters. The RAF pioneered many concepts that the U.S. Air Force still uses today, including aircraft early warning, identification friend-or-foe, track management, aircraft vectoring, and operational research. The Battle of Britain is also one of the most well-documented episodes in military history.[3] Open archives, abundant data, and the electromechanical vintage of information technology make this case an accessible illustration of information practice in action.

German air power enabled Hitler to defeat Poland, Denmark, Norway, the Netherlands, Belgium, and France in rapid succession. The Luftwaffe then nearly destroyed the British Army before it escaped from Dunkirk. It seemed improbable, therefore, that the RAF would soon defeat a Luftwaffe bombing campaign, preclude a German invasion, and inspire hope in the ability of a free people to resist Nazi aggression. Contemporaries compared the miraculous victory to Britain's defeat of the Spanish Armada in 1588 and its victory at Trafalgar in 1805. Later accounts stressed that the Battle of Britain was a close-run affair. The RAF seemed to be enthralled by the ideology of strategic bombing and disdainful of air defense. Luckily, civilian politicians intervened in the nick of time to empower Air Chief Marshal Hugh Dowding to reorient RAF Fighter Command around the scientific novelty of the Chain

Home radar system. Churchill's dauntless "few" in Spitfires and Hurricanes went on to win a decisive victory against narrow odds.

We now know that the outcome of the battle was never seriously in doubt, even though the combatants did not realize it at the time. The Luftwaffe was designed to support the German Army in the field, not to conduct an independent strategic bombing campaign. It had poor intelligence and was riven by disagreements about targeting objectives. German commanders largely ignored British sources of strength and never understood how their own campaign was failing due to inadequate German intelligence and motivated biases in its interpretation. The Luftwaffe also faced an intrinsically difficult mission in attacking the fortified homeland of an advanced industrial adversary. By contrast, Britain's island geography simplified the RAF's information problem while complicating the Luftwaffe's. The RAF made the most of a favorable situation by methodically preparing its defenses throughout the interwar years. Despite all the sound and fury about strategic bombing, there was an enduring consensus in the RAF about the importance of developing and exercising an air defense system. When the battle started, therefore, tens of thousands of military and civilian personnel were ready to support "the few." As C. P. Snow observes,

> The introduction of radar into Fighter Command . . . is a textbook example of the successful application of science to war. It ought to be studied in staff colleges: and it ought to be studied by anyone who still thinks that scientific war is an affair of bright ideas. . . . By 1940 German radar was technically more sophisticated than the British: but the combination of [Henry] Tizard and the officers of Fighter Command had ensured that the British knew better how to use it as a weapon of war.[4]

Britain won the battle because it put together a well-managed solution to the well-constrained problem of air defense. Germany, by contrast, met the inherently harder problem of offensive coercion with a more insular solution. For the sake of narrative chronology, I will first describe the historical development of the British air defense system, or the "internal solution" in the terms of this book. I will then describe the "external problem" that Fighter Command faced during the battle and show how the interaction produced "managed practice" that improved RAF performance.

## The Internal Solution

Scholars of military innovation debate the origins of Fighter Command's C2 system. Some argue that the growing German threat in the 1930s prompted British politicians to intervene in an RAF that was obsessed with offensive bombing.[5] Civilians thus empowered Robert Watson-Watt to invent radar and Hugh Dowding to build an operational system around it. Others argue that the RAF leadership drove peacetime reforms that resulted in "the Dowding system."[6] Both interpretations are misleading. The RAF ideology of strategic bombing, ironically enough, encouraged rather than impeded the improvement of air defense. The RAF prepared deliberately for two decades, but this was neither a purely military undertaking nor the work of heroic inventors.

Air defense provided the impetus for the establishment of the RAF. Britain created an independent air service in 1918 in response to German Gotha bomber raids that killed over two hundred Londoners. The London Air Defence Area (LADA) pioneered basic reporting and interception techniques that would be used two decades later. As the official historians point out, "Nothing essential started in 1936; rather it was a recovery that started. Dowding was the principal builder rather than the architect."[7] The French occupation of the Ruhr in 1923 had raised new fears of strategic bombing (by France rather than Germany), which prompted the resurrection of LADA as the Air Defence of Great Britain command (ADGB). Every annual air defense exercise between 1928 and 1933 experimented with radio direction finding (DF) to track friendly aircraft. Every after-action report from 1931 to 1933 concluded that fighters must have a radio to coordinate defenses, even at the cost of climb and speed or the potential of micromanagement by headquarters (HQ). A key test in 1923 demonstrated that acoustic mirrors could detect an aircraft at eighteen miles and register its bearing at ten miles. These large concrete dishes, twenty feet in diameter, focused sound on a microphone or a human listener wearing a stethoscope. In 1933 the Air Ministry approved the Thames Estuary system, which was to have two massive two-hundred-foot mirrors and eight thirty-foot mirrors, all connected to a central plotting room. The project was suspended in 1935 and canceled in 1937 only because radar was showing far greater promise. Yet the mirror system established the precedent for scientific collaboration with the RAF. John Ferris points out that the "system was ideally preadapted to radar," a conclusion echoed by David Zimmerman in his authoritative history of British radar.[8]

That an air defense scheme could be pioneered against Germany, planned against France, and used against Germany is a testament to the stability of the British air defense problem, as discussed further below. It also speaks to a general consensus in Britain about how to solve it. Air defense of the Home Islands in 1940 was a structurally similar problem to that of 1918. Improvement in bomber technology made it more dangerous, however, by increasing the size of the area that needed to be defended and decreasing warning times. Indeed, the institutional stability of the internal solution was challenged at several points by technical and political developments. The institutionalization of British air defense in 1940 was thus not a given but an effortful achievement.

OVERCOMING TECHNOLOGICAL THREATS

The commander of the ADGB observed in 1933 that air defense "is not merely a problem of relative speeds but of absolute factors, some of which are invariable. The speed of the bombers is steadily increasing; on the other hand two other factors remain constant; firstly, the distance of London from the coast and secondly, the time that must elapse between the aircraft being seen by observers and the defending aeroplanes leaving the ground."[9] The military importance of geography is conditioned by technological possibility. The increasing speed and altitude of bombers compressed the time available between the detection of aircraft over the English Channel and fighter interception and increased the volume of the search area. One influential contemporary writer on airpower in the 1930s despaired, "To find the enemy in this immensity, in any but the clearest weather, may be as difficult as finding a needle in a haystack."[10] Acoustic mirrors could provide a twenty-mile offshore buffer given clear weather and no friendly fighters in the soundscape. Yet it was not until the advent of radar that the RAF was really able to address the so-called Channel Gap.[11]

The technical state of the art in the means of control places an upper bound on the potential of an information system to illuminate dead space and effectively tether friendly forces in space and time. The ADGB had an institutionalized internal solution, to use the terms of this book's framework, but the increasing speed and altitude of enemy aircraft were changing the external problem. The 84 percent interception rate achieved in 1928 dwindled to 40 percent in 1934. As a result of this alarming exercise, the Air Ministry created the Committee for the Scientific Survey of Air Defence, led by chief scientist Henry Tizard and known informally as the Tizard Committee, to investigate the potential of reflecting radio waves, as well as more crackpot schemes for radio death rays.[12]

Information practice theory expects adaptive practice to become more pronounced as the external problem destabilizes. Bottom-up exploitation, if supported and valued by organizational leadership, can lay the groundwork for top-down reform in the institution. A cycle of exploitation and reform enables an organization to reestablish the managed character of its information practice. In this case, British radar progressed from Watson-Watt's jury-rigged proof of concept in early 1935 to integrated operational readiness in five years. Because the ADGB had primed the RAF to think about air defense as an integrated "information system," work on integrating radar into the system proceeded simultaneously along with the development of the revolutionary sensor itself. Watson-Watt described his philosophy of experimental research and development as "the Cult of the Imperfect."[13] Fighter Command, according to David Zimmerman, "was simultaneously undergoing construction, upgrading, expansion, experimentation, training, operational testing and operations."[14] Cutting-edge science was conducted with an eye toward its practical application. Indeed, the very first document known to use the British term for radar—"RDF"—did not actually delve into the details of the novel device, but instead considered how it would be integrated into the operations of the ADGB successor organization (Fighter Command).[15] The Air Ministry's decision in September 1935 to halt the acoustic mirror project and go ahead with five Chain Home stations for £200,000 was a big gamble on an unproven technology that literally could not yet find its bearings.[16]

Methodical reforms to doctrine and technology enabled the RAF to adjust its institutional solutions to the changed structure of the military problem. Experiments at Biggin Hill simulated radar contacts with DF transmitters on bombers that flew progressively more complicated flight paths in order to exercise the Ops Room in constructing tracks and vectoring interceptors. These experiments encouraged Dowding to do away with the concentric rings that LADA had used to coordinate fighters and artillery and instead to vector fighters through the entire airspace. Watson-Watt's technical progress at Bawdsey picked up considerably in 1937 with three Chain Home prototypes up and running and a Filter Room to work out specialized plotting techniques and jamming countermeasures. Chain Home received its first major test in the summer 1938 exercise, which revealed problems in processing a high volume of unpredictable raids, too many radar echoes, difficulties classifying tracks as friend or foe, trouble with teleprinters, and oversaturation of the Filter Room when confronted by more than 141 tracks. With operation of the radar chain transferred to military commanders, scientists turned their attention to "operational research" (OR), a term coined by Watson-Watt.[17]

OVERCOMING SCIENTIFIC CONTROVERSY

The scientific achievement of radar was not a foregone conclusion. Tizard, a talented administrator with a scientific and military background who had credibility with the British establishment, played a key role in spanning different communities. Tizard engaged in what John Law calls "heterogeneous engineering" to procure funding, protect scientific research, and operationalize its results.[18] However, a quarrel between Tizard and Frederick Lindemann nearly derailed radar research. Lindemann was the eccentric science adviser to backbench MP Winston Churchill. He repeatedly questioned the efficacy of the new committee led by his rival Tizard in 1935. A grandstanding Churchill then publicly revealed the committee in order to criticize government inaction in defense. Lindemann was allowed to join the Tizard Committee, where he advocated his half-baked ideas about wire barriers suspended from balloons and curtains of parachuting bomblets. The air minister feared that radar would be compromised and abruptly dissolved the committee, only to reconstitute it days later without Lindemann. Churchill then threatened publicly to reveal the radar secret.

Tizard was deeply shaken by the sense that the government had lost confidence in him. Lindemann was finally muzzled when he was given an opportunity to test some aerial mine concepts, which failed, and Churchill was briefed on the full extent of radar, which greatly impressed him.[19]

The Lindemann controversy was only ever a minor threat. The ruckus calmed once Churchill was read in to the radar secret. The fact of the matter was that there was a growing consensus on the desirability of scientific air defense. Yet if Lindemann had succeeded in wresting control of the air defense research agenda from Tizard, it is possible that the development of radar, and thus the outcome of the Battle of Britain, might have been in doubt. Compromise of the secret of radar to the Germans through Churchill's antics could also have undermined the effectiveness of Chain Home. As discussed further below, however, German intelligence was not very well organized to exploit this counterfactual revelation. The press, remarkably, also agreed to keep the Chain Home system secret even though it was composed of thirty-six extremely conspicuous radar stations. It is perhaps still possible that Fighter Command could have run out the clock on Germany's planned invasion even without radar. Acoustic and visual sensors would have provided some warning, and erratic German targeting might have spared key infrastructure. Britain also probably would have been able to defeat a German amphibious invasion had the RAF been defeated.[20] The costs, however, would have been enormously greater. Protecting the radar secret helped the RAF to substitute information for mass, thereby improving both efficiency and effectiveness.

Lindemann later became Lord Cherwell, science adviser to Prime Minister Churchill. In 1942 Tizard challenged his advocacy for area bombing of Germany and his inflated bomb damage assessments, but this time Lindemann carried the day. Lindemann's last contribution to British air defense would be to doubt the existence of German V weapons, which then set London afire in 1944.

OVERCOMING POLITICAL RESISTANCE

A more serious political challenge to the viability of Fighter Command emerged from within the RAF itself. The RAF had a complicated attitude toward air defense. Even Hugh Dowding, the champion of air defense and the namesake of "the Dowding system," refused at first to fund radar when he first learned about Watson-Watt's idea in February 1935. Dowding asked Tizard to instead consider research to improve "offensive methods."[21] Hugh Trenchard, the chief of the Air Staff in the 1920s, summed up the prevailing offensive bias: "The aeroplane is not a defense against the aeroplane, but . . . as a weapon of attack, cannot be too highly estimated."[22] Yet Trenchard himself also chaired the Joint Committee on Anti-Aircraft Defence that launched the ADGB. In 1923 the RAF agreed to a Home

Defence Air Force plan that fielded twice as many fighters, and 20 percent fewer bombers, than Trenchard wanted.[23] A decade later, the RAF had a ratio of twelve fighters to fifteen bomber squadrons. Bomber Command resisted rigorous testing of its ideas and remained uninterested in OR (at least until 1941 when its loss rates became too terrible to ignore). As a result, Fighter Command did not have to compete for scientific expertise. John Ferris notes that the "RAF could have not done much better had it explicitly aimed for defence rather than offence."[24] While the RAF leadership was publicly enamored with strategic bombing, Fighter Command was quietly empowered to improve defenses.[25] There was a general consensus in Britain about the possibility and desirability of defense. For the same reason it lacked the dramatic fanfare of strategic bombing, the supporters of which had to drum up public fear and support.

The acrimony that emerged regarding British rearmament in 1937–38 highlighted tension between the strategic bombing and air defense programs. The Chamberlain government was terrified of German air power and outraged about weaknesses of Bomber Command that were only then coming to light. The cabinet had listened to years of the Air Staff's ideological attacks on air defense and pleas in favor of strategic bombing, yet Zimmerman finds that "lower level internal memoranda reveal near unanimity of opinion in the Air Ministry and in the cabinet about the influence of the new technology."[26] Air Minister Lord Swinton, a public advocate of buying more bombers, later observed, "As soon as radar was discovered and proved, the theory that the only defense was counter-attack was dead."[27] Bitter negotiations dragged out over the proper mix of fighters and bombers until the civilian cabinet pressed the RAF to adopt Scheme M, which had an emphasis on fighters, on November 7, 1938.[28] The doctrinal controversy in the RAF did indeed threaten Fighter Command by delaying fighter procurement, but it could have been a whole lot worse had the institutional foundations of air defense not already been laid in the years before Dowding and radar arrived on the scene.

Dowding did play an important role in ensuring that the fighters that the government procured would in fact be employed for the defense of the Home Islands. The RAF deployed fighters to support the British Army in France, but on the Continent it had no prepared bases or radar guidance. RAF losses were very high, with 477 fighters shot down in May and June. In a critical meeting, Dowding used data prepared by OR scientists to argue that fighter wastage in France was unsustainable and that no more aircraft should be diverted there. Dowding was willing to leave the British Expeditionary Force exposed to meet this goal, which generated some controversy. By summer 1940, however, the Luftwaffe had begun bombing British soil and the situation appeared too desperate to consider upending Fighter Command. The RAF information system would have been useless without any fighters to direct. Dowding thus sacrificed performance in the Battle of

France to ensure adequate supplies of fighters for the Battle of Britain. As he later reflected, "I am profoundly convinced that this was one of the great turning points of the war."[29]

## UNITY OF COMMAND

In the years before the war, as we have seen, consensus about war aims was threatened by episodes of political disagreement and scientific controversy. During the battle itself, however, Fighter Command enjoyed remarkable unity of command. This elusive principle of war was embodied in the hierarchical architecture of the RAF information system itself. Fighter Command was organized into groups to divide the computational burden of control into relatively independent subproblems. Keith Park's Eleventh Group was the most important given its location between London and the Channel Coast. Tenth Group in the Southwest and Twelfth Group in central England could be called on to provide support to Eleventh Group during large raids. Thirteenth Group covered everything in the North, which the Luftwaffe mostly ignored, making it ideal for training. Each group was further divided into sectors that managed fighter squadrons at dispersed airfields as well as other forms of defense (e.g., artillery, searchlights, and balloons). The apex of information processing was the Ops Room at Dowding's headquarters in Bentley Priory mansion on the northern outskirts of London, which gathered up radar and observer reports into a panoptic representation of the entire airspace.

Dowding understood his primary mission to be the survival of Fighter Command rather than the immediate protection of London. Destruction of the RAF would have exposed the Royal Navy to German air attack. Defeat of the navy in turn would have exposed England to amphibious invasion. To survive, therefore, Fighter Command had to avoid unnecessary fighter duels with Messerschmitt 109 fighters (Me-109s) and shoot down enough bombers to convince Germany to abandon its plans for invasion. The RAF faced a battle of attrition, so it had to be able to endure attacks indefinitely while inflicting losses at a higher rate than the enemy could compensate for. The RAF information system operationalized Dowding's goals by placing British interceptors in the path of only the most threatening German raids, which enabled the RAF to maximize the efficient use of its limited supplies of fighters and pilots.

Dowding was a micromanager, to be sure, but he was willing to delegate authority to group commanders, especially Keith Park, who had to control a very complex battle over southern England. Dowding ensured that Park received the planes, pilots, and information he needed. Senior RAF commanders, as well as radar scientists like Henry Tizard, all spent a lot of time circulating through the organization to socialize common goals and values before and during the battle. During the most critical weeks of the battle there was a high level of cooperative coordination throughout the information system. Unfortunately, the relationship between Park at Eleventh Group and Trafford Leigh-Mallory at Twelfth Group did begin to fray in mid-September because of a disagreement about fighter formation tactics (i.e., the "big wing" debate). Dowding did not intervene to referee. Fortunately, the Luftwaffe offensive was already sputtering out by then.

Finally, the institutionalization of Fighter Command was enhanced by Churchill's unwavering support for Dowding's mission. The prime minister encouraged the nation to resist Nazi aggression at all costs, which helped stiffen the resolve of RAF personnel during the battle. The focused and determined leadership of men like Churchill, Dowding, and Tizard articulated and reinforced the intentionality that guided Fighter Command. For all the structural advantages of defense, had Churchill not resolved to exploit them to the hilt or prevailed over the appeasement faction in government, the battle might never have been fought at all.[30]

## The External Problem

The nature of air defense in general and Britain's geographical position in particular conferred important structural advantages. There are things that Germany could have done to destabilize the problem, but instead the adversary inadvertently contributed to its stability.

### THE STABLE ONTOLOGY OF AIR DEFENSE

Clausewitz points out that "defense has a passive purpose: *preservation*; and attack a positive one: *conquest*."[31] Any Continental adversary, whether France or Germany, would have to send raids over the British Isles while Britain only had to deflect them. Germany also had to destroy Fighter Command by the end of September 1940 before bad weather in the Channel precluded an invasion attempt. Invasion, moreover, would have been difficult even if Germany had succeeded in defeating the RAF: the conquest of France had wrecked the Channel ports; the Royal Navy still commanded the seas; Britons would be defending their home soil. While the German Navy had prepared sketchy plans the year before, Admiral Erich Raeder remained pessimistic about the prospects for success. Fighter Command, conversely, did not have to totally destroy the Luftwaffe but merely had to survive long enough to persuade Hitler to relent. As Clausewitz observes, "Time which is allowed to pass unused accumulates to the credit of the defender. He reaps where he did not sow. Any omission of attack—whether from bad judgment, fear, or indolence—accrues to the defenders' benefit."[32]

Clausewitz further highlights "the advantage of position, which tends to favor the defense."[33] Bombers from the Continent had to cross the Channel and deliver their payloads directly over their targets in England, thereby exposing themselves to detection by an array of fixed sensors—radar and visual observers—on English soil. With the advantage of interior lines, Dowding could shift squadrons from different groups and sectors to meet a raid and could relocate to different airfields if attacked. The Luftwaffe largely avoided Thirteenth Group in Scotland because the longer distances and early warning times created suicidal conditions, which in turn created a training sanctuary for RAF replacement pilots. Because Fighter Command was defending at home, it could draw on the resources of civilian society for communications and emergency response services, scientists to adjust radar and tracking techniques, and fresh supplies of airplanes and pilots. Luftwaffe airfleets, meanwhile, operated from unfamiliar airfields spread out from France to Norway at the end of a long and inefficient logistics chain. British pilots who bailed out and were recovered could be sent back up to the fight, but every German pilot who bailed out over Britain was lost to the Luftwaffe.[34]

Island geography also affected the relative balance of information between the combatants. To calculate relative losses, the Germans relied mainly on the exaggerated claims of pilots who returned to France, but the British could count aircraft downed over England. The RAF could also recover both friendly and enemy pilots, receiving combat reports from the former and interrogation reports from the latter. The British received clear feedback when raids were not intercepted because bombs detonated on British soil. By contrast, German bomb damage assessment relied on irregular photographic reconnaissance flights and exaggerated pilot reporting. British interrogation of German pilots corroborated intercepts of Enigma communications and other sources, which improved confidence in British intelligence and betrayed the dysfunctional decision making of the enemy organization. British estimates of relative power thus became more accurate as the battle progressed, while German estimates became more fanciful. British commanders thus became more confident in their cautious and methodical strategy, while German commanders became more reckless and made more mistakes.

Fighter Command needed to worry about representing only a few types of things. Airplanes flew through a mostly homogenous sky, although weather could create noise. They obeyed rate-time-distance and other known physical constraints that were amenable to mathematical formalization. Aircraft didn't change their allegiance or turn into new types of things, although large formations might split into sections. Fighter Command did not have to attend to messier ontologies on the Continent, such as units on the ground and economic infrastructure, with the same precision as it did the air plot. It was thus feasible for Fighter Command to maintain a small set of discrete symbolic tokens in the space of a plotting room that stood in a structurally corresponding relationship with distant aircraft in an actual sky. This straightforward modeling relationship made it possible to use basic trigonometric concepts to triangulate enemy detections and vector friendly interceptors.

## A COOPERATIVE ENEMY

Germany made the hard problem of strategic bombing even harder for itself through its institutional choices. The Luftwaffe was designed from whole cloth to support the German Army in the field. It did not develop a four-engine heavy bomber, in part to produce higher numbers of aircraft to meet the Führer's political targets. It suffered from a disorganized aircraft industry, had no bombsight until 1939, and neglected essential support technology like communications. While the Stuka dive bomber inspired fear in its close air support role on the Continent, it suffered such grievous losses during the Battle of Britain that German commanders stopped sending it altogether. Germany's fast and agile Me-109s, meanwhile, had to sacrifice performance to escort bomber formations, which improved the prospects of British detection and interception. As Germany's newest armed service, commanded by Gestapo founder Hermann Göring, the Luftwaffe was also its most politicized. Göring's airfleet commanders, Albert Kesselring and Hugo Sperrle, developed independent campaign plans with little coordination with each other or Göring. Their target lists included RAF airfields as well as Royal Navy ports, merchant shipping, aircraft industry, and transportation infrastructure. These disparate lists reflected confusion over whether the bombing campaign was supposed to support the army's landing as part of Hitler's planned invasion (Operation Sea Lion), defeat Fighter Command or other RAF commands, destroy the aircraft industry, disrupt the whole economic system, lower British morale, or some combination thereof. The plan to defeat the RAF came down to little more than hope that British fighters would come up so that German aces could shoot them down. Decisive Luftwaffe victories in Poland and France—catching hundreds of enemy aircraft on the ground—gave German commanders little reason to reconsider their tactics and encouraged foolish optimism in considering war against Britain. On the other side, by contrast, the Phony War gave British commanders time and incentive to redouble their preparations. Sporadic German patrols and insignificant raids during the summer of 1940—a far cry from the blitzkrieg tactics that made the Luftwaffe famous—gave Fighter Command plenty of realistic practice and the opportunity to make vital adjustments.[35]

Intelligence is vital in an air campaign, but Luftwaffe intelligence had little prestige within the organization. It told its customers what they wanted to hear, chronically underestimated RAF strengths, and exaggerated its losses.[36] The assessment before the battle by the Luftwaffe chief of intelligence was wrong on almost every point:

> The command at high level is inflexible in its organization and strategy. As formations are rigidly attached to their home bases, command at medium level suffers mainly from operations being controlled in most cases by officers no longer accustomed to flying (station commanders). Command at low level is generally energetic, but lacks tactical skill. . . . [Thus] owing to the inadequate air defense of the island . . . the Luftwaffe, unlike the RAF, will be in a position in every respect to achieve a decisive effect.[37]

In a classic case of mirror imaging, Germany assumed before the battle that British radar would naturally use short wavelengths like German sets rather than the twelve-meter waves used by Chain Home. When the airship *Graf Zeppelin* flew over the Channel to investigate the possibility of British radar before the war, it failed to recognize Chain Home emissions even though the sky was full of the floodlight radiation it produced. The Germans dismissed the noise as an artifact of inefficient British communication, which made them doubly wrong. Luftwaffe signal troops did finally learn about Chain Home by late July, but the Germans did not attempt any deliberate jamming until late September, even as British operators learned to work through it.[38] Yet the Luftwaffe still failed to understand how British radar fit into Fighter Command's organizational system of group and sector headquarters, in part because the technically more sophisticated German radars had not yet been integrated into an equally sophisticated integrated air defense system. For Germany, bad institutions undermined good technologies. Even as it became apparent via German monitoring stations in France that the RAF was using radar and vectoring fighters, the Luftwaffe opted to interpret this as evidence of C2 rigidity. This can be understood as yet another instance of mirror imaging, because German pilots were themselves resistant to adopting ground-controlled interception tactics.[39] The Germans didn't understand the extent and flexibility of British ground control all the way up to visual engagement, so they underestimated the level of resistance that Luftwaffe raids would meet. Most importantly, they did not realize that their best hope of winning would have been to systematically attack British C2.[40]

The Luftwaffe did not attack C2, nor any other target set, coherently or repeatedly. German targeteers lacked the mathematical expertise of RAF scientists and did not understand the functions of Britain's many airfields, which were parceled out to Fighter, Bomber, and Coastal Commands. The Germans believed, wrongly, that all RAF Operations Rooms were in hardened bunkers, so they didn't focus on them. In fact, somewhat inexplicably, the British had left most of them exposed. Sector Ops Rooms were knocked offline only three times during the war, and the British quickly reconstituted them through repair or relocation. The Germans strafed observer stations occasionally, but not systematically. Chain Home stations were attacked only six times. Tall lattice radar towers and guy wires proved to be intimidating targets for German pilots, and the towers weathered blasts when hit. Luftwaffe damage assessments were thrown off when Chain Home sites appeared to keep working even after being bombed; the Germans didn't realize that the British moved in mobile replacements to fill the gap or broadcast dummy signals to cover downed equipment. The Luftwaffe concluded that the stations were too hard to damage, and of only marginal importance anyway.[41]

The case of Operational Trials Wing 210 shows what might have been possible had the Luftwaffe recognized the importance of the RAF's radar system. OTW-210 had a mix of Me-109 and Me-110 fighters trained to deliver reduced bombloads onto small targets at low altitude with high precision. On August 12, 1940, OTW-210 knocked out four Chain Home stations and threw tracking plots into temporary confusion. Stations at Dover, Pevensey, and Rye were back up within six hours, but Ventnor was down for three days without power. Had the Luftwaffe exploited these temporary windows by following up with bombing raids, it might have been harder for Fighter Command to adequately identify and intercept incoming raids. The same effect might have been achieved with commandos infiltrated on fast German S-boats; Chain Home stations, which were all on or near the coast, had no meaningful ground defenses. The Luftwaffe did not systematically develop or attempt tactics for the coordinated suppression of enemy air defenses. OTW-210 was the exception that proved the rule.[42]

The British Isles were a fortuitous defensive position. Fighter Command's external problem was further stabilized by the insulated nature of its adversary's information practice. While enemy attacks on friendly systems are typically a dangerous source of friction in war, the Luftwaffe's own misunderstanding and disorganization greatly attenuated this source. By comparison with the German bombing campaign, British air defense was a paragon of methodical preparation and enterprise integration.

## Information Practice

Fighter Command was able to manage the battle well because both its operational environment and its own organization were relatively well constrained. The external and internal poles of the representational relationship were solidly anchored. During the battle, therefore, Fighter Command was able to measure the state of the world, coordinate all of its subordinate parts, and enforce its commander's intentions without suffering major breakdowns in the process. This does not mean that the information system was free of friction, or that representation was a wholly top-down affair.

I will point out a number of minor glitches and bottom-up adaptations that affected Fighter Command's performance along the way. Yet centralized management with steady leadership was the dominant theme. Adaptive practice simply greased the gears of an institutionalized machine that was already in place.

The RAF appears to be the first organization to explicitly use the term "information system" to describe a rationalized assemblage of people and machines for data processing.[43] A diagram of the system in an Air Ministry pamphlet (figure 3.1) even resembles a software architecture schematic.[44] In this case, computational functions such as input, comparison, calculation, storage, and output were performed by an organization of human beings working with electromechanical instruments and paper charts. Intelligence came in through the Chain Home radar system, visual observers, and signals intelligence from Bletchley Park. They produced track data that filtered up through a hierarchical chain of command. A central Operations Room compiled an authoritative depiction of the air situation. Instructions flowed back down to Group and Sector Ops Rooms, which vectored fighters, directed antiaircraft artillery, launched barrage balloons, and alerted emergency responders. The time between raid detection and the scrambling of interceptors was usually only a couple of minutes. RAF officers, radar operators, and OR scientists constantly monitored and adjusted the workings of the system in near-real time to ensure that the system could carry out the commander's intent.

MEASUREMENT

Fighter Command tracked enemy and friendly aircraft via radar, observers, signals intelligence, and radio direction finding. To provide an illustration of distributed cognition in action, I will describe in some detail the ways in which radar technology mediated human experience, physical contacts were transformed into data, and the relationship between representations and actual aircraft was systematically constrained. I will provide more summary accounts of measurement via other types of sensors.

*Radar.* Radar operates on the physical principle that solid objects reflect electromagnetic radiation. Wavelengths longer than visible light, which have lower attenuation and absorption rates, can be detected at great distances and through weather. Basic radar concepts were widely understood before the war. Germany had even developed individual radar sets with better technical characteristics than British models. Chain Home, however, was the first radar to be incorporated into an operational system.

Chain Home detected the range, bearing, and height of aircraft out to eighty miles. Greater (or lesser) ranges were possible based on atmospheric

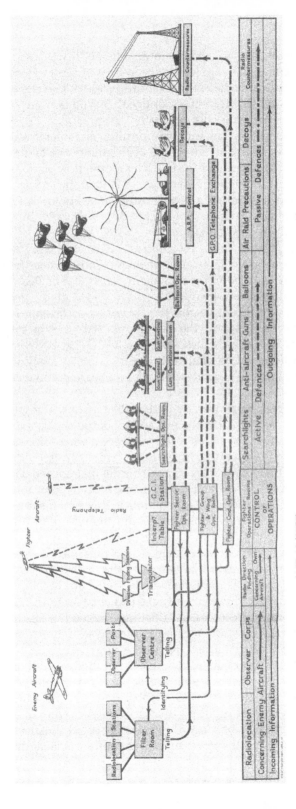

Figure 3.1. Air defense information system (AIR 10/3757)

conditions, target altitude, and operator skill. Yet it was unable to detect low-flying aircraft, so a supplementary system of Chain Home Low stations was built to detect the range and bearing, but not height, of aircraft as low as fifty feet out to fifteen miles, or one thousand feet out to thirty-four miles. Chain Home Low was originally designed as a Coastal Defense radar for ship detection, but scientists fortuitously found it was also effective against low-flying aircraft. At the beginning of the battle there were twenty-one Chain Home and thirty Chain Home Low stations providing coverage of most of the South and East of Britain, including most of the English Channel and some Luftwaffe airfields in France. A Chain Home station transmitted a pulsed "floodlight" of horizontally polarized long-wavelength (12-meter) radiation from wires tensioned between steel towers. Chain Home Low, by contrast, had steerable "searchlight" antennas and a shorter 1.5-meter wavelength. Radio echoes from aircraft, which generally traveled in the horizontal plane, illuminated the massive Chain Home towers and energized sets of crossed dipole antennas that were mounted at different heights on a set of wood towers. This spacing created a slight time delay in the energy reflected from the same aircraft, which enabled height estimates. Given only 120 feet between antennas, however, height estimates of aircraft dozens of miles away were notoriously less accurate than estimates of range and bearing, and tended toward underestimates.[45]

Human beings were physically part of the circuit that established and maintained contact with remote aircraft. All of the dipole antennas were wired to huts at the base of the receiving towers where signals were processed. Many of the operators in the huts were members of the Women's Auxiliary Air Force. They worked at a console that housed a cathode ray tube (CRT), a goniometer dial ("gonio"), and buttons controlling a motorized switchbox. The x-axis deflection of the CRT beam measured the time elapsed between the transmitted pulse and the receipt of the echo. The signal spike was displayed in units of miles, which was a convenience enabled by the constant speed of light. This represented the slant range to the aircraft. The y-axis deflection measured the ratio of the signal strength of two different antennas connected via the switchbox to the stator coils of the goniometer. With the gonio switched to connect to the North-South and East-West crossed dipoles, the operator rotated the dial to minimize y-deflection on the CRT, at which point the physical direction of the dial represented the horizontal bearing to the target. For weak signals the operator could maximize deflection, which required an additional ninety-degree correction for the horizontal measurement. Minimization of deflection provided the cleanest CRT display, but this measurement was ambiguous; the crossed dipoles would give the same measurement of an aircraft on a reciprocal bearing. The operator would thus have to toggle reflectors mounted on the tower one-quarter of a wavelength behind the antennas and observe whether the signal increased or decreased, which would indicate whether the aircraft was in front of or behind the tower, respectively.

This arrangement effectively established a continuous analog circuit between the remote aircraft and the human operator. Thus a physical circuit running through the radar tower and CRT connected the operator's eyes to the target in real time, enabling her to modulate the signal to amplify relevant information. The energized antenna, simultaneously, was coupled to the target via reflected energy. The deflection of the CRT beam and the angular setting of the gonio amplified specific features of the electrical activation of the antenna. When the gonio was switched to connect to dipoles mounted at 215 feet and at 95 feet on the tower, respectively, the operator could measure the vertical azimuth to the target by a similar process. The long afterglow display of the CRT enabled operators to distinguish the reflection signal from background noise and jamming because unsynchronized transient noise did not build up a signal. Experienced operators could count aircraft in formation by observing the beat rate of the echoes on the display. A button on the console could momentarily shorten the transmitted pulse from twenty to six microseconds, which improved range resolution. The operator also attempted to classify the contact as friendly or not based on a distinctive return pattern generated by a transmitter that most RAF airplanes carried, known as Identification Friend-or-Foe (IFF). Furthermore, the operator had to continually tune the apparatus as conditions changed. For example, she had to switch between measuring echoes from the main transmitter (centered at 5.2 degrees) and a gap-filler transmitter (centered at 5.9 degrees) if the aircraft climbed out of the main lobe of the former. The operators' engaged attention and active intervention enabled the instruments to represent features of the world on their displays.[46]

As the operator read off the settings of her instruments, other personnel in the hut logged the measurements in a book. Their representations progressively became more discrete and less attached to a particular physical locale as constraints propagated from device settings, through minds and sound waves, to pen on paper. Once contacts were logged and plotted, these representations were no longer part of a continuous physical circuit with the remote target; that is, the connection to the target became purely referential. With each subsequent step of the process, personnel re-represented the constraints of the radar contact into different formats and media. Once operators had obtained a complete set of measurements from a contact, they then performed trigonometric calculations using several correction parameters specific to the particular radar station, finally plotting the result on a map table. Extensive calibration of the station precomputed part of the representational problem prior to runtime. Once radar personnel in the station had calculated target coordinates, altitude, heading, and IFF reading, another operator known as a "teller" used a landline voice telephone to "tell on" these data to the Filter Room in Dowding's headquarters.

Later in the war, OR scientists automated some of these procedures. At first the calculations to convert radar ranges and angles into grid coordinates and altitudes were performed manually on paper and a plotting board. This tedious cognitive task was eventually simplified by the introduction of a purpose-built tabulating device and automated plotting table called "the fruit machine." OR scientists later connected radar console and fruit machine outputs directly to teletype printers in the Filter Room. Humans thus offloaded data-processing tasks onto electromechanical machines, but only after they were able to formally define their tasks as discrete operations. Human labor at that point shifted to programming and maintaining devices like the fruit machine. Many of the operators' situated, intuitive, interpretive performances were not so well defined, however, and thus radar stations continued to rely heavily on the connective tissue of human interpretation. A hybrid system of humans and machines thus propagated the constrained state of the real world through a series of representational constraints that maintained the validity of the reference to the actual aircraft. The complexity of implementation at even this primary stage of measurement suggests that popular distinctions between data, information, and knowledge are profoundly misleading. A lot of knowledge in the radar huts was needed to produce mere data for the Filter Room.[47]

*Visual Observation.* Once incoming aircraft crossed the coastline, a network of thirty thousand observers at one thousand observation posts tracked them visually. The Observer Corps had evolved from a similar organization employed in World War I. The enthusiastic members of this civilian volunteer organization endured countless hours of boredom and exposure on various English hilltops. Observers used binoculars to sight targets and ear horns to amplify the acoustic signature of passing aircraft. Neighboring observation posts could also provide cuing. Observers estimated bearing and altitude with a sextant-like instrument mounted at the center of a ten-mile radius map. They set a bar representing the height, rotated the instrument, and adjusted the sight angle to point at the aircraft, which in turn moved a pointer to a two-kilometer grid square on the map. The precision of the instrument and its structural correspondence to a physical relationship in the real world enabled observers to manage more relationships more quickly than they could with their minds alone. Yet results varied greatly by operator. Skilled use of the sextant transformed a hard cognitive task (trigonometric calculation) into simpler perceptual task (reading a map).

The observer post then logged its observation and called in to its reporting center via telephone to "tell on" the number of aircraft in a formation, their direction of flight and their height, and the map grid square read off the instrument. Many observers also attempted to identify the aircraft, but they tended to err on the side of designating them as hostile. For sound-only contacts, the observers passed along only the bearing and apparent direction. Over a million observation reports per day flowed via intermediate observer centers up to Fighter Command headquarters, with transmission times as fast as forty seconds. As with radar, the stereotyped and repetitive nature of these tasks made them ideally suited to mechanical assistance. Nevertheless, active and mindful human intervention was still needed to enable the observation post to behave as a black-box sensor from the perspective of upstream C2 nodes.[48] Knowledge is needed to create data as much as the reverse.

*Signals Intelligence.* Like radar and visual observation, radio listening stations exploited physical constraints in the world to transduce radio contact into detachable symbolic records. A further source of environmental structure in this case was the organizational routines of the Luftwaffe. Signals intelligence essentially made copies of the representations generated by the enemy's own representational practice. Intelligence practice can thus be understood as a second-order form of information practice. Bletchley Park, or "station X," broke high-grade encryption produced by the German Enigma machine, which became "Ultra" intelligence. Ultra was sanitized before being sent to Fighter Command in order to disguise its origin and thereby protect the Allies' most secret source of intelligence. While Ultra provided some information on the Luftwaffe's order of battle and readiness, it proved to be of little tactical use because it took too much time to decrypt and figure out who had the need to know.[49] By contrast, the RAF "Y" service intercepted and processed low-grade German ciphers and clear-voice transmissions.[50] Y was more tactically useful because it had less onerous controls on dissemination. Y was reported from Cheadle station to Fighter Command HQ and often provided warning that a raid was being launched, and sometimes was even helpful for vectoring interceptors. The British also exploited a German navigational beacon called Knickebein (bent leg), which guided Luftwaffe bombers to within five hundred feet of their target. German pilots soon lost confidence in it as a result of British spoofing and jamming.[51]

Signals intelligence was structured as much by the target's bureaucracy as by the physics of interception. To the degree that the Luftwaffe's control loops were stable and thus meaningful for German airmen, the British could incorporate the stability of enemy routines into the British control loops that exploited them. For example, Y benefited from de facto cooperation from the target because Luftwaffe security discipline was especially lax. Before the war, some units actually painted their names and radio codes on their aircraft, which was observed and collected by British air intelligence. When the Luftwaffe finally repainted and changed codes at the start of the war, British traffic analysis was easily able to recover the organization. Yet for the same reason, the enabling structure of target signals was more delicate than physical entities like aircraft. Intelligence successes had to be kept secret lest compromise prompt the enemy to alter the external structure on which the internal intelligence solution relied. So long as an intelligence collector wants to keep collecting in the future, it must exercise caution and restraint in how that intelligence is used.

*Tracking Friendly Fighters.* Fighter Command needed to detect and track not only enemy raids but also its own fighters. This function is sometimes described today as "blue force tracking." RAF aircraft were fitted with high-frequency radio sets that automatically transmitted a 1,000 Hz note every minute, earning it the name "pip-squeak." This signal was detected by three antenna towers at a direction finding (DF) station located in each sector. Each DF station had a plotting table and a map with elastic strings attached to the corresponding location of each tower on the map. An operator for each antenna measured the bearing to the aircraft by tuning a goniometer to the strongest pip-squeak and then pulling the string along the map, guided by protractor marks along the edge of the table. A fourth operator plotted the intersection of the three strings as the triangulated position of the fighter on the map. The triangular fix was known as a "cocked hat." The process took trained operators only fourteen seconds, so it was possible for one DF station to track four aircraft at once with only one minute latency. Generally this meant that DF stations tracked four squadrons at once, on the assumption that each squadron flew in formation.[52]

The DF station thus acted as another data-generating black box from the perspective of higher echelons, much like the other sources of measurement described above. While the three signals from the DF antennas contained all the information necessary to fix the position of the squawking aircraft, this solution was not actually available to the organization until each airman physically moved representations of this bearing onto the map that satisfied specific constraints. The constrained propagation of multiple representations onto the structured space of the map in the Sector Ops Center (which itself was precomputed by cartographers in advance) effectively constructed a new fused representation in the form of plotted coordinates. The systematic arrangement of this compound inscription guaranteed that the new representation stood in a reliably constrained relationship to the state of the world at the time that the bearings were read. Any unconstrained degrees of freedom in these transformations—distortions in the signal, friction in the goniometer, cartographic errors, operator incompetence, squadrons not in formation, too much area in the "cocked hat," and so on—would undercut that guarantee.

COORDINATION

The Fighter Command Ops Room was perched at the apex of multiple streams of data flowing up and down the organization. It consolidated the single authoritative track plot from all sources of information—what today would be called a "common operational picture." Women's Auxiliary Air Force members wearing telephone headsets used croupier rakes to move wooden tokens around on a large map table that represented tracks of enemy raids and friendly squadrons. A slotted blackboard called the "total-isator" or "tote board" displayed squadron availability and assignments to particular raids. This enabled operators to keep the plotting table clear except for active raid and interceptor tracks. Officers on a raised balcony had a panoptic view of the island's airspace and telephone links to Group and Sector Ops Rooms.[53] The gendered aspect of this arrangement is striking. The women performed all the hidden work to maintain the integrity of the air plot, while the men in command gazed down from their balcony to gain a sense of mastery over the air situation.[54]

The Ops Room was one of the first true "all source fusion" centers, which combined multiple types of intelligence reporting into an authoritative graphical assessment. The Admiralty created something similar for tracking submarines. Yet sensor data from radar and observers did not flow directly into the Ops Room. Its authoritative representation relied on a series of intermediate steps that turned messy records into cleaner records that could be combined into a synoptic picture of the airspace. Standardized doctrine and representational protocols enabled the efficient propagation of representational constraints throughout the system. Even so, operators still had to compensate for some information friction that emerged in real time.

*Information Friction in the Filter Room.*   Chain Home radar stations reported to a central Filter Room in Bentley Priory. A "teller" at each station reported by telephone to a "filterer" in the Filter Room who plotted every report as passed. Reports from individual stations were unreliable due to the vagaries of station calibrations, operator skills, and atmospheric conditions. Uncertainty persisted about the numbers and types of incoming aircraft. The search-light design of Chain Home guaranteed a lot of overlap among the stations, which were sited about twenty miles apart. This enabled triangulation in principle, but different stations varied in whether and how they reported the same contact. The Filter Room acted as a buffer between the particular mess of radar reporting and the cleaner abstraction of the Ops Room plot.

As filterers plotted reports from individual stations, a messy spider web of plots emerged. Many of them traced a zigzag path because of errors reported in each successive target position and heading. Chain Home height measurements were especially unreliable. Filterers discarded the plots they believed to be erroneous and joined and straightened those they believed to be reliable. This judgment required considerable technical knowledge of how radar worked, familiarity with the idiosyncrasies of each individual station, and experience filtering radar tracks in combat conditions. RAF officers with more advanced education fulfilled the job. They produced an officially recognized track marked by a wooden token placed on the map. Filterers then gave the track a number and assigned an *X* for unidentified, *H* for hostile, and *F* for friendly. The judgment of attribution depended on reports gathered from IFF transponders, the origin and course of the track, and data on the locations of friendly squadrons that were passed over from the Ops Room next door. Filterers passed the official track to the headquarters Ops Room and "told on" its location to Group Ops Rooms by phone.[55]

Figure 3.2 is an illustration of a Filter Room plot from an Air Ministry pamphlet.[56] In this example there are two radar stations that have each made two reports, represented by the circular tokens. The filterer has determined that they correspond to a single incoming hostile raid of twelve aircraft at twenty thousand feet and assigned track number 314. Figure 3.3 illustrates the same track (H314) on the "raid information display" of the Ops Room plot (along with the track of friendly squadron 605 and an unidentified track X316, to which a flight of friendly interceptors 266 has been assigned). Note that the arrows depicting the progress of the track in figure 3.3 are in different colors; as discussed below, Fighter Command plots used colored tokens synchronized with a clock to represent the time latency of plot data. The shape of the arrows indicates whether the contact is tracked by radar over water or by visual observers over land. Much of the Filter Room's work involved in creating the track (figure 3.2) has been

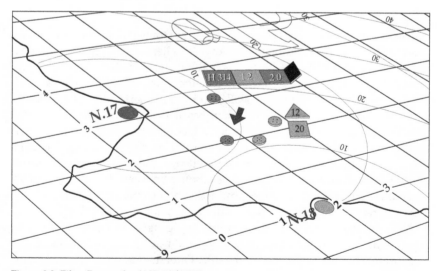

Figure 3.2. Filter Room plot (AIR 10/3758)

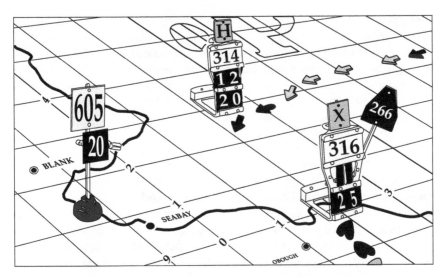

Figure 3.3. Ops Room plot (AIR 10/3760)

filtered out of the cleaner Ops Room view of the world (figure 3.3). Any equipment glitches or operator gaffes might introduce noise into the process, but such equivocation is not visible in the discrete symbolic tracks that emerged from the Filter Room.

The bizarre "Battle of Barking Creek" illustrates the consequences of information friction. On September 6, 1939, a friendly aircraft approaching from the direction of France had failed to file a flight plan. Chain Home operators did not get an IFF reading, so they classified it as hostile and Hurricanes were sent to intercept. Inexperienced operators confused the reciprocal bearing of the outgoing fighters as IFF failed again, so they classified them as a second incoming raid. Another squadron of Hurricanes were scrambled, and these were also classified as incoming hostiles, compounded by observer post misidentifications. After several iterations, there were eventually twenty unidentified X tracks in the Filter Room, prompting the launch of a Spitfire squadron, which eventually opened fire on what the pilots thought were Me-109s. The Spitfires shot down two Hurricanes, and antiaircraft artillery downed one Spitfire. Personnel perceived the breakdown only when they were suddenly confronted with graphic evidence that was at odds with their symbolic track data. OR scientists immediately set about debugging the procedures and equipment to repair and improve performance.[57]

Filtering became more difficult with the rollout of Chain Home Low in early 1940, because it had different imaging characteristics than Chain Home. The rapid expansion of Chain Home from fifty-four to seventy-six stations between July 1 and September 30 resulted in more operators with uneven skills. When IFF systems malfunctioned or operators couldn't read them, filterers usually assumed they were hostile. Counting the numbers of aircraft in a raid was one of the hardest measurements for radar operators. Radar operators often assumed there was only one aircraft when they saw multiple height returns, which made sense in British training runs, but the Luftwaffe used the tactic of vertically stacking fighters and bombers. Chain Home tended to underestimate height, so controllers and pilots each tended to add a few thousand feet to the estimate they received to improve their interception. Once radar and operator skill had improved, however, this habit had the effect of placing fighters too high to do any good. Dowding noted that "the training of the technical personnel and the maintenance of the elaborate scientific apparatus presented great difficulties."[58] No Chain Home station conformed to a standard design, and each required complicated calibration based on local landforms, weather patterns, and the irregular vintage of its equipment. Continual upgrades took individual stations offline, and local efforts to improve station performance sometimes interfered with other stations. For example, each station was allocated a specific area on the 50 Hz waveform and was not allowed to deviate without headquarters' permission, but they often did in order to deal with local ionospheric scattering, which inadvertently degraded the signal quality of neighboring stations.[59]

Radar filterers and plotters were sometimes unable to keep up with the high volume of contact reports. The single air picture sometimes had missing and bad tracks, but there was no way to know from the vantage point of the Ops Room gallery. Sometimes the radar stations were ordered to stop reporting because the Filter Room couldn't deal with what it already had. The Air Staff pressured Dowding to decentralize by also establishing Filter Rooms at the group level, as well as providing greater autonomy to sector commanders. Ever the micromanager, Dowding sought to maintain control of a single air picture at Bentley Priory, lest squadrons scramble prematurely before it could be verified that a track was indeed a hostile raid bound for their sector. Given his focus on husbanding fighters, this may have been reasonable. Dowding's concession to the Air Staff in January 1940 was to move his Filter Room to an underground bunker to protect it and to "tell" tracks to the groups directly from the Filter Room. The saturation problem did not go away, however, and in fact became worse once the real battle got rolling. On October 1, the Air Staff directly ordered Dowding to decentralize, and so a Filter Room was built at each of the group HQs to process radar data from stations in their territories.[60]

Despite the continual glitches, British C2 continued to perform adequately enough. Continual tinkering made this possible. For instance, the X designation for unidentified tracks had been introduced to avoid a confusing and potentially dangerous tendency of filterers to default to hostile. The invention of "macroscopic reporting" enabled the Filter Room to deal with the problems of large formations. Rather than trying to track every single aircraft, which was becoming impossible because of saturation, the raid track would include an estimate of the number of aircraft in the formation. It would then be up to the intercepting squadrons to disaggregate this symbol. Another process improvement was the creation of a "Lost Property Office" to maintain a separate plot of tracks that had dropped off the main Ops Room plot. The main plot never displayed any information older than fifteen minutes according to the color-coded clock, but sometimes raids happened to elude detection more than once or twice. Tracks were most likely to be dropped inadvertently when they crossed sector boundaries or were handed off from radar to visual observers, a problem of crossing the "seams" of an "area of operations" that continues to bedevil C2 today. The Lost Property Office was a messy plot, to be sure, but it enabled the reacquisition of some feral tracks in sufficient time to vector fighters to intercept their real-world correlates.[61] A continuous stream of bottom-up adaptations thus helped to restabilize a system that was constantly subject to wartime destabilization.

*Friction in the Observer Corps.*   Each sector had Observer Centers, which were basically Filter Rooms for the Observer Posts in the sector. Within an Observer Center, via the Ops Room, a "Sea Plotter" accepted radar tracks that were approaching the coast, since these would have to be tracked over land by the Observer Corps.[62] Twelve plotters sat around a map table, each connected via a head-and-breast telephone set to a couple of remote observation posts. They alerted the posts about tracks arriving within their field of view (and hearing). As the Observer Center received reports from its posts, operators placed markers on the map that pointed in the direction of flight and displayed the number of aircraft in red, the track number assigned by the Filter Room (passed via the Ops Room) in white, and the altitude in thousands of feet (or "angels") in blue. They marked flights missed by radar with a special code for their territory. When new reports came in they moved the markers in five-minute intervals, placing an arrow marker at the previous location to "remember" its path. A floor supervisor checked the plots, and tellers on a raised platform phoned information to group and sector Ops Rooms, while a recorder traced out more permanent records. The track thus became visible only as inscriptions percolated in from visual observers to a central location where they could be combined with other observations.[63]

The Observer Corps channel was noisier than the radar channel. Visual observation was unreliable due to low clouds, rain, or darkness. Height estimation was an acquired skill that few mastered. Height errors translated into position errors through the observer's sextant, which in turn created zigzagging and misleading tracks on observer center plots. Posts were easily overwhelmed by too many aircraft, and observer center plots became cluttered. Many aircraft were misidentified as hostile because the Observer

Corps was worried that otherwise their tracks would be ignored. There were many uncontrolled degrees of freedom in the creation and propagation of inscriptions, so the correlation of internal and external structure tended to drift apart. Confronted with too many bad observer tracks cluttering up the Ops Room plot early in the war, Dowding ordered that only radar could be used to identify tracks, using the Observer Corps only to follow the track inland. This self-imposed insulation resulted in some raids being missed by radar but not by observers, yet never tracked by HQ.[64]

*Standardization and Adjustment.* Fighter Command relied on standardized formats to pass data between the Filter Room, Observer Centers, headquarters, and the Group and Sector Ops Rooms. All stations used the same gridded map scheme so they had to pass only a short code for a given two-kilometer square. They all had clocks on the wall that color-coded each five-minute segment. When plotters updated a track, they looked to the clock and chose a token that matched the current color. Only three colors were allowed on the air picture at any one time to ensure that the information displayed was never older than fifteen minutes. Track markers were standardized, with different-shaped tokens for different types of aircraft. Telling procedures were strictly specified with a standard reporting order and pace of speech. Ops Rooms used this data to measure the distance between hostile and friendly aircraft tracks, and also to compare their actual and tolerable wastage rates. The panoptic layout of the Ops Room—the air situation and the tote board—was designed for commanders to be able to visually make these measurements from their seats in the gallery and to make decisions about which gaps to close by issuing orders.[65]

All of this standardization reflected advance agreement among system designers about what categories and properties of things would matter at runtime—that is, in battle. The resulting representations amplified particular aspects of the world (raids and interceptor squadrons composed of multiple aircraft moving along a given vector and altitude, etc.) and kept anything else off the board. Standardization was essentially an advance agreement on how exactly each transformation of one type of representation into another would be constrained so as to preserve referential integrity to the aspects that mattered. This enabled a great deal of speed and flexibility in passing details on changing tracks across all the different C2 nodes. It also enabled commanders to have some confidence that plots were coordinated with what was actually happening in the sky. The price was that the Ops Room was not designed to make anything else legible—for instance, to pick targets for Bomber Command, assess the damage in London, or track civilian passenger flights to their destinations. The shadow of the world cast on Dowding's cave was a carefully curated world of entities that he cared to act on. So long as Fighter Command's defensive strategy fit the problem that the Luftwaffe posed for it, this constrained form of experience enabled the organization to fight effectively.

On the whole, RAF information practice could be characterized as well managed. Yet the exceptions are revealing, demonstrating how ideal types are usually mixed in real systems. In this case, adaptive practice made many beneficial adjustments at the margins to keep the Dowding system running like a well-oiled machine. Just as interpretive work within the radar stations and Filter Room produced a clear plot for the Ops Room that covered up the friction in its construction, continuous bottom-up adjustments enabled Dowding's top-down management to fight the war. Fighter Command can be likened to a man climbing a tall ladder, making hundreds of microadjustments with his muscles to keep his balance at every step, yet repeating the same macromovements over and over again. The importance of the guiding role of intentionality in any control system—whether the goal is climbing a ladder or preserving the RAF—cannot be emphasized enough. A sense of common purpose throughout Fighter Command guided idiosyncratic adjustments by personnel into a coherent and deliberate fighting effort.

While Fighter Command was a hierarchical organization, its peculiar history of pulling radar up by its bootstraps created a legitimate space for adaptive practice by OR scientists and operators. Dowding empowered them to make adjustments to the C2 system that would more efficiently realize his command intentions. The mechanical tabulator known as the fruit machine emerged only in November 1939, and numerous other contraptions were improvised at the last minute.[66] The civilian and military personnel who were building Fighter Command's system worked side by side with, and sometimes were the same as, the people who were operating it. Because the barriers between operational and technical expertise were remarkably low, technical adjustments to operations were less likely to break the system. Fighter Command functioned, in effect, as both a warfighting command and a scientific laboratory. The center of scientific research at Bawdsey was itself an operational Chain Home station. The stations that had a scientist in residence were always the top performers in the chain. OR scientists studied filtering processes and analyzed every raid that escaped radar detection, discovering various problems and recommending solutions to improve filtering, plotting, and telling. Military officers sought scientific advice on force-employment problems, operational planning, staff information processing, and interception tactics. A 1948 study of Fighter Command estimated that radar improved the probability of interception tenfold and OR contributions doubled this probability again. The radar scientists who began by experimenting with a new technology in the Orfordness laboratory ended up reprogramming a large-scale sociotechnical machine while it was running in the middle of a war.[67]

ENFORCEMENT

The Ops Room plot provided commanders with a sense of mastery, but to actually influence the battle they also had to be able to direct friendly forces. The HQ Ops Room "told on" its air picture to the groups, and the groups passed data to the sectors. The groups and sectors each had Ops Rooms organized similarly, with a plot, "tote," and radios. Just as measurement circuits moving up to HQ involved several loops within loops and intermediate integration centers, so too did enforcement circuits. Yet in this case inscriptions moved in the opposite direction, from greater symbolic abstraction to more physical particularity. While a Sector Ops Room resembled the HQ Ops Room at a smaller scale, it actually contained more fine-grained information. For example, in addition to the tote board, the Sector Ops Room also had electrical "state panels" with four-foot-high boxes for each squadron displaying six colored lights representing different states of readiness. As representations percolated down the chain of command, they became more grounded in particular situations, locales, and tactical missions.[68]

*Anticipating Contact.* The Sector Ops Rooms had to compensate for the drifting apart of the state of the plot and the state of the world. According to an RAF pamphlet,

> It takes time to read, report, filter, tell and plot the plan position of an aircraft. During that time the aircraft flies on. A Plotting Counter indicates, therefore, the position of the aircraft when it was read, *not* when it was plotted . . . ; in addition, human inaccuracies in positioning the counters must be allowed for. In practice it is found (a) that the point on a plot is on the average 1½ miles from the actual track that the aircraft has followed; (b) that the aircraft is probably—allowing that it does not continue a straight course—within a lemon-shaped space 16 miles long and 12 miles wide.[69]

To estimate future locations for interception, the sectors extended out past plots through dead reckoning. The trigonometric problem of calculating an intercept path for two nonmaneuvering aircraft is a complicated two-vector problem. Sector Ops Room personnel used a heuristic to simplify the problem. The "Principle of Equal Angles" was affectionately known as the "Tizzy Angle," after Tizard who suggested it. A controller would plot (or simply visualize if he was experienced enough) an isosceles triangle with the baseline linking the interceptor (located by DF plot) and the hostile raid (located by radar or observer plot), lining up the raid's track as one of the triangle's edges. By extending the interception course out at the same angle to the baseline, the controller could provide a heading to the apex of the triangle where the interception should occur. Since fighters were at least as fast as the bombers, they could get there first and take a favorable attack position above their targets. If another fix came in the meantime that indicated that the raid had changed course, the controller simply visualized a new triangle and revectored the interceptors. The Tizzy Angle is yet another example, along with the fruit machine and the observer's sextant, of turning hard cognitive problems into simpler visual problems. These simplifying representations resided partly in the controller's brain, especially if he dispensed with actually plotting triangles. The operation worked because the constrained transformation of one type of representation (the current track plot) into another (the new heading) recapitulated the physical constraints on the trajectory of an aircraft flying through the sky. This mapping failed if the target maneuvered, but a structured feedback loop (running through the Chain Home huts and the Filter Room) enabled the controllers to correct their mapping.[70]

The sector controller maintained authority over the squadron, vectoring its bearing and altitude, until the flight leader gained visual contact with the raid. Prior to visual contact, the sector HQ actually had a better idea of exactly where the fighters were because of the DF plot (so long as it did not have to control more than four squadrons). DF tracking thus offloaded some of the cognitive load of pilot navigation onto an external organization. This allowed pilots to focus on other things, such as the combat merge. Controllers, furthermore, were pilots themselves, which helped them to create trust over the radio and detect nuances of jargon and tone of voice, enabling them to better simulate in their minds what the fighters were experiencing. The common experience shared between the pilot and the controller helped to stabilize this vital information channel. Once the interceptors had visual contact, the controller then transitioned to a monitoring role.[71]

In order to keep its radios up and running, the RAF had launched a selective draft of civilian ham radio operators, of which there were many in Britain, to be radar technicians and DF operators. Already enthusiastic about tinkering with fickle radio technology in their free time, they were able to continually tweak the radar and radio sets that they worked on in an official capacity. Civilian skills and commercial technologies promoted adaptive practice, foreshadowing the explosion of military user innovation in the digital era.[72]

The sectors' primary preoccupation was fighters, but they also activated supporting defenses. Searchlights lit the night sky for fighters and artillery. Artillery barrages placed flak in the path of raids prior to the arrival of interceptors, hence the need for coordination of DF plotting and synchronization by the sector controller to avoid fratricide. Listening stations helped to aim searchlights and artillery. Balloon barrages lifted to force bombers up to higher altitudes in order to interfere with their targeting accuracy. Dowding thought balloons were of limited tactical utility, but speculated that "they exercise a very salutary moral effect upon the

Germans." Fighter Command also used decoy airplanes, hangars, and factories to further confuse German targeting. Lastly, sectors alerted (and de-alerted) civilian emergency services to prepare to deal with the consequences of bombing. All of these tactical functions had their own miniature Ops Rooms to manage their particular translation of symbols into effects.[73]

*Verification in Combat.* The control system could place fighters in a position to intercept a raid, but it was up to squadrons in the air to finally close the loop. The tactical movements involved in intercepting German bombers and dogfighting with their escorts were only faintly legible to radio controllers and DF plotters on the ground. Aerial combat itself was a fast-moving and highly localized control problem. By pushing pilots closer into an uncertain problem, Fighter Command effectively decomposed combat into control problems of a much smaller scale, in which units could connect to the changing structure of the problem through organic sensors and controls. Pilots' eyes and ears acted as measurement sensors, their brains coordinated incoming data, and the aircraft's control surfaces and tactical radios enforced their intentions. The process of ever more particularized enforcement and localized measurement culminated in the merger of distributed cognition in the cockpit with the physical contact of gunfire. The squadron thereby gained some traction on a highly dynamic situation.

As in the Ops Rooms, prior design-time standardization enabled speed, flexibility, and accuracy in runtime combat. Pilots used radio brevity codes to refer to "bandits" at so many "angels" (thousands of feet) and announced "tallyho" at the merge or "pancake" to return to base. Pilot jargon, keyed to standardized tactical routines linked to recurrent patterns of air combat, enabled the squadrons to quickly combine tactical moves in novel ways as the fight developed. Prior coordination could also standardize new interpretations that were intended to deceive German intelligence. For example, pilots might agree that "angels eighteen" actually meant "go to twenty-one thousand feet" in order to lure a listening enemy to a disadvantageous lower relative altitude where the fighter could "bounce" its prey. Fighters were also enabled by a prior logistics effort; while this was not strictly precomputational, it was critical for maintaining the machinery of representation. Dowding noted that "refueling, rearming, engine checking, including oil and glycol coolant, replacing oxygen cylinders, and testing the [radiotelephone] set would all go on simultaneously," realizing an eight-to-ten-minute turnaround day and night, out in the open to avoid Luftwaffe attacks on dispersed hangars in blackout conditions. A great deal of prior preparation helped to structure the rapid interactions of combat. Fighter Command thus used time to structure space.[74]

*Maintaining Contact through Feedback.* The feedback or remeasurement phase of Fighter Command's control loop was already underway during an engagement. Controllers monitored the radio in case it became necessary to order an end to the attack, send in reinforcements, or prepare for German raids that made it through the screen of interceptors. Later when pilots returned to base, they filled out combat reports describing friendly and enemy losses (usually exaggerated), the effectiveness of searchlights and artillery in illuminating or breaking up a raid, and so forth. Fighter-mounted movie cameras provided a gun's-eye view of the interception, which was invaluable for debriefing and training aircrew. Every contact with the enemy provided an opportunity for further measurement and further tweaking of the architecture of control.[75]

## Military Performance

Dowding's after-action report from August 1941 divides the battle into four phases. It began indistinctly in July 1940 as Germany intensified sporadic raids and probing patrols along the Channel coast and ports. The RAF did not rise in force to take the bait because Dowding prioritized the conservation of his limited fighter force. The second and most dangerous phase began in early August with a concerted Luftwaffe effort to destroy Fighter Command by targeting aerodromes and trying to force more British fighters into the air. By the first week of September the Germans believed, based on exaggerated pilot reporting, that the RAF had been defeated. They launched a coercive bombing campaign against London proper in a third phase known as the Blitz. This decision inadvertently relieved pressure against RAF infrastructure, freeing up British fighters to punish the large formations of bombers sent over London operating at the range limit of their Me-109 escorts. The Luftwaffe suffered a sobering 25 percent loss rate during its last major daylight raid on September 15, commemorated as Battle of Britain Day. Two days later, having failed to achieve air superiority, Hitler canceled Operation Sea Lion. The fourth phase was the Night Blitz on London, which continued into October and then petered out through the winter as Hitler's attention turned to the Soviet Union. The Night Blitz aimed to apply political pressure on an isolated Britain rather than destroy the RAF or prepare for invasion. Night bombing created a real challenge for Britain by changing the information problem the RAF system was designed to solve. Fighter Command put in its most dismal performance during this final phase, but fortunately Germany sent only small, poorly targeted raids during this phase. Luftwaffe leaders were only too happy to get back into their comfort zone by supporting the Wehrmacht in a major ground offensive for Operation Barbarossa in the East. Dowding had surmised correctly that the RAF could win by not losing.[76]

EXCHANGE RATIOS

The outcome of the battle turned on the cold accounting of fighter casualty and regeneration rates. Fighter Command needed its fighters to defend Britain from German bombers and invasion. Germany needed its fighters to protect its bombers. Germany thus had to shoot down British fighters faster than they could be replaced, and it had to do so within a short period of two months if invasion was to be viable. Because the RAF was able to husband its fighters and still damage the Luftwaffe, however, Hitler failed to destroy the RAF and opted to cancel Sea Lion.

Figure 3.4 plots the average number of serviceable aircraft available throughout the battle. Each line charts the base inventory, plus replacements generated, less losses to accidents and enemy fire. Britain began with only 752 serviceable fighters on July 10, 1940. The Luftwaffe had 2,462 serviceable aircraft, composed of 1,283 bombers and 1,179 fighters. This appears to be a major advantage. However, the twin-engine Me-110 proved vulnerable to more maneuverable Spitfires and Hurricanes. Thus the most relevant number is the 899 serviceable single-engine Me-109s available to engage British fighters. Furthermore, Germany held back 10–20 percent of its fighters to defend Germany against RAF Bomber Command, and some additional fighters were retained to defend airfields in France. This left approximately 700 Me-109s available to escort bombers to Britain. This number is more comparable to the 752 fighters available to the RAF. Assuming that 2 Me-109s were needed to escort each bomber, moreover, the Germans were constrained to using only 350 of their 1,283 bombers. The Luftwaffe was thus simply not big enough to deliver enough sustained and punishing attacks to create the knockout blow it desperately needed, *unless the RAF was reckless with its fighters*. As it turned out, the inventory of RAF fighters remained steady at around 800 throughout the battle. Luftwaffe totals, meanwhile, declined after major combat operations began in mid-August. Even worse, the Luftwaffe lost fighters three times faster than bombers. From August 12 to September 30, the Luftwaffe lost 1,760 aircraft of all types, but it replaced only 855. It lost 565 Me-109s but only produced 350—a 62 percent replacement rate. By contrast, the RAF lost 1,240 fighters, yet it also received 1,330 new ones from production lines—a replacement rate greater than 100 percent. Britain routinely produced over 400 new fighters each month, while Germany usually produced less than 200.[77] Britain's advantages in mass were thus not inconsiderable, although it still needed information to know how to use them.

In order to impose steady attrition and overcome British production rates, the Luftwaffe would have had to routinely achieve exchange ratios better than four to one. On average nearly every German pilot would have had to be an ace. Yet on only one day, September 28, did the Luftwaffe achieve this ratio (sixteen to four). On only six days did it ever achieve

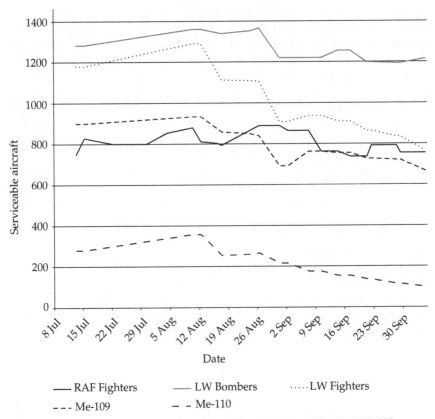

Figure 3.4. Serviceable aircraft. Data from Prior, *When Britain Saved the West*, 228, 230.

something better than parity. Two of these were in the ambiguous first phase of the battle (when the Luftwaffe was inadvertently training the RAF) and one was in the fourth phase (after the RAF had basically won). On all but three days in the critical middle phases, therefore, the Luftwaffe was always losing. By contrast, the RAF achieved equal or better exchange ratios on nine of the ten most intensive days of combat, inflicting higher absolute losses than on the Luftwaffe's best days and routinely better relative losses.[78]

While a lost aircraft could be replaced with a new one, experienced pilots were a rarer commodity. At the beginning of the battle, pilot experience was one of the few advantages the Luftwaffe had over the RAF. Yet experienced pilots who bailed out over Britain were lost to the Luftwaffe forever. Britain, by contrast, could recover its survivors and send them back up. Luftwaffe pilots, moreover, remained in the front lines on a more or less permanent basis, while the British implemented a system of rotation to give pilots a rest. Both sides actively trained replacements throughout the battle, but Britain was more successful in generating higher numbers of pilots. The RAF had 1,259 fighter pilots available on July 8 and 1,581 on September 29— a 15 percent increase. The Luftwaffe had 1,126 Me-109 pilots in July but only 990 in September—a 12 percent decrease. Britain lost 537 pilots killed, compared to Germany's losses of 2,262 aircrew (in part because bombers carried multiple crew members). Half of German casualties after September were caused by weather and ice rather than the RAF, which is another environmental condition that improved the British defensive advantage.[79]

Luftwaffe wastage of men and material was simply unsustainable. German commanders had anticipated losing 10 percent of their aircraft *per month*, which would have been affordable given German production rates. Yet their losses *per week* routinely exceeded that. The Luftwaffe lost 10 percent of its bombers and fighters from August 12 to 18, 10 percent of its bombers and 15 percent of its fighters from August 24 to 31, and 6 percent of its bombers and 15 percent of its fighters from September 1 to 7. Although British fighters sometimes fought at numerical disadvantage against formations of two to eight times as many German aircraft, they usually had the tactical advantage of being vectored into a favorable position where they could "bounce" German formations from above.[80]

## THE CHALLENGE OF THE NIGHT BLITZ

The first major destabilization of Britain's information problem, as discussed previously, was the improvement of bomber performance that created the Channel Gap crisis in the 1930s. The innovation of radar (which expanded the technical art of the possible) and the reform of organizational processes (which made the possible actual) restored stability. The second was the German switch to night bombing during the battle itself. Fighter Command performed admirably against the Luftwaffe during daylight attacks, but it was unable to deal effectively with the night raids that began in earnest in mid-September. This German innovation blinded British measurement and blunted enforcement. The Dowding system drifted toward insulation as the information problem became less constrained. The Night Blitz reveals how a well-regulated control system depends on particular assumptions about its environment.

Of the 6,135 Luftwaffe night sorties over Britain in September, the RAF shot down only four aircraft. Night interception was harder for three reasons. First, the British needed a larger two-crew aircraft to fit the navigational apparatus and more powerful canons for fleeting encounters with the enemy. The Bristol Blenheim, an adapted light bomber, proved barely adequate in this role. The two-seat Bristol Beaufighter, built for this purpose, was still months away. Second, a controlled interception required the elaboration of ground-controlled intercept (GCI) tactics with a radar that could closely track the target and the interceptor overland, rather than the broad-search early-warning methods used over the ocean. Third, and most important, a fighter needed airborne intercept (AI) radar to close the gap between the twenty-thousand-foot limit of GCI and five-hundred-foot visual acquisition. Searchlights were defeated by high altitude because of the sound lag, so GCI and AI were the only intercept options.[81]

GCI became feasible with the introduction of plan position indicator (PPI) radar displays. PPI, which would become a standard for decades to come, represented range and bearing on a polar plot with a sweeping beam, leaving an afterglow to represent the intensity of the radar return signal. PPI is able to display both the target and interceptor locations on a two-dimensional display. When plotting the two tracks independently using the Tizzy Angle method described above, any accumulated errors in position or heading of either contact could botch an interception. When they were displayed in relation to one another on the same PPI image, however, then the errors in the representation would be the same for both contacts, thus preserving the meaningful relative relation between them. This is yet another of many instances of how automation can transform cognitive work into visual problems. Height still remained a difficult problem for PPI, as it had with previous methods.[82]

AI was the long pole in the tent. Tizard recognized Fighter Command's nighttime vulnerability as early as 1936. Indeed, it was one of Lindemann's only legitimate complaints in their quarrel. AI was simply a hard scientific problem at the time, as the transmitter and receiver had to be quite compact to fit in an airplane, and ground clutter swamped the display for targets closer than one thousand feet. In comparison to the methodical research and development effort that produced Chain Home, AI was plagued by chaos and mismanagement. The project was too secret to bring in industrial partners who might have helped. Instead firms received cryptic orders— for instance, to design alternators that looked like an aircraft DC generator.[83] When Chain Home went operational and its management was turned directly over to RAF officers in 1939, Watson-Watt's team then relocated from Bawdsey to Dundee. This move severely disrupted AI research. The demoralized scientists got caught up in installing the inadequate Mark 3 AI, which had been rushed to production, rather than researching improved AI designs. Both Tizard and Dowding understood the night-fighting limitations of their system, but without a workable AI, there was little that they could do.

Dowding was the hero of the Battle of Britain for institutionalizing a winning control system, but he became a scapegoat when the Germans changed the game. The Air Staff informed Dowding on November 13, 1940, that he was being sent to the United States to improve air forces liaison, effectively relieving him of Fighter Command. Controversy has surrounded the ignominious dismissal of Dowding and his key lieutenant Keith Park ever since, but it is clear that the failure to stop the Night Blitz was one of the major reasons. The irony is that Dowding understood how to solve the night-fighting problem better than his detractors, but the AI research effort was too fragmented in 1940. An improved Mark 4 AI set and a new Bristol Beaufighter would not be available until 1941. From March 2 to April 13, 1941, Fighter Command shot down at least thirty-two aircraft and damaged a dozen more. Before the Mark 4, there had been only six successful engagements. Dowding's system proved unable to connect with the Luftwaffe at night, but there was little he could do about this unfortunate insulation. This unfortunate denouement to an otherwise brilliantly fought battle highlights the importance of *both* structure and organization for the performance of C2. When the problem changed and the organization was unable to adapt, Fighter Command could not close control loops on the enemy. It is fortunate for Britain that Hitler had already given up on Sea Lion by then and turned his attention toward the Soviet Union. German targeting during the Night Blitz, moreover, was as erratic as ever. Germany was still even more insulated than Britain, so it was unable to exploit the promising operational possibilities that it had unlocked. The RAF thus had time to renegotiate a coordinated solution for a more complicated air defense problem for the remainder of the war.[84]

## Explaining British Victory

We now know that the odds were stacked in favor of the RAF, even as this was not well understood by contemporaries. No one had ever fought an air campaign on this scale, and strategic bombing was widely feared, thanks in part to the ideology of Bomber Command. Fighting the Battle of Britain helped to clarify the true strength of air defense. Germany could have done a lot to destabilize Britain's information problem, but it had too many of its own self-inflicted pathologies of insulation to deal with. The RAF outfought the Luftwaffe because it knew more, and it knew more because the external and internal poles of its vital informational relationships were better stabilized. The fixed pelagic geography of British air defense, improvements in the art of the possible for measurement with radar, and the insular confusion of German leadership combined to create a well-constrained and thus solvable information problem. The RAF had worked for years to improve the efficiency and effectiveness of its solution via a network of sensors, standardized protocols for data sharing, and ground-vectored interception. This system enabled Britain to fight with fewer fighters than it might otherwise have needed (and a greater number were not available anyway). Indeed, the RAF won the battle with only half of the 120 squadrons that it had estimated would be required before the war.

The Dowding system as designed was hardly free from information friction. Continuous low-level adaptation helped to maintain high-level management because it was guided by a clear sense of shared mission. Stable structure was an active achievement, not a simple fact. Yet it enabled Fighter Command to employ a limited supply of interceptors to maintain a favorable exchange ratio long enough for Germany to call off its invasion. When the external constraints that enabled managed practice loosened, however, performance suffered. The Night Blitz underscores the dangers of a mismatch between institutionalized solutions and strategic problems.

C2 improved the efficiency and effectiveness of Fighter Command, but British victory was hardly a simple substitution of information for mass. Many raids got through, and the battle still required a mass of fighters to oppose the Germans. Five thousand military personnel died, including fighter and bomber aircrew and ground-based personnel on both sides. Still, this is a relatively low number by the horrific standards of World War II. The civilian costs were much higher, with more than forty thousand civilians killed in the Battle of Britain and the Blitz. The asymmetric lethal exposure of civilians relative to military personnel anticipated the information-intensive wars of the future, which is yet another way in which the Battle of Britain is a prototype for modern information practice. In a further tragedy for the civilian population, the RAF's success in resisting the German bid for military conquest led Germany to settle on a strategy of indefinite political coercion via terror raids by manned bombers and unmanned weapons (i.e., V-1 and V-2).[85] Similarly in modern-day drone campaigns, which I discuss in chapter 6, the terrorist organizations that are unable to fight drone warriors directly are incentivized to redirect their efforts toward civilian populations.

Militaries are often accused of preparing for the last war. Sometimes this is a recipe for disaster, but not always. Both the RAF and the Luftwaffe planned to fight a previous war, but this worked out to the benefit of one and the detriment of the other. Great Britain created its air force to keep potential invaders far offshore after the wake-up call of German bombing in World War I. The success of the RAF in the Battle of Britain enabled the Royal Navy to continue deterring invasion and protecting lines of communication, as it had for centuries. Germany, by contrast, bound its air force to its army to facilitate rapid conquest. This worked out very well in Poland and France. Yet the Wehrmacht could not conquer Britain unless the Luftwaffe first won alone. Germany's Stukas were useless for strategic bombing, and it failed in its bid for rapid conquest.

Afterward the Luftwaffe continued its harassing attacks at night, and it later employed V-weapons. Yet the German air threat never again posed an existential danger to Britain after 1940. Germany instead settled into a long-term coercive strategy that attempted to bypass the RAF altogether and negate the strength of the Royal Navy through unrestricted submarine warfare. This change of domain, from air to undersea, altered the structural conditions of British information practice, much as the bomber speedup and the Night Blitz had done within the air domain. The German Navy enjoyed two years of success in pursuit of a goal of naval blockade and coercion, which it could do at relatively low cost by evading the Royal Navy. Britain was not as prepared for the Battle of the Atlantic as it was for the Battle of Britain, in part because it had neglected the lessons of convoy defense learned in World War I. The Battle of the Atlantic lasted much longer than the Battle of Britain, in part because the operational problem was far more difficult, spanning a much larger area and requiring coordination with the United States. Moreover, the British were initially less prepared to solve it. Once again, Britain fought a defensive, attritional battle against Germany, and it would rely on a centralized information system to identify, track, and intercept enemy targets, but at the beginning of the war the antisubmarine warfare system was nowhere near as mature as the air defense system had been. Furthermore, while the German air force fought the Battle of Britain with an ineffective and insulated information system, the German Navy had adopted a more efficient and coordinated information system to fight the Battle of the Atlantic. It was not until 1943 that technological and organizational innovations by the Allies finally closed the North Atlantic gap with long-range aircraft, aggressive surface escorts, and persistent access to encrypted German communications.[86] This reformed information system was finally able to break the naval Enigma and identify the location of German U-boats so that the Allies could sink them faster than Germany could put them to sea. The British system of intelligence fusion and intercept cuing, interestingly enough, bore more than a passing resemblance to Fighter Command's system, albeit at a much grander scale and without the benefit of the focused, prior preparation that distinguished the Battle of Britain.

Britain won in 1940 because Churchill's famous "few" in the air were backed up by thousands more on the ground. As Dowding observed, "The system operated effectively, and it is not too much to say that the warnings which it gave could have been obtained by no other means and constituted a vital factor in the Air Defense of Great Britain."[87] The Battle of Britain is an exemplary illustration of how stable problems and institutionalized solutions can improve information practice, and thereby military performance. It is also an outlier in many ways. So favorable a combination of external and internal conditions tends to be elusive. In the eight decades since Spitfires dueled with Messerschmitts, military organizations have adopted far more sophisticated technologies for measurement, coordination, and enforcement. Yet they have also tried to use them in more ambiguous and complicated environments. As both sides of the relationship between representation and reality become more complex, the collective action problems of control become more difficult.

# User Innovation and System Management

## Aviation Mission Planning Software

The analogue to the Fighter Command Ops Room in the modern U.S. Air Force is a Combined Air Operations Center (CAOC). The air force formally designates the CAOC as a weapon system (AN/USQ-163 Falconer), even as it is basically just a large office space with hundreds of computer workstations, conference rooms, and display screens. The CAOC is an informational weapon system that coordinates all of the other weapon systems that actually conduct air defense, strategic attack, close air support, air mobility and logistics, and intelligence, surveillance, and reconnaissance (ISR). As the central knot in a global skein of informational relationships, a CAOC copes with far more internal and external complexity than anything Fighter Command ever faced. One might be tempted to describe the CAOC as "a center of calculation," in Bruno Latour's phrase, but modern digital technology tends to decenter information practice. Representations of all the relevant entities and events in a modern air campaign reside in digital data files rather than a central plotting table. The relevant information is fragmented across collection platforms, classified networks, and software systems that are managed by different services and agencies. It is up to military personnel to cope with the information friction that results.

### Working the Seams

In each of the four major U.S. air campaigns from 1991 to 2003, CAOC personnel struggled with friction. They rarely used the mission planning systems that were produced by defense contractors as planned, and they improvised to address emerging warfighting requirements. A study of air command and control by Lieutenant Colonel Michael Kometer notes that air planners in Operation Desert Storm (Iraq, 1991) "used markers, pens, and pencils to mark the target locations on charts," which they then input into five different systems without a common database. There was no common "picture of the way all the missions fit together to accomplish the objectives." As a result, "information did not come over data link[s] in a systematic way, but rather from the informal links [the planner] was able to assemble from the sources within his reach."[1]

Over the next decade, automated planning systems began to provide alternatives to markers and maps, but they also created interoperability problems. During Operation Allied Force (Kosovo, 1999), Kometer writes, planning data were "produced in a message format that was readable only by special parsers. As the war dragged on, planners developed their own Microsoft Excel spreadsheets, Word documents, and other tools to perform their own functions." Because these tools were often incompatible and not all planners worked in the same buildings or on the same classified networks, the air commander's staff "took the information, converted much of it to [Microsoft] Access format for manipulation, and created PowerPoint briefings. . . . This enormous task was the digital equivalent of collecting everyone's yellow stickies to create napkin-sketches of the progress of the entire air campaign." System interoperability improved for Operation Enduring Freedom (Afghanistan, 2001), but with a high tempo of operations and many new intelligence sensors to integrate, informal workarounds were ubiquitous. According to Kometer, "Planners said that [Master Air Attack Plan] briefing slides represented their view of the world. . . . It was much easier to understand than the [Air Tasking Order], which was sent out in a confusing message format. They looked at the . . . briefing to see the planned flow of aircraft and an application called FalconView to see where the aircraft were in real time." Personnel improvised digital expedients and physically transported data to bridge gaps in the organization: "Those in the [targeting] Cell coordinated with others in the CAOC by walking around to get signatures on a routing spreadsheet, e-mailing, or telephoning. Since the Judge Advocate General was in another part of the building and the point mensurators were in another building," as Kometer laconically reports, operators "walked around a lot."[2]

General John Jumper, the air component commander during the Kosovo campaign, was frustrated by the fictions revealed in "time critical targeting" during the war. In his capacity as the commander of Air Combat Command after Kosovo, Jumper stated that the F2T2EA "kill chain"—find, fix, track, target, engage, assess—should be "our bumper sticker going into this century." Jumper aimed to realize a 1996 prediction by air force chief of staff General Ronald Fogleman that "in the first quarter of the 21st century, it will become possible to find, fix or track, and target anything that moves on the surface of the Earth."[3] Fogleman's statement, made the same year that *Joint Vision 2010* appeared, came out of the heady RMA milieu of the 1990s. The notion of a kill chain was analogized from the corporate supply chains of firms like FedEx and Walmart. As Vice Admiral Arthur Cebrowski and

John Garstka argued, the RMA should model a supposed "Revolution in Business Affairs" because "nations make war the same way they make wealth."[4] Jumper aimed to operationalize the kill chain as F2T2EA. The goal seemed tantalizingly close.

A March 2002 slide deck by Brigadier General Jim Morehouse, director of command and control and deputy chief of staff for air and space operations at Headquarters U.S. Air Force, discusses the challenges of implementing Jumper's vision. One slide, entitled "Working the Seams" (figure 4.1), depicts all of the different types of assets that are employed in each phase of the F2T2EA kill chain. At the center is a picture of a CAOC, a room full of people in front of computers. The CAOC plays, or is meant to play, the central coordinating role in this system. This image might be compared to the 1942 image in the previous chapter of the RAF air defense system, with its streams of incoming and outgoing information culminating in a central Ops Room. This image is more cartoonish than the RAF schematic, even as the CAOC system is far more complex. Morehouse's slide, in effect, "breaks down" the kill chain to highlight all the inadvertent breakdowns that emerge across its seams—that is, Morehouse attempts to shift the viewer's attention from what information means to how information works. Yet his

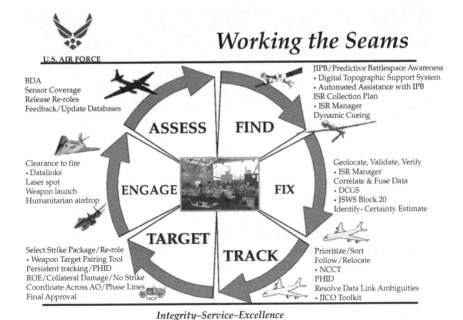

Figure 4.1. "Working the Seams" of the kill chain. Jim Morehouse, "Time Critical Targeting" (briefing to National Defense Industrial Association DOD Interoperability Conference, March 26, 2002), 5.

slide remains a top-down view of the problem from a headquarters per-spective. Working the seams in practice has usually been a more bottom-up affair.

Operation Iraqi Freedom (Iraq, 2003) stressed U.S. information systems further with the requirement to coordinate air operations with a fast-paced land invasion. "Never before," write Lieutenant Colonel Thomas Kelly and Lieutenant Colonel John Andreasen about their experience with the XVIII Airborne Corps, "have the air and land forces supported each other as effectively." Yet they were also alarmed by the effort it took to achieve these results: "The processes and systems designed to support joint targeting and operational fires interfaces between the land and air components proved unwieldy, ineffective and inefficient." Data exchange required "tremen-dous human intervention" and "Herculean efforts of joint and Coalition warfighters throughout the force." As contractor-procured systems proved unwieldy, "most headquarters defaulted to [Microsoft] Office software to create decision products or to communicate ideas most effectively."[5] Kometer, similarly, notes that personnel began using systems in very dif-ferent ways than designers had intended: "The Army worked out a way to send all [Air Support Requests] as interdiction requests and then send code in the remarks section that would indicate the mission was other than inter-diction." The repurposing of systems for local needs was able to solve some local problems, but the organization had many local problems. "There were well over 100 colonels in the CAOC," Kometer recalls, "and each seemed to have another problem like [Air Support Requests] to solve. The result was a lot of customized information formats." All the expedient solutions created confusion in the aggregate. The deputy "could not even tell [the com-mander] what the air component had done, not to mention how [the battle] had gone."[6] Adaptive practice in this case was both a response to and a source of information friction. More generally, user innovation has been a critical, and underappreciated, emollient for the peculiar frictions of modern war.

## Organic and Institutionalized Modes of Innovation

It is widely known that amateur invention produced many start-ups in the hothouse culture of Silicon Valley in the 1980s and 1990s. Modern tech giants like Apple, Microsoft, and Google started off in garages and dorm rooms. It is less appreciated that amateur innovation in the far more bureau-cratized realm of the U.S. military also produced the de facto standard for graphical mission planning and intelligence analysis across the services. As new commercial technologies became affordable, tech-savvy personnel found ways to use them for new types of missions. They also had to struggle through an uneasy symbiosis of institutionalized and organic practice.

During the personal computer (PC) boom of the 1980s, air force pilots wrote flight-planning software using privately purchased computers and shared it freely with their comrades. These expedients anticipated and often suggested features that the defense acquisitions community later incorporated into official systems. In the early 1990s, U.S. Air National Guard fighter pilots teamed up with programmers at the Georgia Institute of Technology to write an application called FalconView to support planning for the F-16 Fighting Falcon. Strictly speaking, FalconView was the mapping interface for a suite of software tools known as Portable Flight Planning Software (PFPS), whose successor was known as Execution Planner (XPlan). Many users simply referred to the entire bundle as Falcon-View, and I will do so here for narrative convenience. The PC/Windows-based application could display a variety of chart and image formats, create custom overlays, perform planning calculations, interface with Global Positioning System (GPS) receivers, and display real-time data from tactical radio feeds. FalconView was a bit like an earlier era's Google Maps.

Just a decade after FalconView emerged as a user-designed prototype in the Air National Guard, many personnel in all four services and other U.S. government agencies had become regular users. The application supported an astonishing variety of intelligence, planning, and operational tasks, often in conjunction with other user-developed extensions of the Microsoft Office platform. Throughout the 2000s, an enthusiastic user community drove its functional evolution and diffusion across the U.S. military services and beyond, all for a total cost of about $20 million. By contrast, official acquisition programs that were inspired by and designed to replace user innovations like FalconView went millions of dollars over budget, totaled in the billions, lagged years behind schedule, and lacked broad appeal across the services. The acquisition bureaucracy attempted to build reliable, interoperable, efficient mission planning systems, but operational requirements kept changing as the U.S. military found itself involved in a string of conflicts after the Cold War. Commercial information technology developed rapidly at the same time. In the framework of this book, the quest for managed practice in the mission planning world gave way to insulated practice. Acquisition programs ended up prioritizing bureaucratic process over operational responsiveness. Frustrated service members defected from official solutions to engage instead in adaptive practice. They took advantage of local discretionary resources and their own technical know-how to adapt information technologies for their emerging operational reality.

From 1998 to 2008, I used FalconView myself as a naval intelligence officer. I also developed a few software tools to extend its functionality. During that time I dealt with countless minor frictions in military networks and observed the challenges of the official procurement process firsthand. This experience facilitated my access to and credibility with the

FalconView user community. This narrative is based on interviews with key participants, supplemented with primary documents that are available in the public domain or were provided to me by participants. Several officers I interviewed were actively involved in both informal projects and official programs, which put them in a position to report on both sides of the debate. This mitigates a potential pro-FalconView bias in my informant sample. Indeed, the interdependence between bottom-up (organic) and top-down (institutionalized) efforts in this case is an important finding. Despite the stark contrast between user enthusiasm for FalconView and user antipathy for official systems, grassroots initiatives often worked as complements to rather than strict substitutes for the official systems. Furthermore, one of the reasons that FalconView survived where many other military user innovations did not is that it gradually took on some of the bureaucratic characteristics of the very systems it challenged.

## THE FIRST WAVE OF USER INNOVATION

Aviation mission planning includes all of the information gathering and processing that aircrew must complete before takeoff. A simple cross-country flight requires them to acquire charts, plot a flight route, gather weather data, update airfield and navigational aid information, calculate fuel consumption data, and file a flight plan with airspace managers. An attack mission additionally requires parsing the air tasking order, studying intelligence about threats and targets, calculating weapons settings, deconflicting operations with friendly air and ground forces, and developing search-and-rescue contingency plans. Different missions have different requirements, such as slow-down and air-release points for cargo airdrops. Aircrew typically print this information on kneeboard cards for easy reference inside the cockpit. "Mission planning is not rocket science," one pilot explained, "but it's complicated with lots of moving parts."[7] Prior to automated mission planning systems, these calculations were time-consuming activities involving paper charts, protractors, dividers, plastic "whiz wheel" calculators, grease pencils, flight manuals, and paper orders and updates. Aircrew pulled strings across charts to measure distances, traced turns around coins, and cut strip charts with scissors and glued them together. Manual planning was not uncommon in some squadrons well into the 1990s.

FalconView emerged during a period of history that was peculiarly conducive to user innovation. New types of air-delivered weapons appeared in the late Cold War, and new missions emerged after it, which together increased the computational burden on operational units. The commercial information technology sector exploded at the same time, with the arrival of inexpensive PC technology, programmable software platforms, and the commercial internet. Military personnel thus gained access to "commercial off-the-shelf" tools that could help with computational tasks old and new. Aviators, with their engineering training and intensive planning responsibilities, were especially well positioned to take advantage of the new possibilities. By the late 1970s, aviators were already experimenting with automating some of these tasks using microcomputers and engineering calculators. The nascent software market was experiencing something of a gold rush, with the appearance of mass-produced microcomputers like the Apple II and TRS-80 in 1977 and the IBM PC with Microsoft DOS in 1981. Most successful software programs that were written *for* microcomputers— the VisiCalc spreadsheet, WordStar word processor, and dBase II database—were also written *on* microcomputers by groups of just one or two programmers.[8] This technological situation created similar opportunities for military aviators who were willing to buy their own computers and program the stereotyped calculations needed for flight planning. Officers were already responsible for their own planning and navigation; therefore, as one pilot points out, they "could fly with beta software because if they screwed up it was their wings anyway."[9] The mission planning space had not yet been colonized by the military procurement bureaucracy, so users had free rein to experiment.

Captains Jake Thorn and Jerry Fleming, both A-10 fighter pilots and avid electronics hobbyists, began playing with a TRS-80 Model 1 in 1981. They happened to meet the commander of Air Force Tactical Air Command, General Wilbur Creech, during a routine visit. Creech looked at the rudimentary machine and asked, "Can you flight plan on that thing?" The captains were invited to give a demonstration at a meeting of air force leadership, and their squadron commander agreed to take them off the flight schedule so they could get to work. The result was "Jake and Jerry's Two Week Flight Planner." Creech was impressed enough that he created the Small Computer Program to provide every squadron with a microcomputer and thus begin "squadron automation." Aircrew started writing and running software on Cromemco System 2 (CS-2) machines.[10] Various aviator-written programs percolated through the community, passed around on floppy disks from one aviator to another as they commuted in their jets to air bases around the country. One program that became widely used was FPLAN ("Flight Planner"), first written in 1983 by Thorn, Fleming, and another pilot at Eglin Air Force Base in Florida. FPLAN performed basic route and fuel calculations, but its real novelty was an updatable database of navigation coordinates and frequencies from the monthly printed instrument flight rules supplement. As if running a start-up company in a garage, the officers persuaded their wives to manually type in data from each update so they could make copies on floppy disks and mail them out to other aviators. Over the next several years, they released new FPLAN updates on the side while working their primary jobs, and their distribution list grew considerably. FPLAN found its way onto the computers of squadrons across the air force and Air National Guard.[11]

The innovation of military aviation mission planning followed the general logic of user innovation.[12] As software production costs declined with the advent of low-cost personal computers, tech-savvy users like Thorn and Fleming innovated for their own benefit. Dramatic changes in the commercial information technology sector provided them with low-cost, flexible tools to solve local problems. Military users innovated to improve their own workflows, and then they freely revealed their programs to other pilots, who then made further improvements. The emergent self-organized user community enjoyed increasing returns from their adaptive practice. Yet these particular users were also members of a military bureaucracy, so their activities also began to create organizational friction.

BUREAUCRATIC RESPONSE

In 1983 Tactical Air Command established a center for testing and evaluating automated mission planning systems at Eglin, which became something of a hub for official and unofficial activity. The first contractor-produced systems, such as the Mission Support System (MSS), also used the CS-2 platform, but they ran a Unix operating system while the squadron machines ran DOS. Thus two paths diverged with the CS-2. One led to officially sanctioned, over-budget, something-for-everyone, difficult-to-manage Unix-based solutions. The other led to informal, run-on-a-shoestring, mission-focused, PC-based solutions. These two paths would characterize mission planning software throughout the 1990s. Another pattern that emerged early on was that the same aviators who were involved with official systems at Eglin were also developing unofficial systems on the side. These individuals had a strong interest in improving mission planning and inside knowledge of problems plaguing official programs. Thorn, for example, became the operations officer of the Eglin Test Squadron conducting the operational test of MSS at the same time he was working on FPLAN.[13]

Aircrew found MSS difficult to use as delivered, which prompted a complete software overhaul in 1983. An even larger version followed in 1986, enclosed in a refrigerator-sized ruggedized container that required four people to lift and a C-130 to transport.[14] Aircrew preferred to use FPLAN and other informal programs whenever possible to avoid using MSS. Thorn once again became a key link between formal Unix and informal PC efforts in 1989 as chief of the MSS Projects Office at Eglin. He was painfully aware of the problems plaguing official development and the importance of ensuring that pilots had something that worked well in the meantime, so he facilitated informal development and distribution efforts even while working to improve MSS. Furthermore, the popularity of PC solutions

among aircrew led to decisions that MSS should mimic the FPLAN interface for better usability and that Eglin should start funding on-site contractors to write small-scale PC utilities as stopgaps for the Unix MSS.
Collectively known as Portable Flight Planning Software (PFPS), this smor-
gasbord of software made it easy for users to mix and match or even add
functionality, rather than wait for upgrades of a monolithic system. Thus
official systems inspired by early user efforts were further shaped by them
through the agency of hybrid user/managers like Thorn. The program
offices also contributed some resources for further informal supplements.[15]

In addition to the first digital map displays, MSS had other important
features that early PC programs lacked. The F-16C, fielded in 1981, had
improved avionics and the first data cartridge, which enabled data to be
transferred from planning computers to the cockpit. The first data cartridge
had only an 8 Kb capacity and could be programmed if needed, but by 1990
cartridge capacity had grown to 128 Kb, which required MSS to load.[16] As
more aircraft were fitted with data cartridges supporting more elaborate
planning, the costs of supporting expanding user requirements increased
for official systems. Unfortunately, not all squadrons with F-16s had MSS,
and those that did found it difficult to use and maintain. Air National
Guard squadrons were particularly hard pressed because they fell well
below active duty squadrons in priority for new equipment. The Guard
also needed to be as portable as possible for short-notice deployments. This
situation led two majors in the Guard, Robert Sandford and Walter Sobczyk, to try to build a device that would hook up directly with a PC and
load a cartridge. Their first prototype was literally carved out of wood at
Hill Air Force Base in Ogden, Utah, and dubbed the Ogden Data Device
(ODD). Soon PCs running FPLAN and ODDs were available to Guard aircrew, which eliminated the need for MSS. ODDs found their way into active
duty squadrons too, despite headquarters protests that they were unsafe.
Reflecting a tradition of aircrew responsibility for their own mission planning, squadron commanders sided with aircrew, who found the ODD/
FPLAN solution to be more reliable and hassle free than MSS.[17]

The 1991 Gulf War was a dramatic deviation from the operational status
quo of the late Cold War. It also provided a major stimulus to automated
mission planning. Many aircrew were exposed to and impressed with digital mapping on MSS for the first time. However, there were never enough
to go around, the machines were frequently down, and manual planning
with paper charts and grease pencils was still common. Mission planning
took at least six to eight hours and sometimes days for complex strikes.
Target imagery and intelligence were often unavailable. Coordination with
widely dispersed members of strike packages was very difficult; coordination with friendly ground forces was sometimes impossible. In response
the air force established a program office at the Electronic Systems Center
in Massachusetts to create a new Air Force Mission Support System

(AFMSS). Its scope was expanded beyond fighters to also encompass bomber, airlift, tanker, and special operations aircraft, as well as headquarters command and control. The effort to make something for everyone at one go would prove frustrating for users and managers alike. The expense and coordination challenges tended to separate engineers from users with many layers of red tape. The AFMSS hardware suite ended up being even bigger than the already large MSS. The entire system was also classified, to include its generic navigation functionality. By contrast, programs like FPLAN had always been unclassified and easy to disseminate.[18]

## THE SECOND WAVE OF USER INNOVATION

Adaptive practice often begins through the defection of personnel from the insulated practice of their own organization. The systems they create, however, can become vulnerable to insulation too. The scientific art of the possible and operational mission requirements continue to change, and the inventors may not be able to keep up. The ODD-FPLAN stopgap was limited by architectural legacies dating back to the earliest microcomputers (e.g., single-character variables) and idiosyncrasies of several ports across different operating systems. Yet Sandford was also pessimistic that the new AFMSS program would ever be viable, especially for the Air National Guard. He thus started to consider a complete PC alternative that would run on a single laptop, display digital maps, support graphical mission planning, and load cartridges via an ODD. Sandford persuaded some National Guard Bureau staff officers to provide him with unallocated year-end money to build a prototype.[19]

In the late 1980s most engineers viewed PCs as too underpowered to support graphics-intensive mapping applications. Sandford presciently believed that this problem would go away as PCs running Microsoft became more capable. Unix machines, meanwhile, would remain scarce assets for squadrons. By the release of Microsoft Windows 3.0 in 1991, basic graphical mapping was finally feasible on the PC. Another challenge was that geospatial software required detailed, updated, compressed map data covering a lot of area. Commercial customers who could afford expensive map data could usually also afford a more powerful Unix machine and Esri software to work with it. A military user like Sandford, however, would be able to freely leverage all the map data that had been developed for Unix systems like AFMSS, even as Unix machines might remain practically out of reach. The commercial maturation of Windows and the availability of free map data for military users thus created the technical conditions for the possibility of FalconView.[20]

In spring 1993, at the Georgia Tech Research Institute (GTRI), Sandford found a sympathetic ally in a research scientist named John Pyles. Pyles had previously worked on army software that displayed photographs of maps on PCs, which showed him how portable and useful laptops were in the field, and on an MSS contract where he became familiar with military map data formats. They signed a contract for a "Flight Plan Simulation Mapping System," with the word "Simulation" added so that the Guard could piggyback on an existing army contract with Georgia Tech for simulation software. Pyles's managers also required him to phrase the contract wording so that deliverables would be "beta"—working prototypes rather than finished product—because they felt the six-month deadline was too ambitious. "We were just too naive to realize this ourselves and just went ahead and finished the initial release anyway," Pyles recalls. "Being fresh out of college, $475K sure sounded like a lot of money to us and as a result we were under a good deal of self-induced pressure to deliver." The beta version was completed in January 1994, and Sandford discreetly provided a tested version 1.01 to F-16 aircrew in June.[21]

Like other engineers, Pyles wanted to do Sandford's project in Unix, even a PC running Unix, but Sandford was adamant. The constraint of programming graphics for the Intel 80x86 processor versus a more powerful Unix minicomputer challenged GTRI engineers to come up with fast and efficient rendering techniques. Although this constraint evaporated as PC hardware performance improved, it paid dividends later on in fast, smooth performance. "I knew technology was on our side and we could ride the hardware wave of hard disks, screen resolution, and PC performance improvement," Pyles says. As Sandford puts it, "PCs are disposable."[22]

Sandford and Pyles consciously based FalconView management on Frederick Brooks's software engineering classic, *The Mythical Man-Month*.[23] Brooks argues that conceptual integrity is the most important and elusive part of software design, which is inherently complex, hard to visualize, changeable, and somewhat arbitrary. He recommends a small "surgical team" of talented programmers, a chief programmer controlling the concepts, and frequent end-user interaction to ensure that the task and solution are coordinated. Pyles played the role of chief programmer while Sandford—the chief user—was closely involved in architecture and interface discussions, sometimes even coding working prototypes to convey a concept. Although he was detailed as a staff officer at the Pentagon, Sandford still managed to get into the cockpit of an F-16 regularly, which kept his understanding of planning requirements fresh.[24]

The team made two key decisions. First, they bet on Microsoft. Lieutenant Colonel Joe Webster, an aviator on the Air Force Reserve staff who helped fund FalconView and would later take over its management when Sandford left to head the F-16 Test Team in 1995, put it simply: "It will be whatever Bill Gates decides." This meant designing for DOS and later Windows, writing in the C++ language rather than ADA (the Department of Defense standard at the time), and adopting the Microsoft Office interface style.[25] They gave a lot of attention to the user interface because they wanted a novice to be productive with daily tasks right away. In stark contrast to the official programs, which required users to memorize cryptic combinations of keyboard commands, FalconView became an application that people liked to play with on first exposure. As one aviator said, "I've learned more in 10 minutes on my own with this software than I did during two weeks of training on [AFMSS]."[26] Second, version 1.0 would be designed only for the F-16, hence the name FalconView, after the F-16 Fighting Falcon. Even though there was interest in Sandford's project elsewhere, especially in the C-130 airlift community, he wanted to avoid the kind of requirement creep that plagued official programs. Sandford and Pyles supposed from the start that a general-purpose, PC-based geospatial package could be broadly useful for other aircraft, simulator instructors, and even nonaviation communities. "We knew it would be a really cool app if anyone would let us build it," says Pyles. Yet they decided to hold these possibilities at bay in order to first achieve success with the basics: "Do only 80 percent, but do it perfect." This set a precedent for small incremental changes in FalconView, to always make sure that basic functionality was solid before trying to add more.[27]

By focusing on the human interface rather than just technical functionality, and by limiting their initial scope, Sandford and Pyles laid the foundation for an application that many others, and not only pilots, would find intuitively attractive. The application was freely available to any government employee, which sidestepped the proprietary licensing associated with most contractor-developed code. It was also unclassified, since any overlays using classified data could simply be treated as separate files rather than part of the application code. Therefore, the program could be disseminated on unclassified CD-ROMs and later loaded onto classified networks, much as an unclassified copy of Microsoft Word could be used to draft a classified memo. Because FalconView was unclassified, personnel could also install it on their personal laptops without asking permission and take it home. Software developers, including users, could then access the public internet to research technical questions if and when they started programming new features for FalconView. All of this helped to hasten diffusion through different user communities.

## BUREAUCRATIC ACCOMMODATION

Military users and program managers are not distinct homogenous groups. Thorn, Sandford, and Webster, like many officers who rotated between operational and staff jobs, could play either role at different times, and they each developed different opinions about the relationship between user-led software efforts and the big programs. Sandford was exasperated with programmatic inefficiency and sought to bypass it altogether with FalconView, an independently funded Guard system. Thorn, by contrast, was involved in AFMSS program management and saw user software efforts as temporary supplements that would eventually help to improve the official effort. Thorn hoped to reform the acquisition programs, but Sandford wanted to forswear them altogether.

The two officers could at least agree that the venerable FPLAN needed to be overhauled. Sandford needed its flight planning capabilities for FalconView and Thorn needed a stopgap until the official AFMSS system was ready. Sandford transferred Guard funding to the AFMSS program office at Eglin, where Jerry Fleming, one of the original FPLAN developers, was now working as a civilian contractor. Thorn achieved reluctant program office acquiescence by promising to keep the FPLAN replacement separate from FalconView, which was perceived as a wasteful distraction from the main Unix effort. Thorn ensured that the new program, dubbed Combat Flight Planning Software (CFPS), shared the interface look and feel of the AFMSS system in order to help train aircrew, thereby achieving program office buy-in and, eventually, some funding. This enabled CFPS to leverage aircraft-specific software modules, also developed at Eglin, to facilitate data cartridge loading. Sandford, in turn, ensured that FalconView, which was under a separate Air National Guard contract with Georgia Tech, would be able to interface with CFPS, which would give him not only flight-planning but data-loading capabilities as well. Ironically, CFPS was separated from FalconView so that the air force program office would not have to fund a PC-based map display; yet in doing so the program office ended up subsidizing more potent capabilities for its upstart competitor.[28]

Thorn and the program office recognized the threat posed by FalconView and tried to slow development. They demanded that route coordinates entered through CFPS remain fixed in FalconView, but Sandford's next version allowed users to graphically construct and modify a route. They demanded that FalconView not duplicate an expensive and classified AFMSS capability to display threat data, but Sandford's next version allowed users to graphically build threat overlays while keeping the application unclassified. The FalconView team could routinely ignore angry AFMSS protests as they added new features because their funding came from the Guard and Air Force Reserve, not the active duty air force. Furthermore, AFMSS development was having numerous problems and receiving extremely negative user feedback. Program managers like Thorn grudgingly recognized that a functional stopgap really was needed.[29]

FalconView version 2.0 was already percolating throughout the air force when a tragedy in Croatia raised its visibility. On April 3, 1996, a CT-43 carrying U.S. commerce secretary Ron Brown and thirty-four others crashed on instrument approach to Dubrovnik, killing all aboard. Investigators opined that had the pilots used FalconView's GPS-enabled moving map in the cockpit, the "controlled flight into ground" might have been avoided.

Later that year, the air force made FalconView use mandatory on all Distinguished Visitor aircraft.[30]

Frustration with AFMSS mounted and the popularity of FalconView grew, spreading informally to an estimated thirteen thousand users by 1997. The issue came to a head at a meeting of air force general officers at the Pentagon. The AFMSS Special Program Office director was fired, and AFMSS was directed to officially incorporate FalconView and related PC applications, although Georgia Tech and Eglin would continue actual development. AFMSS continued funding the Unix systems because some specialized missions or aircraft (such as the F-117 stealth fighter) depended on it, yet FalconView finally had won official endorsement, as well as $4.4 million of AFMSS money over the next two years.[31] By October, GTRI released FalconView 3.0, updated for the 32-bit Windows 95 operating system and including hundred-meter Digital Terrain Elevation Data. That same year FalconView was recognized as one of five finalists in the "core business" category of Microsoft's annual Windows World Open competition. Microsoft CEO Bill Gates later featured a chapter on FalconView in his book *Business @ the Speed of Thought*, which pointed out that reservists had created an avenue for ideas about civilian technology development to influence the military.[32]

Despite these successes, FalconView's future was often uncertain only six months ahead. The AFMSS program office still viewed it as an unsustainable, unmanageable "hobby shop" project. The majority of mission-planning funding continued to go to legacy Unix systems as well as to a new initiative to replace both Unix and FalconView systems with a PC alternative. Before I turn to this new development, it is important to understand how FalconView managed to survive on a fraction of the funds allocated to AFMSS yet appeal to a much broader user community. Ironically, a small program designed for just the F-16 became popular with a diverse user group well beyond the aviation community, while programs of record designed from the outset to support multiple aircraft communities often failed to satisfy even their target users. Through the creative efforts of lead users, the FalconView community survived and enabled successive waves of user defections from official systems that were designed to replace it.

FalconView requirements were informally coordinated with Georgia Tech and Eglin, whose engineers either were former aviators or had developed and maintained close relationships with aircrew. A standing contract with the Ogden Air Logistics Center allowed other military agencies to transfer small amounts of money to support FalconView development. Sympathetic pilots on staff tours with air force, navy, or special operations program offices were often "shaking trees" to keep Pyles's team in business. Each modest contribution funded a specific feature for a specific customer, creating pressure for continuous improvement and aggressive quality control. Every few months Pyles would call his patrons and warn them, "We're about to turn into pumpkins! Do you have more work?" Lacking a guaranteed budget, FalconView had to be responsive, as Webster pointed out, to "the guy flying the airplane right now" to keep the money flowing. Furthermore, adding small sums of money to an existing air force contract allowed officers to avoid a contractor bidding process designed to promote fair competition. Most FalconView contributions were less than $200,000, the outlier being a U.S. Special Operations Command (SOCOM) contribution of $2 million in 2003. By contrast, budgets for the official mission planning systems were congressional line items executed through a more legalistic bureaucratic process that tended to insulate program managers from active duty operators.[33]

As in the fable about stone soup, different organizations that wanted new features contributed small amounts of money or working prototypes, and these improvements then became available for free to all users. Air Force Special Operations funded terrain elevation and threat-masking algorithms to calculate the effect of ground features on radar coverage so that aviators could predict which surface-to-air threats could see them. This feature could then be used by ground forces in reverse to predict what *they* would be able to see from observation points. The navy funded digital nautical charts, a 3-D flight simulator, and image georectification, which appealed to operators at sea, in the air, and on land. Frequent, incremental, modular improvements to the reliable FalconView chassis, together with a basic contracting vehicle, enabled different service components to achieve a degree of interoperability to which much larger "joint" programs only aspired.[34]

FalconView's diffusion to other services featured the same sort of creative bureaucratic maneuvering and frustration with official systems that had enabled the application's initial emergence in the air force. For example, a Marine Corps F/A-18 aviator working at the Naval Mission Planning Systems office, Major John Bennett, saw FalconView as a viable alternative to the navy's own troubled Unix program. The Tactical Air Mission Planning System (TAMPS) had evolved from a Tomahawk cruise missile planning system, passing through three different contractors and routinely failing operational test and evaluation trials because of interface and functionality issues. The physical machine was so large that it had to be hoisted from the aircraft carrier hangar deck and through the floor into the intelligence center above. Navy regulations classified damage to an aircraft data cartridge as a class C mishap (i.e., an event that required grounding an aircraft), so maintenance officers were reticent to let aircrew take the cartridge if they did not absolutely need it. The introduction of the AGM-84E Standoff Land Attack Missile (SLAM) finally made the use of TAMPS unavoidable, but aircrew still preferred to use paper charts and informal techniques for SLAM planning, only at the last step transferring the results to TAMPS.

In effect they used the TAMPS workstation as a $200,000 data cartridge loader. A former A-7 pilot who worked as a TAMPS instructor in the 1990s laments, "When just planning, I was having a hard time coming up with a use for TAMPS. It took more time than with a pile of charts and scissors, which was faster and more accurate for adept planners."[35]

Bennett proactively visited with aircrew and mailed out hundreds of CDs to "beta users" before FalconView was ever certified for navy use. He self-consciously cultivated "disciples" in the fleet to overcome the criticism of engineers at navy labs who, as in the air force story, focused on Unix and were hostile to maverick PC efforts. To build official support, Bennett would ghostwrite messages for units to send to the navy staff with glowing evaluations and endorsements of FalconView. Network managers complained that they would find it on one squadron machine and remove it, only to come back and find it on several more later. Bennet recalls that the software "was somewhere between a cult and a virus." The groundswell of enthusiasm led to the approval in 1998 of a navy channel for distributing and funding small improvements for FalconView. Bennett leveraged this to accommodate requests from Marine units for new features to support ground forces, which is what expanded FalconView's scope beyond aviation for the first time.[36]

To build and maintain FalconView, officers sympathetic to it would opportunistically exploit organizational openings—helping to balance budgets, cutting deals, letting out contracts, and so on—to deliver expedient functionality to users. They also sought ways to lay in more institutionalized support for the growing user community. A dozen aviators held the first Mission Planning User Conference at the Eglin Officers Club in 1993. This became an annual forum to discuss developments in FalconView as well as the official programs, with attendance growing to over 1,100 people in 2000. Program representatives attended the meetings, downplaying the significance of the FalconView 3.0 distribution in 1998. In 2001 the same speakers apologized to the crowd for not having been able to divest themselves of the big Unix systems fast enough. The Guard staff began a detailed monthly newsletter that went out to a growing mailing list with detailed technical, operational, and programmatic advice on both Unix and PC systems. Software traveled with aircrew around the world for exercises and operations. When officers transferred to new assignments, they took FalconView with them to different commands. Some FalconView enthusiasts even took positions within acquisition organizations with the intent to influence mission planning funding. The turnover of user/managers might have been destabilizing but for the complementary cultivation of small groups of engineers at Georgia Tech and Eglin with low turnover and close user interaction, cooperative relationships with program managers, and most importantly, the longevity and perseverance of a handful of officers in key staff positions. The mission planning community was a dynamic hybrid of formal and informal projects and of aviators, managers, and engineers.[37]

## THE THIRD WAVE OF USER INNOVATION

The first waves of user innovation in the 1980s and 1990s launched automated mission planning and culminated in official recognition and support for FalconView. In the process FalconView became a contract-mediated, civilian-managed engineering project, although not quite as bureaucratically encumbered as the larger programs. Institutionalization did not, however, mean an end to the technology's flexibility. Software architectures are layered and modular, so any particular application is really an assemblage of programs and files built on top of other software components and protocols. The same PC that runs software can also be used to make more software. The complexity of the software stack blurs the distinctions between design and use, since designers and users alike might write code, scripts, macros, and data files. Digital technology is combinatorically generative, offering many opportunities for users to intervene or create novel behavior from existing functionality. The stabilization of FalconView as a freely available general-purpose geospatial platform enabled users to leverage it as a toolbox for new applications, thereby catalyzing a new wave of innovation. Closing the black box on the FalconView platform actually fostered an increase in interpretive flexibility for applications that were built on top of it. Institutionalization at one level of abstraction thus facilitated organic adaptation at another.

Some user extensions simply combined existing features for a new purpose, as when bomber crews duct-taped commercial GPS antennas and data loggers into their cockpits to record their flying routes. Many programs generated output files that FalconView could read and display, such as the actual ground footprint of an aircraft's forward-looking sensor, or the expected bomb dispersal pattern. Some low-level utilities, such as automating data sharing between machines, made FalconView more versatile. Contractors also designed FalconView complements when users came up with funding, such as a bird avoidance model, which drew from a database of migratory routes to create overlays depicting the probability of bird strikes at any given time. Another application called TaskView parsed the air tasking order into an interactive FalconView overlay.[38]

Many user applications helped bridge data from noninteroperable "stovepipe" systems, or worked around them altogether. Personnel often wrote Visual Basic scripts within Microsoft Office documents to get around network policy banning uncertified executables. Sophisticated code and operating system calls could then masquerade as "just an Excel spreadsheet." Writing in a language such as C++ or Java would have been a better idea from a purely technical standpoint, but if users had done so then they would not have been able to run their executables on the network. It is ironic that Microsoft products have enabled so much user innovation in a military setting given that the firm has so often been vilified in the open-source software discourse that champions user innovation. For example, a junior officer aboard an aircraft carrier developed a Microsoft Access database named Quiver that automatically generated target and threat overlays for FalconView and target imagery and data on PowerPoint slides. A similar database called Vulture managed intelligence and operational data at the CAOC, sidestepping the unwieldy Theater Battle Management Core System. A versatile tool called Excel2FV, written inside a Word document, simplified the import of other data formats into FalconView. An officer who managed combat information during Operation Anaconda in Afghanistan in 2002 says that Excel2FV "literally saved lives countless times and was the *only* way we could've managed the battle situation."[39]

Recognizing the extent of this activity in 2000 and trying to encourage it, Georgia Tech exposed FalconView application program interfaces and provided a software development kit in 2002. End users/developers gained the ability to add buttons, control the FalconView display from other applications, and respond to user actions in FalconView within their own code. When software designers expose interfaces, provide code libraries and web services, or include scripting/macro languages, they create opportunities for third parties to become experts in customization, which has the benefit of expanding the market for the original application without the original developer having to pay for the work. Explicit third-party programmer support in FalconView led to an explosion of new applications, many created by users on a very small scale for local unit activities. The FalconView team stayed small, never comprising more than twenty people including administrative support. They accepted the fact that they would often have no idea how FalconView was being employed.[40] By facilitating extensibility, they deliberately reinforced the general-purpose nature of FalconView as a reliable and user-friendly geospatial tool rather than just a specialized aviation mission planner. Free, unclassified, general-purpose, and generative, FalconView diffused out to a very diverse community of users.

Georgia Tech estimated that FalconView had more than thirteen thousand aviation and nonaviation users in 2000, twenty thousand in 2004, and over forty-five thousand in 2009, with much of the growth outside aviation. Military interrogators used FalconView with detainees who were unable to read maps but could point out locations on satellite imagery. Special Forces operators strapped FalconView laptops onto their all-terrain vehicles to navigate in Afghanistan. Soldiers in Iraq shared FalconView unclassified satellite imagery with coalition partners to improve coordination. Air force officers developed the ability to broadcast Predator unmanned aerial vehicle data over internet protocol, displaying multiple Predator tracks in real time on FalconView imagery. Other applications aided aircraft crash forensics, tracked whale migrations, or analyzed directed-energy blinding threats. Outside the Department of Defense, the Forest Service used Falcon-View to plan airdrops and track the spread of forest fires, U.S. Customs tracked drug-running aircraft, and dozens of other countries legally acquired FalconView. Colombia used it for intelligence fusion in counterinsurgency operations. China also obtained a copy illicitly when hackers broke into computers at Redstone Arsenal.[41]

## BUREAUCRATIC REPRISE

FalconView catalyzed enthusiastic second-order innovation well beyond military aviation, but the acquisition community continued to criticize it. From 1999 on, FalconView was the de facto mission planning standard for the air force, navy, and SOCOM. AFMSS had been humbled in 1997 when the air force fired the program director and endorsed FalconView. The acquisition community's response was to launch yet another formal program, the Joint Mission Planning System (JMPS), to replicate and replace it. The arguments that FalconView proponents employed early on to secure bureaucratic support (or at least toleration) were now turned against them. To secure Air National Guard funding, Sandford had articulated a vision of Microsoft-based mission planning software running on cheap commercial PCs. Throughout the 1990s, therefore, the mission planning debate was framed in narrowly technical terms as a contest between popular PC laptops and unwieldy Unix boxes, which resonated strongly with frustrated aviators. As long as the program offices were committed to the Unix installed base, this debate played out well for FalconView, the only viable PC mission planner available. By the end of the decade, however, acquisition managers had learned an obvious lesson: a single PC/Windows-based system would be better than struggling with service-specific Unix "stovepipes." That is, they took the problem to be simply a matter of computer architecture, rather than the bureaucratic nature of the acquisitions process and the dynamic evolution of mission planning requirements.

This narrow framing of the issue marginalized FalconView's innovative processes, which closely involved users in development and thereby reduced the costs of linking emerging operational needs to technical expertise. Moreover, the programs were actively critical of FalconView's process, which they characterized as maverick coding unsupported by adequate testing, review, and configuration management, and therefore not scalable, reliable, or interoperable. FalconView's widespread, demand-driven diffusion suggested the contrary, but the argument had some merit. In 1998 the Air Force Electronic Systems Center commissioned an independent Carnegie Mellon study that found FalconView to be too personality dependent and therefore unsustainable. The program offices used the report and austere Defense Department configuration management directives to undermine the renegade FalconView. On top of this, the navy program office did not want to replace its erstwhile Unix program with nominally air force software. The interservice deadlock was resolved by launching a joint program in 1998 to create a new PC-based mission planner for all the services (including special operations). JMPS thus had an even more ambitious scope than the service-wide ambitions that had confounded AFMSS.[42]

Initial hopes for JMPS ran high, even among some officers who had been active proponents of FalconView over AFMSS.[43] Colonel Jake Thorn returned to active duty full-time to take charge of the JMPS project, and some took to calling it "Jake's Mission Planning System." Thorn hoped to use the fresh start to reconstruct FalconView with the sustainability, longevity, and accountability a managed program might provide. Proponents of FalconView worried that a history of mounting expense and inefficiency was about to be repeated, but they were an unfunded minority amid many interested parties. FalconView once again was seen as just a temporary stopgap awaiting the development of a larger program of record. Thorn felt confident that Northrop Grumman, with a $50 million budget and eighteen months, could meet the modest goal of replicating FalconView functionality. He resisted any additional funding for FalconView, which was due for a major overhaul owing to 1990s development legacies and new architectural opportunities in newer versions of Windows.[44]

By 2001, however, JMPS was suffocating with cost and schedule overruns. Northrop Grumman started cutting corners to keep the program afloat. With all of its subcontractors, development problems, and policy compliance concerns, a great deal of effort was expended on coordination within the program, rather than between users and developers. Things that were easy in FalconView became hard in JMPS. Exasperated with the situation, the original three FalconView developers left Georgia Tech in 2000, and Thorn retired from JMPS in 2002. A certified initial operational first version of JMPS still had not been fielded in 2006, six years after the program had planned simply to field a replacement for the 1999 version of FalconView.[45]

Operational requirements did not remain fixed. Throughout the 1990s the United States conducted aerial patrols and strikes in Iraq. In 1999, the U.S. fought the largest air war since Desert Storm over Kosovo. The bulk of mission planning had become digital, with greater emphasis than ever before on time-critical precision strikes. Major combat operations in Afghanistan and Iraq stressed mission planners even further to support special operations, ground forces, and a profusion of unmanned vehicles. FalconView's decentralized mode of development and ease of distribution proved responsive to emerging warfighting challenges as well as to users outside JMPS's aviation-only jurisdiction. Accordingly, customers such as SOCOM and the army made up for FalconView's funding deficit. While JMPS was still struggling to field its first version, four new operationally certified versions of FalconView were fielded. Like the programs before it,

JMPS struggled simply to meet its original requirements even while those requirements changed significantly in the meantime.[46]

All of the versions of FalconView from 1993 through the late 2000s collectively cost about $20 million, while JMPS cost nearly $1 billion in its first decade. Thorn, in retrospect, conceded that it should have been possible to deliver everything JMPS was supposed to do for $200 million using the FalconView model. Advanced aircraft like the F-22, F-35, and F/A-18E/F/G still needed JMPS to fly. Sunk and spiraling costs without a viable alternative led JMPS to be viewed as extraordinarily high-risk. As a result Northrop Grumman lost the JMPS contract in 2008 to BAE, and services other than the air force pulled out of the program, leaving it joint in name only. Once again FalconView came to the rescue, as JMPS incorporated its mapping engine and FalconView implemented some of the planned JMPS functionality. In a fairly radical step for government software development, most of the FalconView code base was moved into a public open-source forum to improve coordination among far-flung developers.[47]

Throughout its first two decades, automated mission planning evolved as a mélange of officially managed and maverick projects that competed and collaborated in surprising ways. Advances in commercial geospatial software—exemplified in applications like Google Maps for mobile devices—continued to lower technical barriers to user innovation. The wars in Iraq and Afghanistan continued to generate more unforeseen data processing needs, and hundreds of thousands of active duty and reserve personnel rotated through combat zones. Mobilized reservists in particular brought with them civilian technologies and mindsets, along with a willingness to adapt commercial devices and software packages for military use. At the same time, official mission planning software, and command and control and intelligence technology more broadly, grew into a multibillion-dollar defense industry. This development brought all of the attendant congressional, corporate, and bureaucratic tussles. The field became more crowded and regulated, which tended to inhibit the emergence of ambitious projects like FalconView. Furthermore, the emergence of cybersecurity as a high-profile national security concern in the wake of Russian and Chinese breaches created additional imperatives for the top-down regulation of military information practice. Security policy tended to raise the barriers to user innovation that commercial technology had lowered. The result, however, was not necessarily better cybersecurity, just as JMPS did not result in a more efficient mission planner.

## War upon the Map

In the early nineteenth century, Antoine Henri Jomini described the management of military campaigns as "war upon the map."[48] Already in the

Napoleonic era, staff officers would position units and plan their movements from afar. Modern military operations take Jomini's metaphor to the extreme. By sharing geospatial data about friendly and enemy forces in a "common operational picture," a networked military organization aspires to see all, strike all, all together. Yet this "network-centric" ideal requires that a common map actually exist in the first place. To fight an external adversary "upon the map," an organization may also have to struggle internally about how to construct and share its maps. FalconView on one hand and AFMSS/JMPS on the other reflected very different understandings of how mission planning systems should be used and designed. While they each relied on the same cartographic data, the former enabled users to create and informally share data overlay files and develop new mapping applications in response to unforeseen warfighting requirements. The latter, by contrast, aspired to provide a sophisticated solution for everyone but became overwhelmed by the complexity of coordinating engineers, administrators, and warfighters.

The popular conception of military organizations as rigid command and control hierarchies obscures the creative ferment inside. User innovation with information technology—a manifestation of adaptive practice—has become ubiquitous in military organizations. Nevertheless, most acquisition program offices tend to treat material improvisation as an aberration that can be eliminated through better specification of requirements and reforms to contracting procedures. Similarly, visions of future war that became popular in the 1990s made little mention of field-expedient digital solutions, by and for warfighters. Proponents of "defense transformation" advocated top-down leadership to enable joint coordination. However, user innovation was often essential for overcoming the frictions that ubiquitous computerization created for U.S. military operations after the Cold War, even as this process was itself not without friction. The RMA in practice had to cope with the enduring tension between institutionalized and organic technology development. It is essential to recognize the strengths and weaknesses of both modes of innovation.

## THE LIMITS OF MANAGEMENT

Militaries fight to impose control on a chaotic battlefield. They are drawn to hierarchical command arrangements to solve difficult internal problems in the face of even more difficult external problems. With their explicit chains of command, visible ranks and uniforms, joint headquarters, standard operating procedures, and mass-produced weaponry, militaries appear to be high-control organizations. Yet these institutions of control are attractive precisely because militaries must operate in environments that are very hard to control. The enemy, the weather, and other external factors can be unpredictable in the extreme. While it can be hard to constrain these factors, an organization can better constrain its own members (via rewards, punishments, socialization, and so on). Indeed, a military *must* constrain its own behavior if it hopes to effectively employ its measurement, coordination, and enforcement capacity to impose constraints on the enemy. A predictable organization not only compensates for the unpredictability of the environment; organizational control is also the means for controlling the environment.

Yet between the absolute hierarchy of the chain of command and the relentless anarchy of the battlefield, more decentralized and organic practices tend to emerge. The discipline and obedience of barracks life or the parade ground stand in sharp contrast to the initiative and creativity often evinced in combat. Nevertheless, there is an important symbiosis between these two modes of social organization. Adaptive practice works to extend the reach of managed institutions into the realm of anarchy. Conversely, problematic practice carries the chaos of the battlefield into the heart of a bureaucratic organization. Indeed, the negative externalities of uncoordinated innovation pose an implicit threat to managed practice. The conflicting imperatives for imagination and regulation in war thus tend to promote mixed modes of control in military organizations. Maneuver warfare (Stephen Biddle's "modern system") relies on common doctrine to enable decentralized units to maneuver independently and support one another to realize the commander's intent.[49] Similarly, the air force principle of "centralized control / decentralized execution," according to Alan Docauer, is intended to enable "an air component commander to plan, coordinate, and control the independent and direct-support actions of air forces in such a way that they meet the intent and objectives of the joint force commander," yet still allow "subordinate commanders the initiative to make decisions based on the best available information, informed by the air component commander's guidance, directives, and rules of engagement."[50]

The RMA doctrine of "network-centric warfare" extends this same idea into an architectural principle that might well be called "centralized design / decentralized use." In their seminal article on "network-centric warfare," Vice Admiral Cebrowski and his coauthor advocate for "a much faster and more effective warfighting style characterized by the new concepts of speed of command and self-synchronization." They write that "excellent sensors, fast and powerful networks, display technology, and sophisticated modeling and simulation capabilities," coupled with a "bottom-up organization," could radically speed up the OODA loop (control cycle) relative to the enemy. While this "bottom-up organization" seems superficially at odds with the centralizing impulse of jointness, they also assume that the entire "Joint Force" will universally adopt common doctrine and systems across all military services.[51] Indeed, the influential 1996 white paper *Joint Vision 2010* champions "the Imperative of Jointness" to realize goals such as "dominant battlespace awareness," "information superiority," and "full-spectrum dominance." To fight faster and more economically, it argued, "we must be fully joint: institutionally, organizationally, intellectually, and technically."[52]

The ideal of jointness married the rationalizing impulses of the industrial era with the millennial excitement of the information revolution. Writing in 1891, Spencer Wilkinson likened the general staff to "the brain of the army," which "can perform its functions only in connection with a body adapted to its control."[53] Similarly, jointness sought to adapt the U.S. military "body" to control by the joint force commander through the universal indoctrination of warfighting concepts and common interoperable systems. Reformers envisioned a future in which all units would have standardized doctrine, C4ISR systems, and common operating pictures. This is less of a novel vision and more of a culmination of a long-term development. As the computational burden of war increased throughout the eighteenth and nineteenth centuries, military staffs grew to meet the demand. Military practice also transformed in the process to make units and operations easier to monitor and control. Hierarchical regimentation and reporting helped to make operational units more legible while standardized battle drills made their movements more predictably controllable. Managed practice, to use the term from this book's framework, thus expanded the scope and scale of controllable warfare. Staff officers became increasingly able to influence battlefield events via plans and reports without themselves engaging in close combat.

Throughout modern military history, the institutionalization of common command and control has been more honored in the breach. Different services and agencies have long pursued different system goals and developed technical standards; the sociotechnical heterogeneity creates friction.[54] Given the dynamic complexity of modern command and control, the realization of something like "centralized design / decentralized use" will most likely remain a chimeric ideal. The process of design will inevitably remain somewhat decentralized too, so long as smart users find themselves with access to flexible technologies in frustrating operational situations. Unfortunately, organic organization can also lead to two different outcomes, which I describe as adaptive and problematic practice, respectively.

THE LIMITS OF ADAPTATION

Throughout the 1990s and 2000s, the information-processing load of U.S. military operations increased because of the availability of data-hungry weapons, the planning challenges of precision targeting, fluid operational environments, and a strategic shift from superpower confrontation to intervention in civil wars. Meanwhile the commercial information technology market delivered flexible, low-cost software tools to an increasingly computer-literate military workforce. The development of FalconView exemplified many features of user innovation anywhere. Individuals on the leading edge of emerging trends in warfare improvised mission planning software solutions using inexpensive PCs, and they freely shared ideas with their comrades. Their expedients anticipated and often suggested features later incorporated into official systems, much as automobile manufacturers incorporated hot-rodders' aftermarket adaptations into muscle cars, and bicycle companies turned user prototypes into the first mountain bikes.[55] Military personnel explored areas of functionality that the official programs thought would be too difficult or too expensive, notably with planning tools that interoperated across service communities. Again and again, the program offices ran into considerable difficulty and expense by separating users from engineers with a morass of formal requirements, federal budget cycles, sharply defined jurisdictions, and security classifications. FalconView, meanwhile, outsourced the definition of requirements and functional prototyping to the user community. By leveraging toolkits such as developer interfaces and the Visual Basic scripting language, personnel were able to produce working applications that helped them to wage war.

At the same time, there were several historical idiosyncrasies that promoted adaptive practice in the case of FalconView: the end of the Cold War, the shock of the Gulf War, an engineering bias for Unix, the PC/Windows opportunity of the 1990s, a chance accident in Croatia, and so on. Yet FalconView was not a unique case of military user innovation. Just as pilots introduced automated aviation mission planning, the first digital aid for naval antiair warfare was a product of a savvy user/developer group and top-cover from senior officers.[56] In the 1980s, similarly, army geospatial software procurement suffered recurring delays and cost increases, while informal prototypes emerged through the West Point computer science department and became popular with users.[57] One U.S. artillery training team that worked with an Afghan National Army unit equipped with Soviet-era weapons developed a Microsoft Access database to translate between NATO and Soviet fusing and aiming schemes.[58] Indeed, it was the rare U.S. unit in the 2000s that did not have a homegrown Access application to pull together all the odd data that did not fit into officially sanctioned data management schemes.[59]

The traditional notions of technology design and use imply a temporal sequence. Engineers make software at design time and users execute it at runtime. A company develops an application and a customer downloads and runs it. The program office collects requirements, and operators use the systems that fulfill them. Yet the reconfiguration of processes or products by the same people who use them might be better described as "runtime design." Military personnel often reconfigure the rules and tools of control in an attempt to improve mission performance as they understand it. In the process, however, they may inadvertently inflict negative externalities on their comrades or neutral parties in the same network. Even successful run-time design can threaten the equities of defense contractors and system management offices. Players in the field will resent managers who seem to not understand the game, and managers will worry that players will forfeit the game if they do not play by the rules.

While FalconView demonstrates that significant user innovation is possible in the military, it also highlights the challenges that user-developers inevitably face in a dense bureaucratic environment. The traditional acquisition and network management regimes viewed military user development activity as inherently unsustainable and illegitimate. FalconView supporters had to turn organizational barriers into opportunities in order to get funding and build institutional support. They leveraged organizational schisms and exploited policy loopholes. User enthusiasm for Falcon-View and antipathy for official systems sometimes seemed quite out of proportion with its mission planning utility. Yet behind the rhetorically useful morality tale, official and unofficial systems formed an alliance of convenience. In a series of mutual accommodations, some of the same officers managed both types of programs, the programs shared common software components and map data, and the formal programs looked to the renegades to cover programmatic shortfalls. FalconView was thus able to afford the luxury of ignoring some high-end computational or operationally sensitive functions, because the formal systems took care of them. Even so, FalconView management eventually took on a more bureaucratic character in order to comply with certification and testing policy. The gradual accommodation of traditional processes also locked in FalconView's 1990s-vintage architecture by foreclosing opportunities for rearchitecting as the mission planning community looked to embrace JMPS instead. Ironically, the emergence of user-centered alternatives to traditional bureaucratic procurement led FalconView enthusiasts to adopt bureaucratic characteristics.

For every story of users developing innovative solutions and building institutional support, there are countless more of inventions that either (1) withered away when their Dr. Frankenstein transferred, or (2) got crushed by venturing into an adverse regulatory jurisdiction. While on active duty, I teamed up with the inventor of Quiver, the naval aviation targeting application mentioned above. In a familiar pattern, Quiver became popular with intelligence personnel and aircrew in many different units. Some junior officers took the initiative to adapt and improve it. Quiver was opposed by the Space and Naval Warfare Systems Command (SPAWAR) because it competed with an ungainly Unix system. SPAWAR could not prevent local network administrators from helping local users to run Quiver since Microsoft Office was an approved application for warship workstations and Quiver was technically "just an Access database." SPAWAR ultimately decided to incorporate Quiver features into a future version of its own system. As with JMPS, the official effort never came to fruition. Unlike FalconView, however, Quiver never gained any powerful advocates who were willing to work in both formal and informal channels to protect it. A version of Quiver lived on in the Naval Special Warfare community as "Aljaba" (Spanish for "Quiver"), adapted to support special operations in Latin America. Another version ("A3") moved into programmatic channels in SOCOM and, with assistance from contractors embedded with working intelligence analysts, adapted to support Naval Special Warfare operations during the 2003 invasion of Iraq. Yet A3 also withered away when the system's supporters transferred or left the service. Informal user innovation occurs spontaneously in most military units, but the survival of prototypes and the flourishing of a user community require the consent and support of formal organizational authority. FalconView and Quiver are, respectively, stories of the success and failure of information system entrepreneurship. FalconView became a platform for adaptation founded on managed practice. Quiver became an aberration stifled by insulated practice. Whenever institutionalized and organic modes of organization become oppositional, the stronger actor usually has the advantage.

The distinction between user innovation and formal acquisition is often wielded rhetorically in procurement battles, but it is a false dichotomy. Both modes of production have strengths and weaknesses. Program offices can maintain a stream of resources and support across personnel rotations. The Siren's song of bureaucratic system integration, unfortunately, grows ever louder as control problems become more complex. User innovation can produce remarkable prototypes that help people to cope with complexity. Unfortunately, many amateur projects are indeed unstable, unsupportable, nonscalable, or insecure. The next chapter will explore the darker side of organic adaptation.

# Irregular Problems and Biased Solutions

## Special Operations in Iraq

The U.S. intervention in Iraq completed a full cycle through the information practice framework between 2003 and 2008. During the invasion and its aftermath, managed practice turned into insulated practice, which prompted both internal and external actors to adapt. During the subsequent occupation, adaptive practice turned into problematic practice, which in turn encouraged the U.S. military to institutionalize doctrinal reforms. This chapter explores the ways in which insulated practice still persisted at the end of this process, curiously enough, even in a tactical unit close to the fight that had ample opportunity to make sense of facts on the ground.

The George W. Bush administration had ambitious plans to use a smaller but smarter military to invade Iraq and overthrow the Baath regime. Initial combat operations appeared, superficially, to vindicate the RMA playbook. U.S. forces employed coordinated fires and joint maneuver to overrun Iraqi forces and capture Baghdad in just three weeks. To be sure, units in the field still had to cope with misunderstandings, interoperability problems, sandstorms, and irregular resistance.[1] Yet for a brief moment the United States appeared to have achieved a lot with a little. The institutionalized solution of joint military operations had worked reasonably well for the defined problem of major combat operations versus Iraqi forces.

When political unrest in Iraq changed the strategic problem into something else entirely, however, the coalition bureaucracy suffered increasingly from insulation. Local resistance networks were joined by an influx of foreign jihadists and Iranian operatives. Adaptive practice in the insurgency was exemplified by the improvised explosive device (IED). The coalition headquarters in Baghdad, meanwhile, mismanaged reconstruction, inflamed resentment, and became consumed with infighting. Coalition units in the field took it upon themselves to adapt tactics, techniques, and procedures to counter insurgent adaptation. Troops improvised countermeasures against IEDs and suicide attacks, and they rediscovered or reinvented many classic counterinsurgency (COIN) techniques. Many battalions physically altered the constraints on their local information problem by creating berms around towns and building concrete barriers and combat outposts. They also made deals with local elites for manpower, intelligence, and protection that altered local balances of power. These efforts sometimes produced a modicum of stability in local areas of responsibility, but the results were often uneven and temporary.

A band of self-described "COINdinisitas" soon emerged within the Department of Defense and advocated for sweeping changes in training and operations. Whereas Secretary of Defense Donald Rumsfeld's "defense transformation" program envisioned a fast and efficient joint force empowered by the RMA, the counterrevolution in military affairs envisioned years of boots on the ground and intensive interagency coordination. This reform movement culminated in a new doctrine laid out in *Army Field Manual 3–24: Counterinsurgency* and the "surge" of additional U.S. troops under the command of General David Petraeus. The ensuing drop in violence appeared, again superficially, to vindicate the COIN playbook. The historical sequence of surge and pacification encouraged a popular morality tale about the failure of the network-centric RMA and the success of population-centric COIN.[2]

The reality was more complicated. Other developments also contributed to the dramatic drop in violence, including the simultaneous culmination of a sectarian civil war, a temporary ceasefire by Shia militias, and aggressive Sunni vigilantism targeting al-Qaeda jihadists. Technology, furthermore, if it is to be useful in war, always depends on the efforts of people, while "war amongst the people" always relies on a lot of technology.[3] Insurgents took advantage of the internet and cell phones for recruiting and operational coordination. U.S. forces built Orwellian surveillance networks to enforce population control. Tactical units gained access to a hundredfold more systems than were available in 2003, and staff officers stitched together software systems to coordinate operations. General Petraeus, the foremost champion of population-centric COIN, also turned out to be enthusiastic about the network-centric RMA:

> It's definitely here to stay. It's just going to keep getting greater and greater and greater. . . . I was a skeptic of network-centric warfare for years. . . . [But with the ability] to transmit data, full-motion video, still photos, images, information . . . you can more effectively determine who the enemy is, find them and kill or capture, and have a sense of what's going on in the area as you do it, where the friendlies are, and which platform you want to bring to bear. . . . We realized very quickly you could do incredible stuff with this. . . . It was revolutionary. It was.[4]

The most impressive appropriation of both the technology and ideology of network-centric warfare occurred in Joint Special Operations Command

(JSOC).[5] Under the leadership of Lieutenant General Stanley McChrystal, JSOC adapted new reconnaissance-strike networks to the age-old problem of manhunting. McChrystal borrowed a slogan from the RMA theorist John Arquilla: "It takes a network to defeat a network."[6] JSOC's centralized joint operations center at Balad air base, much like an air force CAOC adapted for special operations, combined intelligence reports from multiple agencies and directed multiple special mission units. Operating at a frenetic pace with multiple raids nightly, JSOC used intelligence to target raids and raids to collect intelligence. Networked counterterrorism as practiced by JSOC offered a seductive alternative to the messy ordeal of nation building championed by the COIN school. Some argued that COIN and counterterrorism were complementary: COIN could generate intelligence for JSOC, which could provide breathing room for COIN.[7] Yet conventional army and Marine units were often left to deal with the messy implications of JSOC raids in their local operating areas. As COIN efforts to win hearts and minds became the public face of the war in Iraq, the United States conducted relentless counterterrorism operations in the shadows. Between 2003 and 2008, JSOC removed nearly twelve thousand insurgents from the battlefield, killing perhaps a third.[8]

Other special operations forces (SOF) and conventional units attempted to imitate or reinvent the JSOC innovation.[9] Many of them lacked the experience or resourcing of JSOC, which only exacerbated the tension between counterterrorism and COIN. This chapter tells the story of one such unit. From autumn 2007 to spring 2008, I mobilized to active duty with the Office of Naval Intelligence (ONI) and led a Tactical Intelligence Support Team. My unit deployed with a Navy SEAL Team that formed the core of a Special Operations Task Force (SOTF) in Anbar Province.[10] On the SOTF staff, I served as the "nonlethal effects officer" responsible for managing tribal engagement, information operations, and civil affairs. The concept of "tribal engagement" took on a dual meaning for me as I navigated the military subcultures of the U.S. occupation in an attempt to understand the Sunni tribes of Anbar.

This chapter draws on personal observations at the SOTF of over two hundred missions resulting in the capture of over three hundred detainees over the course of a six-month deployment.[11] In my official capacity I was able to visit operating locations throughout Anbar, which gave me the opportunity to survey the SOTF's information system. I gained firsthand experience of information friction within a strong organizational culture and an ambiguous counterinsurgency environment. As the information practice framework predicts, given these conditions, insulation was the overall theme. Yet the other three types of practice also manifested to varying degrees. As in the previous chapters, friction prompted practitioners to make organic adjustments, but unlike in the previous chapters, adaptation did not necessarily improve performance. Instead, amateur repairs to the SOTF's noisy control system tended to reproduce and amplify the SOTF proclivity for counterterrorism over COIN. After exploring this pathology in detail, I will briefly compare the SOTF to other units that conducted a similar mission (JSOC) or operated in the same environment (Marines) to demonstrate how different institutional choices can generate different qualities of information practice.

## The External Problem

Information practice works best when an external problem is constrained and a complementary internal solution is institutionalized. Irregular warfare, however, is poorly constrained by nature. The breakdown of government institutions erodes social legibility. Insurgents blend in with civilians and subvert security forces. The complexity of the second Gulf War was also self-inflicted. The Bush administration politicized intelligence about weapons of mass destruction, which did not exist, and neglected occupation planning over the misgivings of Iraq experts, who did.[12] Planning slides from August 2002 anticipated an orderly drawdown of U.S. forces from 270,000 troops at the close of major combat operations to only 5,000 troops eighteen months later.[13] The illusion of control persisted through spring 2003. The U.S. military easily defeated an Iraqi military weakened by years of sanctions and containment. Joint maneuver warfare was a reasonable solution to the problem of major combat operations against a regular but ineffective army, but the emergence of multiple irregular insurgencies changed the problem. Even during the invasion, U.S. forces ran into pockets of fierce resistance from Saddam Fedayeen irregulars, a harbinger of the political chaos lurking beyond the PowerPoint slides.

### THE AMBIGUOUS ONTOLOGY OF CIVIL WAR

Iraq under Saddam Hussein had been hollowed out by decades of sanctions and war. Ethnic and sectarian tensions simmered under the weakening grip of the Baath Party, which was dominated by Iraq's Sunni minority. Hussein had tried to burnish his Islamist and Bedouin tribal credentials prior to the invasion, which effectively reversed his earlier suppression of religion and tribalism in the name of Baathism.[14] American analysts failed to realize that while Hussein had dismantled his weapons program to appease the United States, he continued to bluff about its existence to deter Iran, which Hussein viewed as the greater theat.[15] After the U.S. invasion shattered Baathist control, Iraqis were left to fend for themselves. The Coalition Provisional Authority under L. Paul Bremer failed to prevent massive outbreaks of criminality, prematurely disbanded the Iraqi army, banned experienced Baathist officials from serving in government, and refused to countenance a legitimate role for Sunni tribal leaders in northwestern Iraq. The Shia majority asserted its right to govern Baghdad, which was supported by U.S. officials. Iran also saw an opportunity to intervene. Sunni Iraq became a fertile recruiting ground for insurgency and a base for foreign jihadists. The occupation thus excluded and disenfranchised the very people it needed to reestablish order. Army and Marine battalions outside the Green Zone struggled to deal with an increasingly bloody set of overlapping conflicts (Baathist, nationalist, Islamist, criminal) that the U.S. government refused to even describe as "civil war." Coalition forces conducted cordon and search operations and mass roundups of "military-aged males," and units retaliated against mortar attacks with indirect fire into populated areas, which heightened the risks of civilian casualties. Some enterprising commanders began to experiment with different approaches that emphasized protecting the population instead. Some also looked to lessons learned from colonial conflicts.

Classic COIN theory describes a triangular relationship between the underground rebellion, government security forces, and the civilian population.[16] The quintessential task for the counterinsurgent is to separate the guerrillas from the population by encouraging locals to become informants. To accomplish this, COIN doctrine recommends clearing insurgents from a strategic area, holding it against their return in partnership with local security forces, and building new institutions and economic opportunities to reduce the popular grievances assumed to be the causes of insurgency. This "clear, hold, and build" strategy, according to its advocates, requires large numbers of troops (i.e., one soldier for every ten civilians) employed in constabulary roles for many years. COIN is a nation-building project that tries to reform an entire society, thereby imposing order on the unconstrained problem of civil war. Insurgency, meanwhile, uses intimidation and infiltration to exploit the very institutions that COIN builds. The counterinsurgent has more strength but less information, while the insurgent has less strength but more information. The local insurgent is often more resolved than the corrupt government or the foreign occupier, for whom the war may not be a vital interest. Indeed, the U.S. military risked embarrassment in Iraq but never annihilation. As in the Aesop fable where a hare runs for its life while a hound runs for its dinner, the quarry has strong incentives to adapt faster than the hunter.[17]

One alternative to COIN, as practiced by counterterrorism units like JSOC, is to kill or capture insurgent leaders faster than the enemy organization can replace them. This approach has also been described as manhunting, leadership decapitation, or counternetwork operations. Throughout history, military and police counterterrorist forces have relied on human informants, interrogation, and even torture to extract actionable information about insurgent leaders.[18] A captured insurgent can provide information directly through interrogation. In exceptional cases, captors might be able to "flip" a prisoner and send him back as a mole. Administrative files recovered during a raid, similarly, can provide insight into insurgent financing and organization. Counterterrorism thus attempts to substitute actionable information about insurgents for masses of COIN troops that protect the population. JSOC, in effect, attempted to clear without holding or building. Compared to previous counterterrorism efforts that have relied heavily on human informants and coercive interrogation, moreover, JSOC relied relatively more on technical collection such as signals intelligence (SIGINT), satellite imaging, and full-motion video captured by drones. The substitution of technical intelligence (standoff collection) for human intelligence (local informants) is yet another manifestation of the substitution of information for mass, which is the hallmark of the RMA.

Both the triangular struggle of COIN and the bilateral contest of counterterrorism are gross simplifications of the social complexity of civil war.[19] COIN doctrine describes the population as "human terrain," but this is a poor metaphor. Local cultures are not passive geographical structures but rather social networks of active agents. They cannot be terraformed at will, but instead may resist, preempt, or redirect COIN efforts. Local informants can attempt to exploit the combat power of ignorant security forces to liquidate their own private grudges.[20] The same individuals may play friendly, neutral, or hostile roles in different social networks; for instance, a local elite might accept a paycheck for a government job, shelter a family member in the insurgency, provide dispute resolution services to members of his tribe, and attempt to divert toward his cronies the river of development aid flowing into the country.[21] The local groups that protect insurgents from security forces may be motivated more by norms of honor and loyalty to family in the insurgency than by popular grievance or ideological sympathy with the insurgent cause.[22] Some insurgencies may act more like profiteering cartels than political movements.[23] State building in actual civil war, as distinguished from the nation-building ideal of COIN doctrine, is a process of violent negotiation in which government elites may collude with militants or willingly cede territory to militant control.[24] Compared to traditional force-on-force combat, the ontology of foreign intervention in civil war is inherently more ambiguous. Civil war research has come a long way in recent decades, but it remains a far less formalized and more contested intellectual pursuit than the physical sciences. While the characteristics of material weapon systems are amenable to mathematical modeling, predictive modeling of social behavior in civil war is less practical, even with the help of machine learning techniques.

Any political problem for which COIN seems like the answer is already considerably out of control. Armed rebellion does not simply hide within the population but rather is symptomatic of serious weaknesses in state institutions. The emergence of anarchy in civil society erodes the formal legibility that state institutions provide and, instead, heightens the salience of local interactions. Local elites, power brokers, criminal rackets, corruption networks, kinship relations, affiliative ties, and culturally grounded perceptions all take on a new relevance. Foreign troops who do not speak the local language become doubly disadvantaged. Not only are they vulnerable to deception by the population they seek to influence, but they also must rely on interpreters who themselves may have divergent interests. Formidable body armor and force protection requirements create yet more distance between COIN forces and their potential informants. In sum, the pertinent ontology of civil war is nuanced, fluid, and difficult to access.

## UNCOOPERATIVE ADVERSARIES AND OPPORTUNISTIC ALLIES

Anbar was the epicenter of Sunni insurgency in the early years of the occupation, epitomized by the bloody battles of Fallujah in 2004. Then, surprisingly, Anbar became one of the least violent provinces in Iraq. This surprising turnaround occurred before the troop surge of 2008 and differed in important ways from the self-serving narrative of COIN success that emerged in the wake of the surge. Changeable loyalties create problems for referential integrity.

Almost all of Anbar's 1.2 million inhabitants resided in towns like Fallujah and Ramadi along the Western Euphrates River Valley (WERV). The WERV is a longstanding historical corridor for both licit and illicit trade through al-Qaim on the Syrian border to Baghdad. The vast deserts to the north and south are inhospitable, so telephone systems, highways, dwellings, and other infrastructure generally follow the geography of the WERV. This constraint helped slightly to focus U.S. intelligence collection and troop deployments, although the WERV was still a very large "haystack" hiding a few thousand insurgent "needles." Anbar's population was almost exclusively Sunni Arab, whereas the rest of Iraq was fractured by ethnic Arab-Kurd and sectarian Sunni-Shia schisms. This provided another weakly simplifying constraint.

Anbar's superficial homogeneity was complicated by an influx of radical Sunni extremists after the 2003 invasion. The longstanding prominence of rural Bedouin tribes organized around patriarchal social networks also militated against government legibility. Saddam Hussein had attempted to subordinate the tribes to the Baathist state via patronage and coercion, but as international sanctions eroded Baath power throughout the 1990s, Hussein instead catered to the tribes by ceding them de facto control over the WERV's lucrative smuggling economy. The new Baghdad government, which Anbaris derided as "Persian usurpers," had neither Hussein's coercive leverage nor his Sunni identity to manipulate the tribes. Rising Islamic fundamentalism reinforced Sunni resentment of the U.S. occupation and the Shia government, especially in Fallujah ("the city of mosques") where the Marines fought two large battles in 2004 against nationalist militias, former regime elements, and jihadist groups. The most virulent was al-Qaeda in Iraq (AQI), led by a Jordanian veteran of Afghanistan known as Abu Musab al-Zarqawi. AQI fanned sectarian fires with videotaped beheadings, mass-casualty attacks, and the destruction of the Askariyya shrine in Samarra (an important Shia holy site). Al-Zarqawi acted over the objections of al-Qaeda's central leadership, and a nationalist rebellion transmogrified into a sectarian civil war.[25]

In the early years of Operation Iraqi Freedom, the U.S. military made simplistic distinctions between "coalition forces" and "anti-Iraq forces," as if loyalties were limited to supporting or resisting the government or the violence was directed only toward the coalition. Yet Arabs and Kurds, Shia and Sunni Arabs, populist and clerical Shia factions, foreign and Iraqi Sunni insurgents, tribal clans, and organized criminals all used violence instrumentally against coalition forces or one another at various times. Distinct but overlapping conflicts in the Kurdish North, Sunni West, Shia South, and ethnically mixed center affected each other. The WERV fed all of them. Social identities in Anbar could incorporate elements of Sunni sectarianism, Iraqi Arab nationalism, Islamic fundamentalism, Baathist loyalty, or tribal kinship. Identities could shift with the play of local politics or economic opportunity. Individual loyalties could also change surprisingly as neighborhood balances of power shifted, insurgents defected to the side of the government, or Iraqi security forces turned on their American trainers. For example, Muhammad Mahmoud Latif was a Sunni nationalist responsible for killing U.S. troops in Ramadi in 2004, but he organized an anti-AQI coalition in 2005. Another leader of the 1920 Revolution Brigade named Abu Maruf had once been number seven on the targeting list of a U.S. battalion; yet in 2007 he began using his militia to assist the same battalion in locating AQI, even executing an insurgent on another battalion's top-ten list. New ontological types emerged in the course of the war as U.S. forces struggled to represent the situation. As "red on red" conflicts emerged between insurgent factions, or "green on green" property disputes between rival clans turned violent, U.S. forces had to look beyond their traditional focus on "red versus blue" conflict.[26]

U.S. forces killed al-Zarqawi in June 2006 after a lengthy JSOC manhunt. Still AQI violence continued to increase. A classified report two months later by Colonel Peter Devlin, an intelligence officer with the First Marine Expeditionary Force, was flatly pessimistic:

> The social and political situation has deteriorated to a point that [Multi-National Forces] and [Iraqi security forces] are no longer capable of militarily defeating the insurgency in al-Anbar. . . . Underlying this decline in stability is the near complete collapse of social order. . . . Nearly all

government institutions from the village to provincial levels have disintegrated or have been thoroughly corrupted and infiltrated by Al Qaeda in Iraq.[27]

The month after Devlin's report, violence in Anbar peaked at nearly 2,000 recorded incidents, or significant activities (SIGACTS), which was more than in any other province in Iraq.[28] Then, remarkably, SIGACTS plummeted dramatically, registering just 155 in January 2008, the lowest rate since the beginning of the insurgency.

The cause of the drop in violence in Iraq remains a matter of debate. An initial wave of triumphalist accounts, written mainly by American authors focusing on American experiences, credited adoption of population-centric COIN in place of sweeps and raids.[29] The drop in SIGACTS correlated with a decision by the Bush administration to "surge" an additional five combat brigades to Iraq and double down on COIN operations. General Petraeus, the charismatic commander of Multi-National Forces Iraq during the surge and a champion of COIN doctrine, received many kudos for the turnaround.[30] COIN critics argued that correlation is not causation, highlighting factors other than COIN, to include the culmination of ethnic cleansing in Baghdad,[31] a ceasefire by the Shia militia leader Moqtada al-Sadr, a degree of restraint on the part of Iran, the success of JSOC's counterterrorism operations,[32] and the local militias turning on their former allies.[33] Anbar in particular went over the tipping point before the first surge troops arrived in Iraq in mid-2007, and most surge troops were deployed around Baghdad, so the surge itself cannot be credited for the pacification of Anbar. Some scholars have emphasized the synergistic effects of COIN doctrine with other factors or variation of their salience across regions.[34] For example, the brigade commanded by Colonel Sean MacFarland in 2006 did indeed "clear, hold, and build" to win the Battle of Ramadi, yet success also relied on the Anbar Awakening (Sawaha al-Anbar) to mobilize tribal support against AQI. More recent scholarship has drawn on Iraqi perspectives to correct the U.S.-centric bias that marked the early English-language historiography of the war.[35] Religious identity, tribal honor, family safety, revenge, and greed have all emerged as different motives for resisting or assisting U.S. efforts in local areas, or working around them altogether to engage in vigilantism, enrichment, and self-defense.

Iraqi agency was critical for the turnabout in Anbar. Sunni tribesmen made a political calculation about their dismal future with AQI in a Shia-dominated Iraq without American protection, and they decided to take advantage of American firepower while it was still available. By 2005 tribal militias, to include some members who had once aligned with AQI against U.S. forces, had started to attack AQI in reaction to its brutal imposition of Sharia law and encroachment on tribal resources. AQI retaliated aggressively, but, surprisingly, the tribes began soliciting support from U.S.

military units. They won victories together, first in al-Qaim with a tribal militia known as the Desert Protectors, and then in Ramadi and Fallujah with the help of vigilante groups like the Anbar Revolutionaries and more overt political movements like the Anbar Awakening. AQI alienated its Iraqi hosts and pushed them into a marriage of convenience with the U.S. Army and Marines. Just as the United States had defeated the Iraqi military but faltered in the occupation of Iraq, AQI had won the war in Anbar but lost the peace. Central Iraq was riven by even more complex Sunni-Shia and intra-Shia conflicts, but soon after the improvement in Anbar, the violence dropped in other places as well, often for subtly different reasons.

The fact that the causes of the (temporary) pacification of Iraq are still contested a decade later suggests that many contemporaries did not, or could not, understand what kind of war they were in. The invasion of Iraq hastened the breakdown of a corrupt authoritarian regime riven by sectarian fault lines. U.S. troops struggled to understand which Iraqi actors or interactions were most influential in a given locale, why a given group might be willing to help or fight them in different circumstances, or what might be the political effect of killing or supporting a particular person.

## The Internal Solution

It is reasonable to code the operational problem in Anbar as unconstrained, in part as a result of endogenous interaction with the U.S. intervention. Coding the SOTF's internal solution is more complicated. In the previous chapters we have already seen cases of mixed institutionalized and organic organization. Constant adjustments by Fighter Command's radar scientists and operators tended to stabilize management in the Ops Room. Falcon-View was a user innovation that both competed with and benefited from the systems management bureaucracy. So too in Iraq, different parts of the coalition occupation and the SOTF itself could be described as more or less institutionalized or organic. The very concept of an institutionalized solution, moreover, is an aggregate type that bundles together different ideas about organizational centralization, formalization, size, and identity. This case is complicated because some of these components point in very different directions for the SOTF.

THE BUREAUCRATIC ENVIRONMENT

The SOTF was a small unit in a large occupation (figure 5.1). The larger coalition forces solution to the larger problem of civil war, therefore, must itself be considered a part of the SOTF's own external problem. The internal constraints of any large organization are effectively also external constraints for smaller parts of that organization.

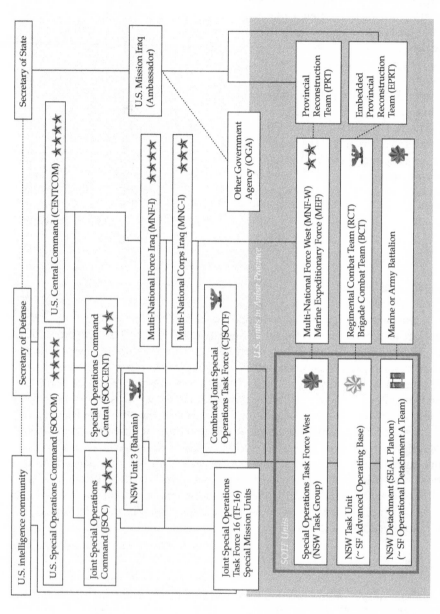

Figure 5.1. U.S. command relationships in Anbar

Nevertheless, weak centralized authority in the occupation ecosystem gave it a somewhat organic quality. In addition to conventional and SOF units in Iraq, there were also administrative and support commands located throughout the theater and back in the United States, as well as State Department provincial reconstruction teams, distinguished visitors, journalists, nongovernmental organizations, private contractors, and the Other Government Agency (a euphemism for the CIA). Many units had their own unique information systems, formats, requirements, and chains of command. This system had too many parts, with too many subcultures, to be consistently and uniformly institutionalized, but that did not prevent eager staff officers from trying. Interagency friction just generated further demand for process reform. Units exchanged liaison officers to improve coordination, but too often the person that a unit was willing to give up would not be the most competent officer, as evidenced by the unit's willingness to let the person go. Action officers tried to visit each other for face-to-face coordination, but a helicopter trip just a few dozen miles away might require hours or days of working through schedule conflicts and sandstorms. The size and complexity of the administrative overhead tended to turn staff officer attention inward.

Even so, all of this churn was more legible and predictable from the SOTF's perspective than was the Iraqi political environment. Other American units still spoke the same language and had a shared professional culture. Even if they had sharp disagreements about policy, they were ultimately on the same team. The occupation bureaucracy thus imposed a degree of stability on the SOTF's representational problem. Moreover, because there was a great deal of bureaucratic friction in the occupation, the administrative superstructure also absorbed a lot of the SOTF's time. Whereas adaptation enabled management in the case of Fighter Command, constant adjustment in this case reinforced insulation. Many Americans in Iraq experienced Iraqi society only indistinctly, therefore, and were only dimly aware of the political dynamics outside the wire.

The SOTF was hosted, but not operationally controlled, by Multi-National Forces West, the "battlespace owner" commanded by a division-sized Marine Expeditionary Force (MEF). SOF typically enjoy considerable responsibility and influence for their rank relative to conventional forces in the same area; for example, the SOTF commander was a navy commander while the MEF commander was a Marine Corps major general. SOF commanders were able, in principle, to set their own campaign objectives apart from their conventional hosts. The SOTF could also leverage dedicated SOF aviation and intelligence assets from its headquarters in Balad (Combined Joint Special Operations Task Force-Arabian Peninsula, or simply CJSOTF) through a separate chain of command. Note that the "white SOF" headquarters (CJSOTF) did not command the "black SOF" units from JSOC; the former conducted "theater" missions in coordination with conventional forces in Iraq while the latter conducted "national" missions with different authorities and a different chain of command.[36] In practice, the four hundred personnel of the SOTF depended on thirty thousand Marines for basing, sustainment, helicopter lift, intelligence, quick response forces, casualty evacuation, and other forms of life support. The MEF was thus another important factor that constrained and enabled SOTF operations.

## MILITARY SUBCULTURES

Subcultures are reservoirs of ideas and practices that shape doctrine and adaption. Austin Long demonstrates that military subcultures are resistant to change and tend to selectively embrace COIN methods that accord with their essential identities.[37] The Marine Corps and Naval Special Warfare (NSW), despite both being part of the U.S. Department of the Navy, approached COIN quite differently. Both the SOTF and the Marines were charged with building up Iraqi security forces and consolidating the political gains of the Anbar Awakening. Both the Marines and the SOTF hunted insurgents who could sabotage the peace, but the SOTF had a strong predilection for JSOC-style raids over COIN methods. The Marines focused more on COIN while the SOTF emphasized counterterrorism, which made for an uneasy symbiosis.

After the violent battles of Fallujah in 2004, the Marines reoriented their focus in Anbar with a tribal engagement strategy. They drew on a cultural heritage of fighting "small wars" in a storied history of overseas expeditions (e.g., "From the Halls of Montezuma, to the Shores of Tripoli"). The First MEF in Anbar developed a COIN-focused campaign plan in 2006. The Second MEF improved on it in 2007. Its commander, Major General Walter Gaskin, added an additional deputy commander, Brigadier General John Allen, to his staff to lead tribal engagement and civil affairs. Allen served as a coequal with Gaskin's maneuver and aviation deputies, leading the non-kinetic "exploitation" of the combat "breakthrough" achieved by the Awakening. Allen met with senior sheikhs throughout Anbar and encouraged those who had fled to Jordan to return, offering moral suasion, material patronage, and military protection to reassure them and bolster their position within their communities. Gaskin directed Marines at the battalion, company, and platoon levels to follow Allen's example with their local counterparts.[38] According to Allen,

> We were faced suddenly with the reality that if we didn't do something quickly about governance and reconstruction . . . we were going to face a pretty serious difficulty, because the people were throwing off al Qaeda

and yearned for some kind of change. . . . We can't afford . . . to have the kinetic phase suddenly be followed by a large, pregnant silence, where there's no governing and nothing being built and no money coming into the province.[39]

By early 2007, after years of frustration with an enemy-centric approach, most army and Marine units in Iraq had embraced a more population-centric approach to COIN. At the same time, ironically enough, SOF developed in the opposite direction. Conventional forces embraced unconventional warfare by necessity while SOF become hyperconventional by choice.[40] JSOC honed "direct action," or combat operations, to a high art. There was no similar institutional champion for "indirect action," or negotiation and persuasion. Even Army Special Forces, which had been created during the Cold War to institutionalize expertise in unconventional warfare, had somewhat shifted away from its founding emphasis on working "by, with, and through" local partners. Military jargon tends to describe informational, advisory, and other soft-power missions as *in*direct, *non*lethal, *non*kinetic, *un*conventional, *ir*regular, or other-than-war. The negative prefix marks the exceptional status of these activities. SOF personnel sometimes described the difference more pejoratively as "pipe hitters" versus "social workers," or "snake eaters" versus "lettuce eaters." Compared to commando direct action, indirect action takes longer to demonstrate progress, is harder to train, lacks clear tactical metrics, and does not readily lend itself to heroic warrior narratives.[41]

The general SOF bias for direct action was particularly pronounced in NSW, and thus also in the SOTF, which was embodied by an NSW Squadron (i.e., SEAL Team). From late 2005 through early 2007, NSW deployments to Anbar had emphasized kinetic operations to kill or capture insurgents. The teams engaged in the most ferocious combat they had experienced since Vietnam, with numerous acts of heroism. Petty Officer Second Class Michael Monsoor, for example, posthumously received the Congressional Medal of Honor for jumping on a grenade to save his teammates. This period has been dramatized in the Clint Eastwood film *American Sniper* and a flattering book by Dick Couch, *The Sheriff of Ramadi*. Yet by Couch's own account, local commanders realized that sniper operations were becoming a magnet for insurgents and had to be curtailed.[42] The SEALs generated excess fighting that became counterproductive for the ostensible mission of protecting U.S. troops. During the same time, unilateral Anbari vigilantism and cooperation between tribal militias and army and Marine battalions resulted in some major setbacks for AQI. By early 2007, the Awakening had pushed AQI out of major towns along the WERV and into rural redoubts in the desert. Still, the SOTF continued to focus on targeting AQI throughout 2007 and 2008, even as its operators spent most of their time training Iraqi partners to carry out the same mission.

## THE CULT OF THE FROGMAN

NSW culture engendered a shared sense of purpose at the SOTF. Preferences and values rooted in the SEAL identity informed every phase of the SOTF control cycle. The NSW community cultivates and reinforces a very particular worldview throughout a frogman's career. Prospective SEALs enter an intensely competitive selection process (Basic Underwater Demolition/SEAL) that emphasizes physical endurance, teamwork, and mission accomplishment at all costs. Graduates who pass a formal review board and another qualification course earn the right to wear the Trident badge, one of the few navy warfare insignia worn by both officers and enlisted personnel. A new SEAL must continue to prove himself in the field and in additional training to acquire sniper, freefall, demolition, and other combat skills. The corridors of NSW buildings are decorated with trophies and lore to reinforce the brotherhood's identity. Popular culture reinforces the frogman swagger by venerating SEALs in action movies, video games, and hagiographic tales.[43]

Most SEALs reluctantly train partner forces and conduct various forms of indirect action on their deployments with "white SOF" units, but they actively lobby for direct action missions whenever possible. As one operator explained to me, "SEALs are like thoroughbred horses, and when you put them on the racetrack they want to run." Most SEALs aspire to be selected to join the NSW Development Group (DEVGRU, popularly known as "SEAL Team 6"), a special mission unit attached to JSOC. DEVGRU is responsible for many famous SEAL missions, including the killing of Osama bin Laden.[44] William H. McRaven, the SEAL who succeeded McChrystal to command JSOC, wrote a "theory of special operations" that asserts, "All special operations are conducted against fortified positions."[45] McRaven's book neglects information operations, civil affairs, combat advising, and foreign military training.

There is a clear totem pole of prestige in the NSW community. Special "operators" are at the top, and supporting "techs" are at the bottom. Techs include not only administrative, intelligence, communications, computer, logistics, and medical personnel, but also explosive ordinance disposal technicians and tactical cryptologists ("ears on target"), whose deployments can be quite dangerous. In the hypermasculine culture of NSW, moreover, this status hierarchy reinforces gender roles; all operators are male (as of this writing), while supporting techs include men and women. Tech work is almost invisible in the image that NSW projects to the world, even as techs outnumbered operators by more than two to one in 2007.[46] Operators do welcome and appreciate the contributions of techs, so long as they work hard to advance the NSW mission. Sailors from other parts of the navy, meanwhile, usually find a tour in NSW to be a refreshing change. Yet skilled professionals in tech communities remain forever outsiders to the frogman fraternity.

This pecking order had unfortunate implications for the quality of information practice at the SOTF. Information specializations like intelligence and network administration were concentrated in the tech community. SOTF operators completed an intensive eighteen-month training cycle, which focused on direct action, prior to their six-month deployment. Most of the techs, by contrast, were gathered together shortly before deployment with uneven training, if any. For instance, some "individual augmentees" from other parts of the navy arrived without being able to read Military Grid Reference System coordinates, a basic skill in land warfare. Sometimes, however, reservists arrived with uniquely useful skills. One junior sailor on our team happened to be a vice president for information technology at a major Manhattan bank; another was a federal lawyer and special agent; others were policemen. Unfortunately, reservist qualifications on paper did not always translate into proficiency in practice. One officer likened the SOTF to deploying an aircraft carrier with a random collection of airplanes and aviators who had never flown together; he implied that this would be insane given all the teamwork required to competently operate a Carrier Air Wing.

Four separate attempts to reform NSW intelligence prior to 2007 had resulted in a hodgepodge of collection and analytical capacity at the SOTF. The first was augmentation from the broader naval intelligence community, which generally preferred to send its most promising officers to sea rather than to NSW. Second, the Mission Support Center in Coronado, California, was established to provide "reach back" support to NSW units in the field. Its remote analysts proved to be too far out of the loop to provide anything more than simple products. Third, the Office of Naval Intelligence stood up another unit known as Trident to field Tactical Intelligence Support Teams like mine. ONI's location in Washington, DC, proved similarly distant from deployed units. Fourth, NSW converted congressional appropriations earmarked for two additional SEAL Teams in 2006 into two new types of units known as NSW Support Activities. They fielded so-called Cross-Functional Teams for Advanced Special Operations (human intelligence), Tactical Information Operations (field cryptography), and Technical Special Reconnaissance (tactical drones and tracking devices).[47] These teams were manned by operators who had been retrained for intelligence work, rather than intelligence professionals trained for special operations. NSW thus finally seemed to find an intelligence concept it could embrace, led by and for operators. This new concept was still embryonic and haphazardly integrated during the 2007–8 SOTF deployment discussed here. As we shall see, the SEAL identity of its collectors influenced the type of information that the SOTF gathered and reported.

A CULTURE OF IMPROVISATION

Although the SOTF's guiding intentionality was strongly institutional- ized, its information system actually evolved in a very organic fashion. For most of its history NSW was the redheaded stepchild of the blue-water navy. Long neglected and underfunded, frogmen historically had to beg, borrow, and steal resources. They largely had to train and equip them- selves. NSW thus developed a meritocratic, upbeat, and informal culture that encouraged personnel to improvise and accomplish their mission by any means necessary. After U.S. Special Operations Command (SOCOM) adopted NSW, and after the SEALs enabled the navy to make a contribu- tion to the Global War on Terror, the resources increased dramatically. Yet in keeping with the historical legacy of the NSW community, operators and techs alike were encouraged to exercise initiative, interpret rules creatively, recruit allies, and generally push limits to get the job done. The nature of any job, of course, was always informed by the commando identity of the SEALs.

Informal, organic experimentation was commonplace at the SOTF, focusing first and foremost on improving the direct action mission. Indirect action was a much lower priority. The SOTF's information specialties were also concentrated in the neglected tech community. Indeed, information prac- tice itself can be considered a form of indirect action. The management of data, even data about direct action, matters only indirectly for organiza- tional outcomes. Data are only meaningful *for* someone when they are *about* something and afford control *of* something that is located elsewhere in space, time, or abstraction. Information technology mediates human expe- rience of entities in the world that cannot be perceived or influenced directly. Because the NSW community devalued indirect action in general, the technical skills needed to improve indirect representations, even those required for direct action, were also in short supply.

Another factor promoting organic adaptation was the SOTF's geographic dispersion. The battalion-sized SOTF had a headquarters element in Fal- lujah and three company-sized NSW Task Units located in Fallujah, Ramadi, and al-Asad. They each had multiple platoon-sized NSW detachments deployed across eight Marine Forward Operating Bases (FOBs) along the WERV. This archipelago of bases complicated coordination and provided space for individual detachments to develop their own information pro- cesses and relationships with local Marine and army units.

Pointing weakly in the other direction toward institutionalization, how- ever, the U.S. military's reliance on the Microsoft Office suite and public internet protocols provided a common platform for communication. Yet the flexibility of the software enabled a great deal of local adaptation and a pro- fusion of local representations, discussed in more depth below. The SOTF, again like many other units, was also gaining access to many new techno- logical capabilities, such as unmanned drones and other sensors. All of these capabilities had the potential to establish connections with friendly and enemy entities far away. Yet any technological potential for reducing the dead space in which insurgents might hide was highly dependent on organizational implementation.

## A MIXED SOLUTION

In sum, the SOTF internal solution had both institutionalized and organic aspects. To tease out the implications for information practice, it is helpful to distinguish between the intentionality and the implementation of control. Intentionality directs the control cycle and determines what an organization considers meaningful. The sociotechnical implementation of measurement, coordination, and enforcement processes embodies that meaning. Intentionality explains why, and implementation explains how. Table 5.1 summarizes the situation at the SOTF. First, the SOTF was situated within a dense administrative ecosystem, so the broader coalition solution shaped the SOTF's information problem. The availability of common commercial technologies and networking protocols was part of this weakly constraining milieu. Second, units within this ecosystem had different subcultures, interests, and degrees of autonomy. The SOTF in particular had very strong preferences for the types of missions it wanted to conduct—namely, direct over indirect action, or counterterrorism over COIN, performed by operators over techs. Third, NSW underinvested in indirect action, intelligence, and other information-intensive skills concentrated in the tech community. Thus SOTF personnel tried to make sense of these missions on an ad hoc basis. Fourth, as a result, the SOTF's actual data management systems and protocols were implemented haphazardly with insufficient coordination.

Information practice theory suggests mixed expectations. The warfighting problem of civil war and reconstruction was unconstrained. A strongly institutionalized solution in this environment should promote insulated practice. The SOTF's ingrained preference for direct action was a poor fit for the irregular situation in Anbar, where indirect action had been and would continue to be critical for the stabilization of the province. At the same time, the geographic dispersal of camps along the Euphrates and the NSW culture of improvisation promoted organic adaptation, which has

Table 5.1 Intentionality and implementation in the SOTF

| Solution | Intentionality | Implementation |
|---|---|---|
| Institutionalized | Strong cultural preference for direct action | Occupation bureaucracy and shared commercial technologies |
| Organic | Neglect of indirect action and intelligence | Ad hoc data management |

the potential to improve fit. Problematic practice becomes more likely, however, when adaptation is not mindful of salient interdependencies and constraints, such as those created by the administrative superstructure of the U.S. occupation and its common technologies. The net result was that the SOTF had a shared consensus on what to do, but not on how to do it. The SOTF's mixed solution thus created the worst of both worlds: problematic practice reinforcing insulation.

## Information Practice

How did the SOTF's information system link organization and environment? Counterterrorism doctrine aims to use targeted raids to generate intelligence that enables new targeting opportunities. JSOC described this process as "find, fix, finish, exploit, analyze" (F3EA).[48] F3EA is the special operations version of the air force "kill chain" (F2T2EA) discussed in the previous chapter. While the assets involved are different, the basic control cycle is similar. In the measurement phase, a diverse array of intelligence platforms identifies and locates human targets. A joint operations center (JOC) plays a coordination role much like the CAOC. Enforcement is performed by special operators rather than strike aircraft. The cycle then repeats as new measurements are fed back into the system. Just as "working the seams" created problems for the "kill chain," moreover, friction could also undermine the "kill or capture chain."

In theory, repetition of the F3EA cycle enables counterterrorists to work their way up from foot soldiers, to lieutenants, to middle managers, to the leaders of the enemy organization. In practice, F3EA can potentially be iterated forever so long as the organizational infrastructure of the target survives. F3EA is, more or less, a recursive search algorithm that uses the results of one raid as input for another to, eventually, produce actionable intelligence about enemy leadership. Alan Turing famously demonstrated that it is impossible to decide whether any arbitrary algorithm and input will eventually produce an output or just keep on running forever.[49] To make a loose analogy, F3EA also has a halting problem. It is impossible to say whether any arbitrary SOF unit hunting any given target will eventually deliver strategic victory or just keep on hunting forever.

The SOTF attempted to implement the JSOC model of F3EA, but it did not perform as well. The SOTF was a much smaller unit in a different chain of command with different authorities, so this is perhaps not surprising. What is surprising, however, is the degree to which the SOTF realization of F3EA was shaped by the culture of Naval Special Warfare. An institutionalized preference for direct action and a fragmented sociotechnical implementation at the SOTF virtually guaranteed that F3EA would go on forever without halt.

MEASUREMENT

All intelligence, whether collected by technical sensors or human informants, must be transformed from some initial physical contact with the environment into mobile inscriptions that analysts and operators can view and combine to make sense of the battlefield. Standardized reporting channels and formats, together with shared habits of interpretation, make it possible to reliably and repeatedly deliver records of contact with the battlefield. Yet representational output is like a black box that hides assumptions built into the implementation. Each processing step has the potential to discard provenance metadata or introduce spurious information. Experienced intelligence professionals usually cultivate a feel for pitfalls in different collection disciplines. Yet experience was uneven in the SOTF. Operators wanted intelligence personnel to find actionable targets. Techs were reticent to question the validity of intelligence processed under this assumption. The SOTF's raw intelligence was cooked from the get-go.[50]

Within the SOTF's sensitive compartmented information facility (SCIF), cryptologic technicians received SIGINT from national, theater, or organic sources. They were gatekeepers more than analysts. "Sanitization" processes were intended to protect sensitive sources by making it plausible to those outside the SCIF that SIGINT reporting could have originated with, say, a human spy rather than an electronic intercept. Some staff members, if they had the right clearances, made regular pilgrimages into the SCIF to hear the latest unsanitized news. Access to highly controlled compartments became an in-group marker of social prestige, and the value of intercepts was often conflated with the height of the barriers protecting them. The utility of SIGINT depends on its fusion with other sources, even as the secrecy and fetishization of SIGINT creates barriers to collaboration.

Technical intelligence, surveillance, and reconnaissance (ISR) was described as an "unblinking eye" with a "persistent stare" at its targets. By one estimate, the amount of intelligence data flowing back from automated surveillance increased by 1,600 percent over the previous decade. Drones collected over two hundred thousand hours of video in 2007 over both Iraq and Afghanistan, but there was no automated way to find and catalog interesting results.[51] The SOTF scheduled collection missions against its targets through its higher headquarters (CJSOTF), and it communicated with drone operators located a continent away using a chat client over the classified internet (Secure Internet Protocol Router Network, or SIPRNET). Sometimes chat room voyeurs inadvertently blocked users who were actually guiding and monitoring the operation from reconnecting after a dropped circuit. SOTF analysts had to watch hours of sensor video (which they often described as "like looking through a soda straw" because the zoomed-in video lacked context). Usually they saw nothing out of the ordinary as Iraqis went about their daily lives. This boring task was often assigned to junior personnel who were expected to store snippets of interesting video (in their judgment) on the Task Unit's share drive. These data, stored away as files in folders for whatever target was being monitored at the time, and lacking a geographically tagged database of historical coverage, were as good as lost for future analysis by any different sorting scheme (e.g., by location or jurisdiction rather than by target). These were solvable engineering problems in principle, but SOTF personnel either lacked technical expertise or did not agree about common cataloging procedures. As a result, the eye in the sky blinked a lot.

As an alternative to technical ISR, human intelligence (HUMINT) was potentially useful for gaining access to underground organizations and insight into their motivations. Spies can elicit information directly in the traditional manner, or they can emplace technical sensors. Yet instability is built into HUMINT because it relies on spies, and spies rely on duplicity. Iraqi agents might finger "terrorists" to liquidate personal grudges, fabricate stories to make money or enhance their status, play different HUMINT organizations off one another, or infiltrate as double agents to lure U.S. troops into an ambush. HUMINT collectors could vet agents by cross-checking their reporting and asking them to perform nontrivial observable tasks, but these safeguards are imperfect, especially if not applied conscientiously. Furthermore, the same skills that were useful for manipulating informants could also be useful for the unscrupulous manipulation of colleagues in one's own organization.[52]

Target-hungry SEALs who had recently been retrained as HUMINT collectors (as members of a Cross-Functional Team) pressed Iraqi informants for the names and locations of insurgent operatives. Their reporting focused narrowly on suspected targets rather than the local political, economic, or tribal dynamics that might have enhanced understanding of local atmospherics or enabled indirect-action alternatives to killing or capturing someone. The irony is that collectors often discussed local politics and motivations for fighting or defecting in the course of vetting and establishing rapport with Iraqi sources, but this information remained segregated in source-management computer systems or was never written down. SEAL collectors believed that outsiders in their own organization had no need or interest to know the political and personal backgrounds of their sources. Their finished reporting was often vague and uncritical (e.g., "The man with a mustache wearing a *dishdasha* is a terrorist" could indict almost any Iraqi man). Yet collectors bristled at the suggestion that their informants might be out of their control ("My sources don't lie!") or that they might mislead U.S. collectors to settle private grudges. Skeptical questions, especially from second-class "techs," were discouraged. Reports were informally emailed as Microsoft Word documents (as a "draft" intelligence information report) to avoid the cumbersome reporting bureaucracy of official HUMINT filing. Some collectors simply passed information verbally to target-hungry operators. These informal practices left little trace for further analysis or analysts working outside the organization who might have benefited from the information, let alone audited it. The collectors' relaxed grooming standards, civilian clothes, segregated workspaces, isolated computer systems, and improvisational ethos ostensibly facilitated clandestine HUMINT collection and source protection. These same factors also created cultural barriers to collaboration and inhibited performance inquiry.[53]

Five years into the war in Iraq, most U.S. personnel had little familiarity with the Arabic language, let alone the Iraqi dialect. The constant rotation of personnel, many of them reservists, made language proficiency an uneconomical investment compared to improving tactical proficiency. Iraqi interpreters ("terps") became the gatekeepers to linguistic access to the population, and thus another source of equivocation in measurement. Some HUMINT collectors, unpracticed in reading Iraqi informants' body language or communicating nuance, even allowed their terps to run the meetings. Many collectors let their terps gloss over informant comments and offer their own opinions, so a terp might summarize a large amount of speech from a detainee with something like, "He says he doesn't know anything, but he's a lying piece of crap." Relevant interrogation and other reporting might be lost in databases as Iraqi names were transliterated inconsistently (e.g., a name could be rendered variously as "Mohammad," "Muhammad," "Mohamed," etc.), or family and tribal names were concatenated differently. Intelligence organizations might have addressed these problems by investing in Arabic language competency among analysts. Instead they offloaded fluency onto automated translation systems and sophisticated pattern-matching algorithms, which themselves required expertise to reduce false positives. Perversely, this created increasing returns to illiteracy.

Records from the measurement phase of the control cycle at the SOTF propagated through long and complicated chains of transformation. There were plenty of opportunities for human selection, or suppression, of some features over others. Only the simplified results propagated downstream into further representations. The multiple degrees of freedom in all these computational transformations, to include pervasive cut-and-paste operations, heightened the potential for data equivocation and loss of provenance. Yet throughout, the NSW tendency to construct a world full of targets exerted a steady influence. Thus NSW measured its own signal through the noise. Channels for gathering and translating targeting information were actively stabilized by humans and machines, and equivocation that might have been relevant to understanding the social milieu for indirect action was buffered out in the construction of dossiers about "bad guys."

COORDINATION

In contrast to the materially substantial plotting boards in Fighter Command's Ops Room, the SOTF had no one centralized representation of its world. Across its multiple computer workstations, and in multiple applications on the very same workstation, computer users made and copied digital files with abandon to organize their small slices of operational reality. Usually they did not, or could not, articulate all the assumptions that went into their constructions, and they discarded or lost track of data provenance along the way. The SOTF's distributed infrastructure had a decentering effect on representations across the archipelago of camps along the WERV, and even among personnel working in the same room. The control system was shaped by a curious combination of fragmentation (of implementation) and consensus (of intentionality) that called into question the referential integrity of targeting representations. To use a computer metaphor, the SOTF's sociotechnical operating system (the assemblage of tools used for coordination) was full of bugs, but the organization relied on it to run its critical targeting application (the substantive data they coordinated).

*The Sociotechnical Operating System.* The SOTF operated in a classified environment. Information, far from being virtual or weightless, was protected by an elaborate physical infrastructure of bases, barbed wire, and controlled entry. Computer users had multiple noninterconnected machines on the same desk in order to access networks rated for different levels of classified data. Users resorted to USB thumb drives to move data across "air gaps" between nominally disconnected networks, risking infection by malware or "spillage" of classified data. Because it was easy to create new digital documents but hard to verify the proper level of classification, most users defaulted to the highest level on their systems, resulting in the routine overclassification of mundane emails and PowerPoint slides. Other personnel had to make contingent judgments about how to handle or release information to neighboring units, let alone Iraqi partners. Second-guessing undermined the protection that classification marks were intended to provide.

Each of the SOTF's eight camps had its own tactical local area network (TACLAN), which was mostly inaccessible from the others. TACLANs had been designed to provide a small SOF unit with local computing capability and connectivity to home base as it went "downrange" for a brief special mission. The Marines, by comparison, implemented a more robust SIPRNET that was a historical product of large-scale occupation of the same bases in Iraq year after year since 2004. This difference was reflected in the types of computers in use: SOTF personnel relied on small ruggedized laptops designed for field deployment, while Marines used desktop machines and larger monitors more ergonomically suited for a long occupation. The

SOTF conducted the bulk of its business over TACLAN, so transfer of ubiquitous PowerPoint files to a Marine compound right next door routed slowly through satellites and the continental United States (CONUS). Voice networks were similarly segmented by classification and institutional provenance, preventing many routine communications for simple want of interoperability.

The SOTF's file server alone held over a million files, accumulating on average 526 new files per day. Half of them duplicated existing files, and dozens of files were duplicated thousands of times. All were saved in a byzantine labyrinth of folders that were nested ten deep on average. These warrens of data were functional for users who lived in them but inscrutable to others spelunking through the share drive. Search engines were nearly useless amid all the duplicate copies, divergent versions, and strange naming conventions. Users relied on other users to find things. Email, the primary means of staff communication, further amplified the clutter by providing users with additional data warrens and more capacity for duplicating data. As a new SEAL Team rotated in every six months, the SOTF had a lot of orphaned data. A newly arrived staff officer typically dumped data into folders named "old shit" or "Joe's stuff" and began building the same files and folders anew. There were instances of archived folders nested several levels deep after several turnovers, containing exactly the same files that had been rediscovered and resaved by different personnel during each rotation. When asked where to find some particular data, personnel often said, "It's on the share drive!" Yet data were effectively forgotten once the people who saved it departed, which can be likened to deleting the pointers in a file system that index memory blocks where data are stored.

In this disjointed setting, runtime redesign was the only emollient for information friction, but it also created friction anew. Ubiquitous commercial software, notably Microsoft Windows and Office, provided the SOTF with a common set of tools and enough flexibility for adaptation to unforeseen requirements. PowerPoint became a general-purpose graphics editor and ersatz database for recurrent intelligence products, target folders, and mission reports. Ad hoc Excel spreadsheets tracked the recurring flights, missions, targets, logistical tasks, and reports in the routinized military universe. FalconView, Google Earth, and ArcGIS provided geospatial graphics for intelligence data and tactical coordination graphics. Local expedients such as homegrown databases, improvised data formats, and other software experiments could be functional for their local author; yet marginally satisfied users had little incentive to contribute to public goods like organized files and common data management schemes. PowerPoint slides proliferated without provenance. Key data were squirreled away in the fields of structured databases named "miscellaneous." Excel spreadsheets encoded important data in the visual layout (i.e., location, color, size, and font of data), which was illegible for automated analysis without manual preprocessing. The meaning of icons and drawings in shape files was elusive without any supporting documentation. Personnel throughout the enterprise adjusted on the fly to deal with information friction, but a lot of the friction resulted from other users making similar adjustments. SOTF staff officers became preoccupied with debugging representations along well-worn paths rather than taking in the more panoptic view that flexible software applications could have afforded, if implemented differently for different intentions.

Digital media did not replace paper and other hard copy, but in many ways made them more important.[54] Wall maps, whiteboards, hand-annotated printouts, bureaucratic stamps, mission-support documents, and yellow sticky notes retained important advantages in their durability and ergonomic interface. Yet hard copy also tended to localize important information. Paper representations, together with the tacit knowledge embodied in local personnel that was needed for locating relevant data, reinforced the importance of physical place.

Digital communication did not replace travel, likewise, but rather increased the demand for it. Frequent travel, or "battlefield circulation," was necessary for commanders and staff to understand the situation at each camp. Digital technology made travel easier by facilitating more efficient scheduling of helicopter flights and passengers. Yet the pragmatic context of digital information was not easy to transmit. The daily routines, conversations, and interactions that were not explicitly recorded in reporting became crucial for understanding what role any specific inscription played in a unit's operations. Remote communication via email or video teleconference tacitly advertised the fact that there was even more to learn in person. Action officers sought personal contact with their counterparts up and down the chain of command in order to orient to and influence one another's intentions. Face-to-face interaction suggested questions and answers that would not come to mind through virtual interactions alone. At the same time, travelers felt compelled to return to home base as soon as possible to access the mounting pile of emails trapped in their own TACLANs. The SOTF's air operations shop spent the majority of its effort coordinating the routine movement of all of these travelers rather than reconnaissance, raids, and fires. Hardly a static collection of people and functions, the SOTF was a hub in a circulatory system of people and data, but information friction created clots that impeded the flow.

The usual solution to coordination failure is more hierarchical control, better configuration management, or an enterprise-wide architecture. SOTF officers did recognize the importance of "knowledge management" and sought to impose regularity, but geographical and technological variety frustrated standardization. Institutionalization efforts perversely added more potential for delays, misunderstanding, and bureaucratic insulation. The "battle rhythm" of daily, weekly, and monthly briefings, teleconferences, and reporting tended to channel the attention of SOTF personnel inward toward the preparation and coordination of products required by their chain(s) of command. The administrative load was reproduced at each echelon as slides and text were aggregated and sent forward, each requiring slightly different formatting and content to appease different superiors and their staff surrogates. The SOTF's three Task Units had twenty different kinds of daily and weekly regular reports due to the SOTF, as well as endless one-off requests for information, which the units usually pushed further down to their platoons: a daily situation report, draft and final concepts of operation, target intelligence packages, video-teleconferencing "quad charts," after-action "quick looks" and "storyboards," Commander's Update Brief slides, training statistics and photographs, intelligence reports, engagement reports, civil affairs packages, equipment and medical casualty reports, and so on.

Forward units that had regular contact with Iraqis were constantly hit with new data calls to provide additional information, or to repackage the same information in a different format, each of which required manual production and took time away from the interactions with Iraqis that the reports were ostensibly about. CONUS "reach-back" agencies and administrative commands tried to provide intelligence and other support to forward units. Reach-back intelligence was optimized for stereotyped tactical support products (for example, a remote analyst with little local knowledge could obtain a satellite image of a target building and measure and annotate its dimensions to create a target graphic) rather than providing insight into dynamic sociopolitical networks, yet another way in which the direct action bias was institutionalized. Furthermore, reach-back units were also prone to "reach forward" for juicy data to advertise their "support to the warfighter" to vicarious audiences in the Washington Beltway, which added to the coordination burden on forward units. Finally, the imminent redeployment of SOTF units back to CONUS tended to capture staff officer attention as the deployment went on. The SOTF began requesting inputs for end-of-deployment awards and performance evaluations when its members were only four months into a six-month tour of duty, even as it took a new SEAL Team about a month to get into an effective battle rhythm. If awards were meant to incentivize productivity, therefore, it was possible for them to do so only three months out of six.

This convoluted sociotechnical operating system ran the SOTF's targeting applications. Intelligence fusion was meant to combine multiple sources to identify a target and discover enough "pattern of life" to support a tactical interception of the "high-value individual" (HVI). SOTF briefings to visitors regularly invoked the phrase "ops-intel fusion," a term that really just underlined the persistent boundaries and confusion between work centers. The SOTF did not and could not have one master fusion system because there was no one common representational center.

*The Targeting Application.* In the panoptic fantasies of counterterrorism professionals, a common database would be constantly updated with new intelligence from an interagency dragnet, making it possible to generate visualizations of the entire terrorist network and clearly identify the HVIs. In reality, network diagrams obscured as much as they revealed. Working analysts defined links and nodes incommensurably across different applications; their partial diagrams (subgraphs) parsed the world into inconsistent sets of entities, relationships, and characteristics. One analyst included farm animals in the "person" category so that he could track sheep that a shepherd was using to smuggle weapons. As with Excel spreadsheets, analysts encoded information in the visual layout of network diagrams by drawing shapes around icons to indicate organizational affiliation (e.g., the icons representing people would be enclosed in a rectangle to indicate that they were all in the same insurgent cell); icons clustered closer or further apart might indicate the strength of a relationship, or the analyst might include clusters of icons representing people in Iraq and icons representing intelligence reports about them. Meaningful social network data became inaccessible to graph-theoretical algorithms operating only on links and nodes on the assumption that they represented all of the relevant relationships and people. Such problems were less severe in cases where durable infrastructure such as engineered roads, telephone networks, and electronic transactions helped to standardize link and node ontology, or where technical specialists worked with well-understood formats (e.g., call chain analysis with SIGINT metadata). For analysts at the Task Unit level, the idiosyncratic diagrams served more mnemonic and heuristic roles to keep track of reporting as they read it in relation to some specific, circumscribed problem. The act of building a diagram forced intelligence specialists to read reporting more closely, which made them smarter about their topic. The diagram itself might receive little further use and remain inscrutable to others.[55]

Real insurgent networks continued to evolve at the same time that analysts built their time- and labor-intensive social network diagrams, which they described as "hairballs" or "star charts" (i.e., insurgent celebrities and their tangled entourages). These charts continued to hang as wall decorations long after the analysts who built them had transferred and the insurgency had evolved into something different. Impressive visualizations created an aura of professional expertise and an ego boost for techs in the eyes of operators. The hairballs appeared seductively detailed, but they hid assumptions and transcription errors in their construction. Software databases had similar problems. The reification of a link labeled with the vague catch-all phrase "associated," or reliance on SIGINT communication patterns alone without reference to other social context, might turn innocent delivery boys into nefarious villains ("nefarious" was a word used with surprising frequency at the SOTF).[56]

Target lists—usually names and mug shots arranged on a PowerPoint slide for easy viewing rather than in an Excel spreadsheet that could have been sorted and analyzed—often took the place of a network diagram. One SEAL mentioned that a target list made a good slide because "it's satisfying to cross bad guys off the list when you get one." After-action reports that could claim, "Captured SOTF HVI #2!" looked impressive. Like a good action-movie franchise, target lists were inexhaustible. If the SOTF got two of the ten most wanted, numbers eleven and twelve could get promoted on the list, if not in the insurgent organization. Every American unit had a target list, so even if the SOTF was unable to find its own HVIs, it might get someone else's. The bullet point "Killed 3rd Battalion's HVI #1!" also looked good and demonstrated a commitment to jointness.

While the SOTF lacked any one center of calculation, the representation closest to something like an apex of incoming and outgoing cascades of inscription was the PowerPoint target intelligence package (TIP), or target folder. The TIP consolidated intelligence reporting and fusion products (maps and network diagrams), helped to make the case for targeting a "nefarious" individual, and supported construction of a concept of operations (CONOP) document needed to secure headquarters permission to assault the target. TIPs might have multiple slides on a single HVI depending on the background and intelligence data available. TIP slides were usually compiled into "target decks" depicting multiple HVIs. Intelligence techs at the lowest echelons faced a great deal of pressure from operators to produce targets. They infrequently exercised skepticism about target quality or pushed back against pressure to be a team player. By producing targets, techs could participate in the hunt along with operators venturing outside the wire to catch their prey. Personnel became disinclined to ask whether some "bad guys" were necessary evils in local society, whether some were potential allies if they could be persuaded to defect from the insurgency, or whether some were not actually bad guys at all.

In theory a commander's guidance prompts targeting analysts to identify functional "lines of operation" in the insurgent organization (e.g., logistics, propaganda, combat) and then the key HVIs that would disrupt them. In practice the process worked in reverse. Rather than value-free pictures of the world, network diagrams were seeded around specific personalities of interest, and then targeting guidance provided a retroactive legitimation of the targets that bubbled up along the way. Previous SEAL Teams, conventional battlespace owners, intelligence agencies, indigenous partner forces, tribal allies, and Iraqi informants all held opinions about who the bad guys were, and they kept target lists and TIPs on individual suspects. Marine battalions sometimes handed off actionable targets to SOTF Task Units with little advance notice, enabling a frustrated Marine officer to work around his chain of command and enabling the Task Unit to build up social capital with its Marine hosts. The target pool expanded as working-level analysts or operators took the initiative to follow up leads from a HUMINT source, an engagement event with local tribal elites, or a SIGINT report. The SOTF headquarters held a weekly targeting meeting to gain insight into the targets its three Task Units were pursuing. It maintained a "top ten" list that was balanced evenly across them, not because the insurgency was equally active in all areas but rather to avoid the appearance of headquarters favoritism. Headquarters guidance did not so much direct targeting as organize the ferment of tactical activity into something more coherent and legible for external consumption. Rhetorically, targeting guidance and lists enabled the SOTF to demonstrate to CJSOTF that it was fighting the war in a deliberate fashion.

Once reified on a PowerPoint TIP, the target *as such* became a black box. Questioning assumptions or methodologies behind the initial target designation became harder as the hunt gained momentum, as more collection effort was invested, and as the representational residua of tracking accumulated. More machinery was in place for building up a target than for inspecting the quality of construction. The time and effort spent going after a target reinforced a sense of the target's value. Moreover, while methods for understanding and developing targets were ad hoc, processes became more well-defined and reliably routine as target development moved closer to direct action. Pattern-of-life analysis accumulated representations of the likely whereabouts of HVIs with the goal of identifying tactical "triggers" that placed a target at a particular location within a given timeframe. Operators took more interest in intelligence at this point because they wanted to improve the reliability and frequency of triggers, which provided more targets to assault. Reliability did not refer to the quality of the target per se, an intelligence fusion problem of less interest to operators. The ideological consistency of TIPs with the NSW fixation on bad guys covered up the considerable friction involved in TIP construction. Actionable triggers were the culmination of the coordination phase of targeting. Triggers enabled direct action, the real goal all along.

ENFORCEMENT

SOTF practice took on a more managed quality for the tactical operations closer to the archetypal core of the NSW organization. Digital technology enhanced the performance of tasks closer to the essence of the organization but did not necessarily improve tasks that were perceived as more peripheral. More friction emerged at the back end of the F3EA cycle, where operations were supposed to generate intelligence.

The "full mission profile" is the central ritual in the cult of the frogman: insertion of operators into the target area, infiltration to the objective, and actions on the objective, followed by exfiltration and extraction back to safety. The NSW community generally and the SOTF in microcosm were designed around its reliable performance. The target (a person) and objective (a location) were reified in the TIP. The receipt of an actionable trigger set in motion a well-rehearsed drill. An impromptu planning session with SEAL Platoon leadership determined feasible options for the mission. Intelligence techs scrambled to assemble tactical support products such as maps and imagery of the objective and assessment of the anticipated tactical threat. The platoon's superior Task Unit generated Word and PowerPoint versions of a standardized CONOP detailing communications, maneuver, and emergency plans. Each CONOP had an information operations statement as to the expected effect of the mission and perceptions that Iraqis might have of it. This was almost always a vague pro forma statement about disrupting the insurgency and sending a message, content unspecified. The information operations statement varied little from CONOP to CONOP, a testament to the lack of attention given to COIN effectiveness (indirect action) compared to tactical performance (direct action).

Creation of a CONOP began the negotiation for mission approval with the chain of command. Task Units were prone to play games, such as waiting until the last minute to submit a CONOP in order to restrict headquarters decision time. Headquarters concerns that proposed HVIs might not actually be important insurgents, or that hitting them would be counterproductive for the local community, were construed as second-guessing the unit on the ground, which was seen as bad for morale. When the SOTF did raise concerns about the low quality of intelligence justifying a target, the Task Unit could counter that direct action would serve as a useful "confidence mission" for training Iraqis. Few missions were disapproved because of worries about target validity. Concerns about excessive operational risk to U.S. personnel, by contrast, were more likely to lead to disapproval. To check the tendency of NSW units to indulge in direct action missions and to encourage more attention to the combat advisory role instead, the CJSOTF headquarters (which had more Army Special Forces personnel with greater relative sensitivity to indirect action) levied a requirement that there be a minimum two-to-one ratio of Iraqi personnel to U.S. operators on any given mission. This limited the range of missions that SEALs could conduct, because of the larger force footprint and lower proficiency of Iraqi partners, even though there were benefits, such as putting a less offensive Iraqi face on raids, incorporating native speakers, and training Iraqis to take over their own security. The exception to the SOTF's general reticence to disapprove missions was a concern about "dry holes," the name given to a mission that failed to encounter a target because the HVI either fled the objective in advance or was never there in the first place. The SOTF had to negotiate with CJSOTF for air support and ISR, which meant taking high-demand, low-density assets away from other regional SOTFs. Expending these on dry holes made it harder to get them for future missions. Too many false alarms also looked bad on higher-headquarters statistics. As one exasperated colonel from CJSOTF exclaimed, "The kinetic stuff is fun and easy, but this staff jujitsu stuff wears you the fuck out!"

Once launched, the assault force and the Task Unit maintained coordination through an integrated framework structured by the CONOP, shared maps, overhead drone surveillance, SIPRNET chat rooms, and a communication plan with "prowords" (procedure words) keyed to tactical milestones and contingencies. Prowords were agreed on in advance by mission planners to streamline radio communication and provide a modicum of security. In the movie *Black Hawk Down*, for example, the proword "Irene" signals Task Force Ranger to commence an assault on two of Mohammad Aidid's lieutenants. Prowords, projected on a wall or printed out for ready reference in the Task Unit operations center, provided a shared script for the operation so that the assaulters on the ground could efficiently communicate their progress along predefined waypoints (insertion, rendezvous, security set around the objective, objective secured), or movement to a different branch of the plan (troops in contact, explosives discovered on the objective), which might require additional Task Unit action (activating the Marine quick response force, coordinating close air support, marshaling additional reconnaissance). Real-time tracking of the assault force was provided by "blue force tracking" beacons carried by the NSW unit in the field. The beacons broadcast encrypted GPS coordinates to other U.S. units to facilitate coordination, but the absence of an icon on the digital "common operational picture" did not necessarily mean an absence of friendly troops. A beacon could malfunction, the unit could split up, or highly classified missions might preclude public broadcast. All of these tactical C2 procedures were exercised in advance during the eighteen-month predeployment workup. They served to orient the attention of the Task Unit commander and ISR officer as they monitored the mission unfolding on radio traffic and the ISR video.[57]

Unlike the Task Unit with its active tactical coordination with ISR operators, a larger audience at the SOTF played a more passive role. The headquarters audience watched the action on a video feed from drones circling overhead, which was usually projected right beside a television displaying Fox News or football games. There was a dissonant aura of normalcy in the SOTF office space during combat operations. The televised drone feed, laconically described as "kill TV," provided deskbound personnel a way to vicariously identify with heroic operators on the ground. Combat was actually rare during this period in Anbar's history. Usually only the initial entry onto the objective held the potential for action. Curious SOTF personnel usually wandered into the TOC to watch the video at that moment and then wandered away if there didn't seem to be much going on. Senior officers were hardly immune from the allure of televised tactical operations well below their pay grade.

SOTF headquarters "battle tracking" in practice meant copying Office files emailed from the Task Unit into the share drive folder and updating a "mission tracker" spreadsheet as milestones were met. The mission tracker surprisingly contained no fields directly linking the mission to the target—not even the target's name—and little data concerning mission outcomes. The SOTF's overriding focus was on the safe performance of the full mission profile rather than on connecting missions to targets and targets to COIN effects, let alone analyzing the data in aggregate to make corrections.

After mission completion, the Task Unit would forward a short blurb of any significant activity on the mission, followed later by a more thorough operations summary and a PowerPoint "storyboard." The graphic storyboard, festooned with photographs of any killed or captured Iraqis and weapons material, became the SOTF's trophy of a successful manhunt. If the intended target was captured, he was labeled a "Jackpot" in a large colorful font, while other detainees collected were simply labeled PUCs ("person under control"). The more a mission resonated with commando archetypes, whether heroic or tragic, the more detailed and graphic the storyboard deck would be. Storyboards were emailed out to the wider NSW community back in the United States, and they would go on to form the core of the SEAL Team's after-action briefs. Enforcement practice worked as a sort of translation function to convert a PowerPoint TIP into a storyboard that reinforced the cultural narratives of the frogman community.

FEEDBACK

In F3EA doctrine, a raid is supposed to generate intelligence for the next raid to eventually locate HVIs. Enforcement via raid is thus an opportunity for further measurement via "sensitive site exploitation" (SSE). Yet the SOTF performed its "finish" activity (where intelligence enables operations) with more finesse than the "exploit" effort (where operations enables intelligence). This accorded with the greater prestige of operators relative to techs. SEALs treated SSE as more of a cleanup requirement following a successful raid, rather than as the setup for follow-on raids. Analysis of the take was not well-integrated into active target development. As a result, the F3EA cycle was not truly cyclic at the SOTF, so few raids led to new ones. Instead of adjusting its behavior based on feedback from new measurement cycles, the SOTF "fed forward" its expectations of desirable behavior.

Weapons, equipment, documents, pocket litter, cell phones, computers, and other things found on the objective could potentially have intelligence value for hunting insurgents. They also had evidentiary value for convicting detainees in Iraqi courts. The evidentiary justification of SSE usually held more appeal for operators than intelligence gathering. A guilty prisoner sent to long-term detention justified the risk of the raid and resonated with a heroic narrative, but intelligence analysis seemed like a job for some tech. To collect SSE, operators had to first judge material valuable enough to collect out of all the printed or digital material to be found in any house and then transport media back to base without corrupting it along the way. Forensic and document exploitation was painstaking work because insurgent hard drives were cluttered messes too. Thus SSE was outsourced to reach-back organizations. Products with a long turnaround rarely reached the same SEAL Team that recovered the material in the first place. Organizations that took F3EA more seriously had organic capabilities (i.e., attached to the unit), such as forensic analysts, evidence rooms, and skilled linguists.

Detainees were turned into data through "tactical questioning" on the objective or interrogation in a detention facility. When SEALs burst into a house, the people inside found themselves face down on the ground with their hands zip-tied. Fearing for their lives—a natural by-product of a raid—PUCs were assumed to be more likely to respond to direct questions with veracity than if they were given an opportunity to compose themselves in captivity. SEALs had to be willing to ask the right questions and, of course, care about the answers. Even if direct questioning provided no information of obvious value, the responses or lack of responses could potentially enable the triage of detainees for expected intelligence utility, as well as development of subsequent interrogation approaches. The inclusion of trained interrogators on the assault force, a practice adopted by JSOC, would have improved the effectiveness of the tactical questioning process. But the trained "gators" at the SOTF were techs left behind at camp to await the return of the raiding party and its detainees.

Interrogation, as contrasted with tactical questioning, uses psychological manipulation and specialized approaches. Under strict U.S. military regulations, interrogation can be conducted only in controlled detention facilities by trained and certified interrogators. The notorious abuse at Abu Ghraib prison ushered in reforms and more regulation, but an unintended consequence was that skilled interrogators found it difficult to run the friendlier or seductive approaches that are generally considered to be more effective than fear in eliciting information.[58] HUMINT from detainees is subject to the same problems discussed above, such as untrustworthy and lying sources, communication through an interpreter, collector bias toward targeting intelligence, and the neglect of social and economic detail in reporting, if not in conversation. Furthermore, SOTF gators were techs, whereas the SOTF's HUMINT collectors were SEALs, so gators had less prestige and influence. Working in prison added opprobrium that further distanced gators from others at the SOTF.

If the detainee didn't give up any useful information, SEALs often concluded that the gator had failed to break him. They resisted considering the possibility that they had recovered and traumatized an innocent person who didn't know anything. Some SEALs took the attitude that if they had risked their lives to take an Iraqi off the objective, then he must be guilty: "Why would we go get this guy if he didn't know anything?" The potential innocence of detainees was a sensitive topic for SEALs, for it called into question the judgment of SEAL HUMINT collectors, Task Unit target development, and the operators who captured the detainees. Interrogation metrics, furthermore, framed success in terms of conviction rates rather than targeting effectiveness. They took little account of the radicalizing effects of incarceration; long-term detention was cynically called "Jihad U." The SOTF did not bear these costs directly and thus preferred metrics that could be counted within the six-month timespan of a SEAL Team deployment.

The SOTF tended to perform each iteration of the targeting cycle as a single phased evolution rather than a canonical F3EA loop. A process that should have provided fundamental feedback on the quality of the targeting cycle and new inputs to it instead became an auxiliary protocol to finish off a mission. Once a Task Unit bagged its quarry and filed a storyboard, it turned to the next target at the top of its stack, whatever its quality, rather than looking for follow-on targets or patiently analyzing the underground. There were always more ready leads in internal target development that had been passed from a Marine unit or named by a HUMINT source. By contrast, the exploitation of prior missions might drag on inconclusively for weeks or months. While the SOTF always had multiple targets at different stages of development, they tended to be independent threads rather than part of a systematic effort to unravel the fabric of insurgency. Many SOTF personnel voiced a desire to avoid distraction by "low-hanging fruit" and to exercise "operational patience"—that is, developing intelligence on the target network per the F3EA ideal. Yet such patience was aspirational. The SOTF's failure to connect the enforcement phase of one raid to the measurement phase of another probably provided some breathing room for the insurgency.

## Comparative Performance

In this section I compare SOTF performance with shadow cases that underscore the importance of institutional choices for information practice. The SOTF and JSOC both addressed the counterterrorism problem, but JSOC implemented a somewhat more institutionalized solution. Conventional units in Anbar operated in the same environment as the SOTF, but they adopted a different understanding of the COIN mission and empowered local units to adapt to perform it. This produced different qualities of insulated, managed, and adaptive practice, respectively.

INSULATED PRACTICE IN THE SOTF

There are many questions one might ask about SOTF performance. Did the unit capture the people it targeted? Did it retarget when it missed? Did it capture any people it did not target? Did its detainees provide any useful intelligence about new targets? Did these raids produce any undesirable consequences? Did they help or hurt the broader U.S. COIN mission? How did SOTF counterterrorism performance compare with JSOC, which invented F3EA doctrine? How did it compare with Marine performance of COIN, which was a different approach to the same province?

To query the SOTF's "database" for answers to these questions, one would have to start by gathering together the records the SOTF used to produce targets in the first place. The SOTF improvised a jumble of ad hoc representations to execute the types of direct-action missions it most preferred to perform. Each product was designed to support particular staff functions throughout the F3EA cycle. They resided in different share drive folders, web portals, or email accounts, or on the networks of different organizations altogether. One would thus have to negotiate access with the data owners and understand the nuances of inscription for each product. Figure 5.2 depicts some of the products and their organizational locations. It also shows a simplified ontology of some of the entities that these products represent, as well as the relevant phases of the F3EA cycle.

Asking questions about mission performance in the aggregate cuts against the grain of this structure. SOTF staff officers who lacked the time or ability to analyze their own data stores tended to charge subordinates with new reporting requirements to answer headquarters queries that cut across the grain. Learning the answers to some questions, moreover, might have been uncomfortable if they did not conform to heroic narratives. After managing to access all the data needed for a query, one would then have to reformat the data to be comparable and analyze the results in a program like Excel. In the meantime, the SOTF would keep executing missions and politics in Anbar would go on. The painful manual query would have to be repeated all over again to update any answers to questions.

The SOTF only ever measured a subset of possible contacts with local society. Biases and friction affected choices about which of these were captured in formal representations. Meanwhile, the actual insurgency continued to evolve as it lost members to battlefield attrition or voluntary dissociations and gained new members through recruitment or alliances with other militants. Figure 5.3 illustrates how external and internal worlds might drift apart. In the Battle of Britain, as described in chapter 4, the gap between representation and reality tended to be corrected through constant feedback. The SOTF, by contrast, tended to operate in a feed-forward fashion, always starting new operations rather than systematically following up. As long as there were no spectacular breakdowns that injured

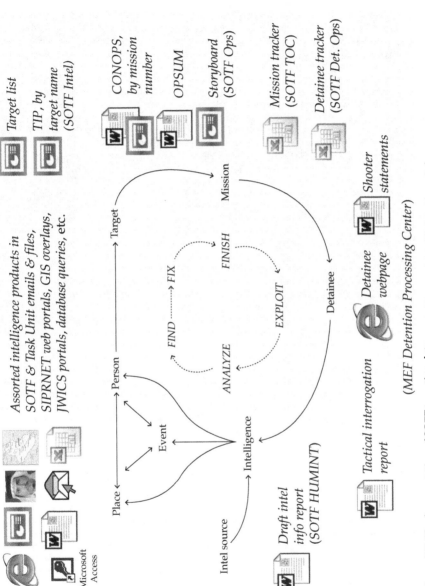

Figure 5.2. The fragmentation of SOTF targeting data

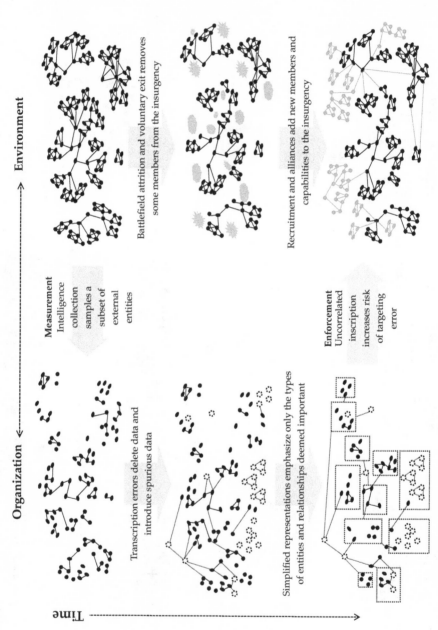

Figure 5.3. Representation and reality become uncoordinated in social network analysis

operators or resulted in excessive collateral damage, it was hard to know how or to what extent the representation might have drifted away from reality.

When measured in terms of completed missions, dramatic storyboards, and detainees in long-term detention, SOTF direct actions seemed successful enough. Yet a narrative of SOTF targeting success based on "Jackpot" storyboards could coexist, in principle, with a thriving yet illegible underground network, as depicted in figure 5.4. By pursuing several disconnected targets without taking the time to follow up and identify connections between them, the SOTF would merely nibble at the edges of AQI without perceiving, let alone defeating, the whole.[59] Since the underground was unmeasured by the SOTF, or perhaps unmeasurable, it is impossible to know how extensive it might have been or what other operations might have, counterfactually, been more effective in defeating AQI without undermining Marine COIN efforts and the Anbar Awakening in the process. Most of the top HVIs on the SOTF's list remained active after this deployment ended until JSOC killed them. AQI remained viable and violent long after this SEAL Team redeployed. Indeed, the so-called Islamic State would arise years later from the AQI underground to dominate Anbar Province by early 2015.[60]

The SOTF's information technology did not create its bias for direct action, but it did amplify it. Assumptions about the world became built into the architecture of distributed cognition. Personnel adjusted and repaired systems to meet emerging needs as they understood them, informed by the worldview of Naval Special Warfare. The system tended to search for opportunities to conduct counterterrorist raids, but it neglected the information needed to evaluate their effectiveness in the broader counterinsurgency mission. SOTF performance became a mystery unto itself.

At a press conference about the war in Iraq, Donald Rumsfeld answered a reporter's question about weapons of mass destruction with a response that gained some notoriety: "Reports that say that something hasn't happened are always interesting to me, because as we know, there are known knowns; there are things we know we know. We also know there are known unknowns; that is to say we know there are some things we do not know. But there are also unknown unknowns—the ones we don't know we don't know."[61] The logical category that Rumsfeld omits is the "unknown knowns"—the things we do not know that we know. The SOTF had a lot of relevant data on hand, but they were not readily accessible for queries that it was not inclined to make. SOTF personnel did not really know what their own information system might have been able to tell them.

## MANAGED PRACTICE IN JSOC

The SOTF was a "white SOF" unit with limited resources for counterterrorism and competing demands due to requirements for training and

| SOTF activity | SOTF representation | Insurgent reality |
|---|---|---|
| Intelligence collection | The SOTF finds several disconnected insurgent cells and develops actionable intelligence on them. 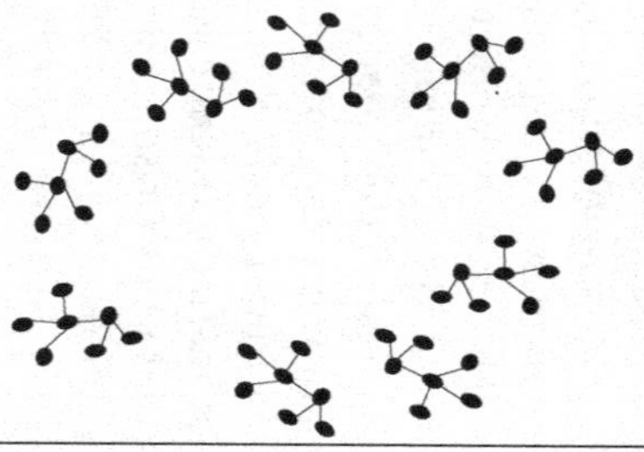 | The SOTF never finds information revealing the insurgency's support infrastructure and leadership. 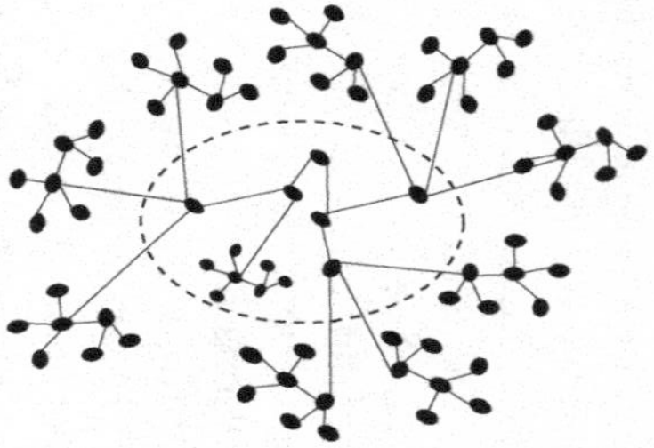 |
| Operational missions | Tactically successful missions generate PowerPoint storyboards, but only one generates follow-on missions. 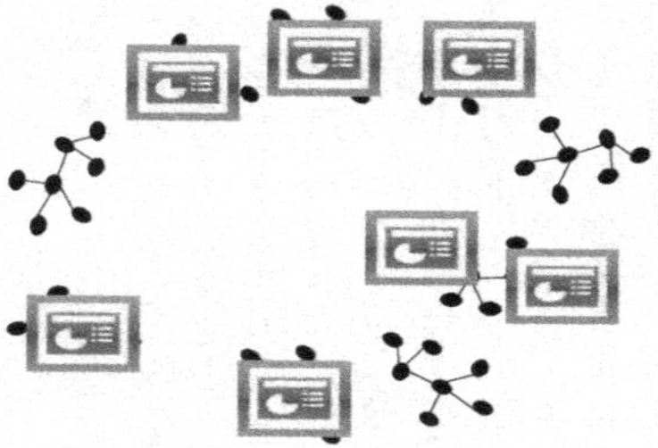 | The uncompromised administrative core can replace lost operatives and continue operations and civilian intimidation. 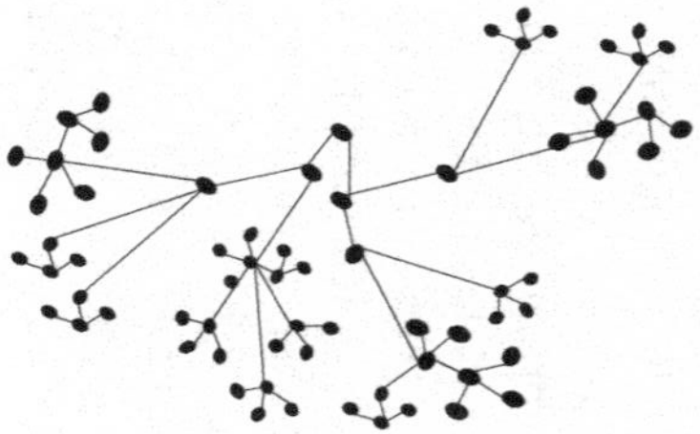 |

Figure 5.4. Successful direct action missions produce attractive storyboards but leave the insurgency intact

engagement with Iraqis. By contrast, JSOC was a well-resourced "black SOF" unit with a single-minded focus on counterterrorism. Under McChrystal's leadership, JSOC grew from 800 people in 2003 to 2,300 in 2008.[62] He flattened the organization, promoted information sharing, brought in representatives from the national intelligence agencies, and dramatically expanded its F3EA capacity. During August 2004, JSOC conducted eighteen raids, but by spring 2006 it was conducting thirty to fifty a week. McChrystal described his juggernaut simply as "the machine."[63] His joint operations center (JOC) had hundreds of workstations and nine giant screens displaying ISR feeds and mission planning data on developing targets and live missions.[64] Participants called it "the Death Star" because of a sense that, as one put it, "you could just reach out with a finger, as it were, and eliminate somebody."[65] McChrystal was willing to use as many as three drones to watch a single target in case one malfunctioned or lost the target.[66] He also had a cyber-operations squadron, a unit of Iraqi operatives known as the Mohawks, clandestine intelligence collectors known as Task Force Orange, professional interrogators, and forensic analysts for rapid exploitation of recovered media.[67] Delta Force raids had once been meticulously planned for weeks, but under McChrystal, Delta began launching raids with little notice for the express purpose of gathering intelligence.[68] The organization allegedly identified over 75 percent of its targeting leads in 2007 from interrogation and document exploitation opportunities generated by its own activity—quite a contrast with the infrequent follow-through at the SOTF.[69]

The contribution of JSOC to the decline of violence in Iraq remains an open historical question. Archives are still closed, and the anecdotal evidence is mixed. General Petraeus asserted that "JSOC played a hugely significant role" in Iraq by killing or capturing "high-value targets."[70] President George W. Bush said simply, "JSOC is awesome."[71] A few shining success stories are well known, such as the 2006 killing of the AQI emir al-Zarqawi, the 2008 raid across the Syrian border to kill the AQI facilitator Abu Ghadiya, and the killing of al-Zarqawi's successor Abu Ayyub al-Masri in 2010. Petraeus also credited SIGINT from the National Security Agency, according to a press release from the NSA, "with being a prime reason for the significant progress made by U.S. troops in the 2008 surge, directly enabling the removal of almost 4,000 insurgents from the battlefield."[72] On June 7, 2005, JSOC recovered an AQI campaign plan that stated that the United States was winning in "the Triangle of Death" (Baghdad-Tikrit-Ramadi), which AQI assessed had become unsafe for senior emirs.[73]

Nonetheless, JSOC targeting ran into some problems. Nouri al-Maliki's government was only too happy to let JSOC's Task Force 16 chew threw the Sunni insurgency, but it was hesitant about letting Task Force 17 go after Shia HVIs with political connections.[74] McChrystal committed his best units to Task Force 16, which he viewed as the top priority in Iraq, but he staffed Task Force 17 with "white SOF" units from CJSOTF that were less familiar with JSOC's processes. "The targeting was a joke," said one member of an NSW Task Unit in Task Force 17.[75] Moreover, even in the face of serious attrition by Task Force 16, AQI demonstrated impressive resilience.[76] AQI violence soared after al-Zarqawi's death until the tribes made a decision to oppose it. Relentless raids also interfered with COIN operations at times. "Sometimes our actions were counterproductive," McChrystal admitted. "We would say, 'We need to go in and kill this guy,' but just the effects of our kinetic action did something negative and [army and Marine units] were left to clean up the mess." Two senior commanders stated in 2011 that JSOC's targeting success rate was only ever about 50 percent.[77] Thus, while JSOC came closer to achieving managed practice than the SOTF in its performance of F3EA, there is reason to suspect that JSOC, with its frenetic pace of operations, was also prone to insulation.

## ADAPTIVE PRACTICE IN THE MARINE CORPS

The turnaround in Anbar relied on the Anbar Awakening and similar realignments, as discussed above. This was a political reaction to AQI gains in the province rather than a military result of U.S. operations. The situation in Anbar would likely have improved even if NSW and the SOTF had not pursued counterterrorism in the province, therefore, but the same cannot be said about the Marine Corps, which recognized and encouraged the growth of the Awakening. The Marines in Anbar hosted the SOTF on the same archipelago of bases along the WERV, yet they implemented a different quality of information practice with a different focus. Leadership empowered Marines in the field to discover and exploit new opportunities as they emerged. They were guided by a shared commitment to COIN, proselytized by General Allen and others, that framed kinetic targeting as only one means among many rather than an end in itself.

In contrast to the SOTF's focus on kinetic raids triggered by technical intelligence, Marine battalions took a policing approach to COIN. They encouraged a "cop on the beat" mentality to gain information about local society. For example, First Battalion, Seventh Marines (1/7) organized a census and vehicle registration program, using color-coded stickers to help patrols identify cars from out of town, and they installed covert and overt surveillance sensors in urban areas with high IED activity. To coordinate the new data it was measuring, 1/7 expanded its intelligence complement from five to over thirty analysts, and each unit was required to provide patrol and intelligence data for input into battalion databases. Important among these was COPLINK, a law enforcement software package to facilitate the analysis of criminal networks and support police-style investigations. Foot patrols performed both routine measurement and enforcement functions, backed up by more consequential intelligence-triggered raids on occasion. Eighty percent of the battalion's own intelligence was generated by 1/7. The unit also worked to turn its enforcement missions over to the Iraqis, building new police stations and a force of 1,400 policemen with the support of the Albu Mahal tribe. Violence dropped significantly during 1/7's rotation.[78] An Army Stryker Brigade Combat Team adjacent to 1/7's area of operations worked along similar lines to couple technical and human intelligence collection to law enforcement techniques; as one officer noted, "It's been more of a police action than combat."[79] The Marines reinforced local policing through a program known as Legacy, which embedded veterans of the Royal Ulster Constabulary within Iraqi stations to build up an indigenous intelligence collection and analysis capacity.[80]

The Marines were exceptional but not unique in this regard. Soldiers under the command of Colonel Sean MacFarland in Ramadi improvised "census patrols" for the express purpose of gathering information, filling out bilingual questionnaires, registering vehicles and weapon serial numbers, and photographing the interior of households and all military-aged males (with the support of tribal sheikhs and neighborhood mukhtars). These data were input into a Microsoft Excel spreadsheet and PowerPoint deck that linked detailed maps and imagery of the neighborhood with a standardized numbering scheme for all buildings and streets in Ramadi. Ramadi went from one of the most violent cities in Iraq in mid-2006 to one of the safest in mid-2007.[81]

Resources are an important part of this story, which is one discrepancy with my generalization that necessity is the mother of invention. Allen and his fellow Marines could dispense more largesse to their tribal clients than SOTF officers could. There were a hundred times more Marines than SOF in the province, so it was possible for Marines to conduct continuous foot patrols and develop close relationships. Without enough mass on the street, the SOTF simply could not generate sufficient information or the right kind of information. Yet this explanation is incomplete. The SOTF and its Task Units had access to most of the information that the Marines collected, had it elected to use it more systematically. The Marines, moreover, were spread thinly throughout the province as well. The COIN literature calls for a ten-to-one ratio of troops to population, but in Anbar it was more like a hundred to one. The Marines gained an informational advantage by using indirect action to get actionable information from tribesmen who shared their goal of vanquishing AQI. Once the sheikhs decided that alliance was both possible and desirable, Marine and army units in Anbar responded by adapting COIN operations. Unlike the SOTF, Marine intentions were better aligned to the nature of the problem in Anbar in 2007.

The decentralized Marine organization thus promoted bottom-up innovation that enhanced control of the province.

### Find, Fix, Unfinished

The SOTF was given, and its command brief advertised to the many visitors who rotated through it, a full-spectrum COIN mission. Yet it only reluctantly embraced the training of Iraqi security forces and haphazardly dabbled with civil affairs or tribal engagement. By contrast, it enthusiastically embraced combating the insurgency. Occasionally I did encounter instances of incompetence, parochialism, or adventurism; however, most SOTF personnel most of the time worked hard to do the right thing as they understood it—get the bad guys and protect the good guys. Intelligence analysts and staff officers toiled sleeplessly to build information products to support their mission. Operators helped Iraqi partners to improve soldiering skills, courageously took personal risks, and conducted themselves professionally in the field (for the most part). They also killed or captured a few dangerous characters. Their earnest performance was at best difficult to evaluate. At worst it was potentially harmful to the broader coalition forces COIN effort. Personnel finished their deployment without clearly affecting politics in Anbar one way or the other.

Insulated practice was the dominant theme in the SOTF. Organic adjustment was also pervasive, but it tended to reinforce the strong preferences that guided SOTF measurement, coordination, and enforcement. The propensity of systems to constantly break down in small annoying ways captured the attention of practitioners and refocused them on the marginal improvement of missions that nearly all agreed were desirable. Persistent friction did not cause any spectacular fratricide or collateral damage during this particular deployment, but this owed as much to good luck as anything. Something like managed practice emerged for the full mission profile at a very tactical level because there was a strong institutional consensus in NSW about the need to get it right. Furthermore, most tactical missions were small-scale, well-constrained tasks such as raiding a house. The broader social ambiguity of Anbar could usually be ignored without immediate tactical consequence, even as angry family members might decide to join the insurgency in the future. The SOTF could sweep away methodological and intelligence gaffes about targets that bubbled up from questionable sources as it constructed target slides and storyboards that supported a simpler narrative about killing and capturing bad guys. Questions about targeting performance were difficult to answer, and the asking was not encouraged. Self-evaluation cut across the grain of inscriptive channels that were grooved for direct action.

Navy SEALs are some of the most proficient tactical operators in the world, and NSW is a valuable national asset for the United States. But before a tactical unit is permitted to both task and evaluate its own assignments, leaders should determine whether its aggressive bias is well aligned with the strategic problem. In a province recovering from a violent paroxysm and struggling to consolidate a newfound peace, the SOTF's strategic preferences were misaligned. One SEAL officer said to me that the entire NSW community should be put under glass with a sign reading, "Break only in case of war." Yet by mid-2007 the strategic problem in Anbar had become something other than war. Some officers at the SOTF were fond of a COIN slogan: "We can't kill our way out of this problem." They were still eager to try. The SOTF information system illuminated a world that NSW was predisposed to see and suppressed the rest.

*Alice in Wonderland* provides a fitting epitaph:

"Herald, read the accusation!" said the King.

On this the White Rabbit blew three blasts on the trumpet, and then unrolled the parchment scroll, and read as follows: "The Queen of Hearts, she made some tarts, All on a summer day: The Knave of Hearts, he stole those tarts, And took them quite away!"

"Consider your verdict," the King said to the jury.

"Not yet, not yet!" the Rabbit hastily interrupted. "There's a great deal to come before that!"

# Increasing Complexity and Uneven Results

## Drone Campaigns

The word "drone" suggests mindless action, but the aircrew who operate drones are mindfully engaged. The U.S. Air Force prefers the term "remotely piloted aircraft" (RPA) to emphasize that the machines are flown by skilled human beings. The other alternative, "unmanned aerial vehicle" (UAV), is not only awkwardly gendered but also destined to sound as anachronistic as "horseless carriage" or "wireless telephone." Whatever one calls them, the drone on patrol and the people in control are part of a sprawling system of military units, intelligence agencies, government policymakers, military and civilian lawyers, uniformed technicians, defense contractors, communications infrastructure, and commercial software and hardware. Just as "the few" in the Battle of Britain depended on many people on the ground, drones rely on a large society of human beings to direct them. As Lieutenant General David Deptula points out, "It takes over 200 people to operate [an] MQ-1 Predator or MQ-9 Reaper RPA orbit for 24 hours."[1] Hundreds more are needed to plan operations and analyze data. The command and control of drone operations spans international, organizational, and intellectual boundaries. Civilians play an important role, too, as contractors, lawyers, administrators, policymakers, and sometimes as targets or victims.

The United States has employed Predator and Reaper aircraft in every phase of the F3EA cycle (find, fix, finish, exploit, and analyze) discussed in the previous chapter. A drone can loiter around a target area for hours collecting full-motion video and electronic signals to enable analysts to conduct "pattern of life" analysis. Video of suspicious behavior or the activation of a known "selector" (i.e., a particular phone number or device identifier) can provide an actionable "trigger" on the location of a "high-value individual." Remote operators can directly attack targets, which drastically shortens the "kill chain" between target detection and execution. Drones can also lase targets for other strike aircraft or provide reconnaissance and strike "overwatch" for ground forces that finish the target. Loitering aircraft then enable analysts to observe the aftermath of the attack by assessing bomb damage, monitoring the behavior and communications of survivors or bystanders in the target area, and continuing the search for more targets. Lethal strikes preclude the interrogation of targets, obviously, which removes an important source of feedback in the F3EA cycle. Yet killing also sidesteps the politically fraught problems of detention.

Targeting strikes on the opposite end of the earth is an inherently difficult problem. In combat zones like Afghanistan and Iraq, drones tended to work as complements to boots on the ground. Troops in forward areas were important participants in the targeting control cycle, both guiding and guided by the eye in the sky. In areas beyond, such as Pakistan and Yemen, by contrast, drones were used more as substitutes for ground forces. The computational burden of gathering and evaluating data in those cases was shifted completely onto remote organizations. The additional complexity complicated information practice, to be sure, but it also brought aspects of the targeting problem into better focus. It is a military truism that troops in the field have better situational awareness than rear echelons. Yet sometimes a telescope planted firmly on a hill can reveal features of enemy forces that soldiers cannot perceive in the heat of battle. The previous chapter raised some doubts about the objectivity of forward units. The SOTF had a systematically biased information system that tended to prioritize the reenactment of warrior archetypes over the understanding of politics in Anbar. Importantly, the SOTF's relatively permissive oversight mechanisms empowered it to find, fix, and finish its own targets in an active combat zone. The SOTF, in effect, proposed and graded its own assignments, and mistakes in the construction of its targeting products were covered over by the same process.

Drone information systems, by contrast, offer unprecedented possibilities for oversight and accountability (and micromanagement too). Tactical data feeds and target support materials can be, and usually must be, shared across a global network. Compared to SOTF data that remained trapped in local servers in camps along the Euphrates, drone networks make it easier for many different actors to access, replay, inspect, and evaluate data. Commanders and lawyers gain the ability to peer through an intercontinental sniper scope as the drone pilot pulls the trigger. The political sensitivity of operations outside defined combat zones also creates strong incentives for practitioners to improve the reliability of systems that could create embarrassment or worse. Drone-enabled counterterrorism never eliminated some of the same insular tendencies evident at the SOTF, to be sure. Targeting errors have generated public controversy and enabled terrorist targets to survive and devise countermeasures. Critics question the strategic efficacy of perpetual counterterrorism, the toll of strikes on civilian populations, the ethics of killing without risking death, and the legitimacy of a secret assassination program.[2] Proponents counter that drone operations enable the penetration of terrorist sanctuaries, limit the footprint of foreign interventions, minimize civilian deaths, save friendly lives, and improve civilian oversight.[3] Over the course of a decade, information practice in the U.S. targeted killing system has indeed realized major improvements in the acuity, effectiveness, and accountability of its operations. These achievements are easy to overlook amid legitimate concerns about the utility and morality of killing by remote control.

The complex information system that directs U.S. drone campaigns developed historically through many iterations of exploitation and reform. This chapter focuses on the dynamic interaction of different types of information practice over time. In a process reminiscent of the bottom-up development of FalconView, practitioners made many improvements to Predator and Reaper drones and to the information systems that controlled them. This long learning process was punctuated with tragic fratricides and civilian deaths. Catalyzed by such errors, an institutional framework gradually evolved to improve the efficiency and effectiveness of drone operations. Practitioners throughout a global command and control system struggled to capture the benefits of managed and adaptive practice while avoiding the pitfalls of insulated and problematic practice. Emergent military solutions encouraged actors to alter the strategic problems, and new solutions increased system complexity over time.

## Sociotechnical Innovation

The drone revolution kicked off by Predator built on nearly a century of evolution in automated aircraft. The U.S. military began experimenting with drones in World War I with a gyroscopically guided "aerial torpedo" known as the Kettering Bug. Nazi Germany inaugurated the jet and missile age with the self-guiding V-1 and V-2 munitions. American drones like Firebee and Lightning Bug emerged early in the Cold War, but they remained a sideshow without a significant constituency. The U.S. services viewed manned aircraft as more practical for airborne reconnaissance and strike against a dangerous foe like the Soviet Union. Unmanned aircraft also challenged aviator identities in an air force dominated by pilots. Other types of automated weapons, such as cruise missiles and space satellites, found important niches where human occupation of the vehicle was impractical. Predator itself emerged through a curious saga of accident and improvisation. Its proponents exploited both technological and bureaucratic opportunities and evaded obstacles. Then al-Qaeda opened up a niche that no other system was ready to fill.[4]

In the 1970s a talented Iraqi-Israeli engineer named Abraham Karem designed aerial decoys for the Israel Defense Forces. After Israel turned down his more ambitious ideas for a reconnaissance drone, he immigrated to the United States. Karem set up shop in a Los Angeles garage with modest support from the Defense Advanced Research Projects Agency and the navy. Throughout the 1980s he worked on aircraft called Amber and Gnat, which had the same iconic v-shaped stabilizers and pusher propeller engine that distinguished the more famous Predator. The money ran out, however, and Karem sold his company in 1991 to General Atomics, a small defense firm in San Diego. Its eccentric owners, Neal and Linden Blue, were obsessed with the idea of selling armed drones to the U.S. government, which they envisioned as cheap cruise missiles that the CIA could use to attack Communist Sandinistas in Nicaragua. The Arnold Schwarzenegger movie *Predator*, which was set in Central America, was released within days of the launch of a new General Atomics project to convert an ultralight aluminum monoplane into a GPS-guided drone. Karem, who was by then facing bankruptcy, pitched his more elegant design. The Blue brothers bought it and simply recycled their favored name. Karem eventually also left General Atomics in frustration, but his aircraft remained.[5]

General Atomics landed a small contract with the CIA, which flew Gnat in 1994 on limited two-hour missions over Sarajevo from an airfield in Albania. The tenuous line-of-sight radio link relayed from a manned aircraft was subject to interference from weather and other electronics, and the CIA obtained meager intelligence results from the grainy black-and-white video, but the proof of concept intrigued officers at the Pentagon. The Predator Advanced Concept Technology Demonstration, an initiative to fast-track acquisition, was set up to experiment with a successor to Gnat that would carry a better reconnaissance payload and a satellite antenna. The RQ-1 Predator first beamed live video to a Pentagon command center during a January 1995 test at Fort Huachuca, Arizona. It used a jury-rigged connection through a Trojan Spirit satellite truck parked next to a ground control trailer, a commercial video codec, and a video-teleconferencing system. That summer Predator made its first combat deployment over Bosnia with an Army Military Intelligence Battalion operating in Albania. It also suffered its first combat losses. NATO commanders in Italy were enchanted by the novelty of live color video over Sarajevo, and they did not have to write any condolence letters for the lost aircraft. Color video in Bosnia was possible, incidentally, only because a lucky gap in the mountains between Gjader air base and Sarajevo enabled the ground control station to communicate with Predator via the line-of-sight C-band radio.[6]

Most air force officers saw drones as a curiosity. The air force chief of staff, General Ronald Fogleman, however, had experience in Vietnam as an F-100F Misty Fast FAC (Forward Air Controller) and RF-4C airborne reconnaissance pilot. Fogleman saw Predator as an opportunity to address the service's looming gap in reconnaissance capabilities. He also sought to renegotiate centralized control of the air campaign by consolidating air force control over army and navy drones. Fogleman reestablished the Eleventh Reconnaissance Squadron, a unit he had flown with in Vietnam, at an auxiliary airfield in Nevada in July 1995. The squadron deployed to Hungary in summer 1996 to fly Predator over Bosnia and later Kosovo. A series of deft bureaucratic maneuvers led the secretary of defense to give the Predator program to the air force in April 1996 and Congress to strengthen air force authority over the program in the 1998 defense budget. During the 1999 war over Kosovo, commanders became further addicted to real-time video from unmanned aircraft, and they also encountered serious difficulties coordinating time-critical targeting through the CAOC.[7]

More alignments of personality and circumstance transformed the reconnaissance drone into a strike platform. General John Jumper had commanded U.S. Air Forces Europe during the Kosovo campaign, where he had been frustrated by the inability to strike enemy targets that he observed through the Predator feed. When Jumper assumed command of Air Combat Command, which was responsible for the readiness of the nonnuclear air force, Jumper also gained authority over the Predator program. Fogleman had given it to the 645th Aeronautical Systems Group, an unconventional unit better known as "Big Safari" that had "rapid acquisition authority" to outfit "special purpose aircraft." Jumper gave additional funds to Big Safari, which then worked with the Eleventh Reconnaissance Squadron to weaponize Predator in less than sixteen months.

Big Safari did a lot of creative hacking to arm Predator. The Wartime Integrated Laser Demonstration (WILD Predator) used the Raytheon AN/AAS-44(V), a gimbal-mounted infrared camera and laser designator employed by navy antisubmarine warfare helicopters. They used part of a navy helicopter rack to attach a single AGM-114 Hellfire (Heliborne-Launched Fire-and-Forget Missile) under each of Predator's delicate wings, which they worried would tear off when Hellfire launched. Hellfire was a low-altitude antitank weapon first used by the army in the 1991 Gulf War that seemed ill suited for medium-altitude antipersonnel purposes. Big Safari experimented with different warhead and fusing combinations, blowing up hundreds of watermelons at the China Lake range to test the effectiveness of Hellfire against human targets. To evade concerns that a weaponized drone violated the Intermediate-Range Nuclear Forces Treaty ban on ground-launched cruise missiles, Big Safari first tested Hellfire Predator with its wings detached from the fuselage. Pressured by White House counterterrorism tsar Richard Clarke, government treaty experts finally decided in December 2000 that Predator did not violate the ban because it did not itself have a warhead, like a cruise missile, but was merely a warhead-carrying platform that could return to base intact, like a traditional warplane. Big Safari successfully tested the whole contraption in flight in February 2001. Predator had been designated RQ-1 for the reconnaissance Advanced Concept Technology Demonstration, but it was redesignated MQ-1 after the incorporation of WILD and Hellfire, receiving the new call sign Wildfire.[8]

More creative effort was needed to stabilize the distributed information system that linked remote pilots to distant targets. The technical difficulty of the initial reconnaissance problem was a function of geography, radio, and changing political requirements. Prior to the summer of 2000, Predator took off and landed via a line-of-sight C-band radio link to the ground station because the higher bandwidth and shorter delay of C-band facilitated responsive control. Operators switched to a Ku-band satellite dish in the aircraft's nose to fly it beyond visual range. The latency of the satellite link became less problematic once in flight, but it still required finesse to fly without crashing. The ground station in this arrangement had to be located within a few hundred miles of the target. Clarke, however, wanted the CIA to hunt Osama bin Laden, who had became America's most wanted terrorist after the 1998 embassy bombings in Africa. The proximity constraint on the ground station was impractical for clandestine reconnaissance in Taliban-governed Afghanistan. To get around it, Big Safari and the Eleventh Reconnaissance Squadron pioneered "split ops," whereby a small team in Uzbekistan launched Predator on C-band and then transferred control to a different ground station in Germany that conducted the mission on Ku-band. This hack relied on an auspiciously borrowed satellite terminal and leased bandwidth on a Dutch communications satellite that had become available after the dotcom crash in March. Predator thus provided, in September 2000, the first remote video of a tall man in white at Tarnak Farms, whom the CIA assessed to be Osama bin Laden. Clarke was as frustrated as Jumper had been in Kosovo by the ability to see but not strike a valuable target, so he redoubled his support for Big Safari's ongoing efforts. Controlling a targeted killing from a manned site in Germany was politically untenable, so Big Safari came up with a workaround. As long as operators pulled the trigger at Langley, not in Germany, a mere communications relay seemed less likely to create controversy. They built a ground control station in the woods outside CIA headquarters, linked via submarine fiber-optic cable to the satellite terminal in Germany. "Remote split ops" thus implemented a control cycle that transited thousands of miles under the ocean, through overland circuits, and beamed through the air, into orbit, and back again in under two seconds.[9]

The full arrangement was first tested on September 10, 2001, auspiciously enough. Both Gerald Ford and Ronald Reagan had signed executive orders banning assassination, and the CIA had been hesitant to reconsider targeted killing after its early Cold War adventurism. The shock of 9/11 abruptly changed attitudes. Policymakers had not expected to employ Predator until the following year, but they accelerated their plans immediately. President Bush signed a Memorandum of Notification on September 17 that specifically granted "exceptional authorities" to the CIA to use lethal covert action to disrupt al-Qaeda and affiliated terrorist networks.[10]

Predator thus evolved from an obscure technology demonstration in 1994 into an intercontinental sniper system. This innovation resulted from a fortuitous combination of engineering resourcefulness, bureaucratic entrepreneurship, and unforeseeable opportunities. As military information problems developed from Bosnia to Kosovo to 9/11, adaptive practice reconfigured the institutional and technical dimensions of the internal solution in search of a better fit. Organic development would continue to be a defining feature of the drone system, but problems would also begin to emerge as drone operations became more embedded in, and constrained by, large-scale combat operations.

## Adaptation and Problems in Afghanistan

Predator's first shot in anger was on October 8, 2001, not even a month after the first test of remote split ops. It targeted what appeared to be armed personnel and a truck parked next to a building where the Taliban leader Mullah Mohammed Omar was supposed to be meeting. The event was eagerly watched not only by drone operators in their control van at Langley but also by senior generals in Florida, Saudi Arabia, and Washington, DC. Drones enabled unprecedented tactical micromanagement by senior leaders, which produced confusion about the chain of command, hesitation over collateral damage concerns, or, alternatively, premature pressure to act. Mullah Omar, if he was ever there, survived the muddled execution of the October 8 strike. A few weeks later, on the first night that Predator operated under a new set of rules that delegated targeting authority to the CAOC in Saudi Arabia, aircrew inadvertently bombed the Al Jazeera news bureau in Kabul. Yet despite the growing pains, Predator soon became invaluable at the CAOC for corroborating intelligence reporting from other sources or providing occasional fires, even though remote pilots in Langley didn't always know what they were watching or bombing. Other platforms found new ways to use Predator too. Big Safari devised a way to share live video with AC-130 gunships in flight so they could aim their cannons without alerting targets on the ground and, similarly, to provide terminals to troops on the ground so they could access real-time video. The Wildfire concept finally proved itself in close air support at a key moment during the battle of Takur Ghar when it eliminated an enemy position "danger close" to a Ranger quick-response team.[11]

RUNTIME DESIGN

The information system that enabled operators to participate in a fight thousands of miles away was not designed according to any technocratic master plan. It was patched together over the years as military personnel and civilian engineers encountered unexpected problems and overcame them with unconventional solutions using discretionary resources. General Atomics initially focused on engineering a light, high-endurance, inexpensive aircraft, not building a remote cockpit for combat. The early interface featured a keyboard, text displays, brightly colored graphics, and intricate menus that obscured functionality that pilots considered vital. There were no standardized manuals or aircrew checklists, and no two aircraft or ground control stations were identical. Prior to 2002, Predator aircrew flew only a preplanned "collection deck" with procedures modeled on the U-2 spy plane, and operators had only two line-of-sight radios and four telephones to connect them to the outside world. There was friction between General Atomics civilian test pilots and military personnel who were "voluntold" to join what they saw as a "leper colony" in Nevada isolated from their peers fighting in theater.[12]

This situation changed dramatically during the next few years of continuous combat in Afghanistan and Iraq. General Atomics delivered a larger version of Predator, the MQ-9 Reaper, with improved speed and payload capacity. Whereas Predator could carry only two Hellfires, Reaper could carry four, plus two additional five-hundred-pound GPS or laser-guided bombs. A complementary ferment of user innovation was perhaps even more significant. Aircrew reconfigured the technologies of their information system by wiring in numerous new communication circuits and data feeds; installing commercial computers in racks to run new automated tasks; mounting interface devices and six additional displays in the remote cockpit to view moving maps, chat conversations, email, and other data; coding custom software tools using FalconView and Microsoft Office; and acquiring a new heads-up display that tailored features of a traditional cockpit to the unique requirements and opportunities of remote-control reconnaissance. Reaper aircrew also adopted the rhetoric of the strike fighter community to cultivate a focus on warfighting. They added or eliminated supporting crew positions to customize mission planning and real-time intelligence support, and they figured out techniques and procedures to improve situational awareness and reduce errors. Aircrew established enabling relationships beyond their local squadrons by developing more flexible and collaborative procedures to support units on the ground. Connections with personnel at Distributed Common Ground System (DCGS) units who monitored Reaper sensor data helped to coordinate the flow of intelligence throughout the operational theater.[13]

Runtime design also flourished in the CAOC to connect Predator to a wider social world. The data architecture developed for Kosovo provided stand-alone video feeds that were disorienting to observe (as mentioned in chapter 5, personnel described the experience of viewing drone video as "like looking through a soda straw"). CAOC personnel wrote a software program that enabled FalconView to plot the Predator sensor field of view over a moving map or baseline imagery display. They also adapted a navy helicopter software program to combine Predator data feeds into an internet stream that could broadcast to any computer in the CAOC or on the wider global classified intranet (SIPRNET). The lead user in these efforts stressed the importance of a rapid prototyping mentality: "Field an 80% solution or even a 50% solution until you can get feedback and determine what people really need," emphasizing that "in Wartime you've got everything *but* time."[14] Flexible commercial software tools, technically savvy personnel, immediate experience with emerging operational requirements, and heedful sensitivity to the needs of other users throughout the network enabled opportunistic adaptation to advance the collective mission. A shared framework of interoperable software platforms and network connections enabled decentralized innovation that enhanced the collective performance of the information system.

EPISTEMIC INFRASTRUCTURE

A careful ethnography of Reaper crews in action by Lieutenant Colonel Timothy Cullen, a U.S. Air Force fighter pilot who later went on to command a Reaper unit, reveals the proactive engagement of human operators with both local and remote events.[15] Aircrew awareness of the battlefield filtered through technical devices, but they had real-time contact with commanders, analysts, advisers, other aircraft, and troops in the field through numerous phone, radio, email, chat, video, and data circuits at multiple levels of classification. Control nodes based in the continental United States provided a stable anchor for one end of all of these inscriptive channels: infrastructure was fixed, personnel were safe, bandwidth was plentiful, networks were reliable, and support was available. The situation of operators thousands of miles away from the battlefield, ironically enough, enhanced their awareness of it.

Although they sat in Creech Air Force Base, Nevada, operators mentally felt as if they were "flying the sensor" in Afghanistan. They kept their eyes focused on the heads-up display, manipulated levers with their hands to adjust the quality of the image, explored the target scene to detect changes and anomalies, and maintained constant dialogue with other aircrew. They were not simply television watchers or video game players. According to Cullen, Reaper "operators negotiated and constructed a constrained environment in the ground control station to coordinate verbal, typed, written, pictorial, and geographical representations of a mission; to identify patterns in scenes from the aircraft's sensors; and to associate those patterns with friendly and enemy activity."[16] In other words, personnel constantly shifted their attention between the battlefield events they sought to control (what information means) and the complicated array of software and hardware interfaces that enabled control (how information works):

> In order to fly the aircraft and control the sensor ball, Reaper and Predator crews had to coordinate the meaning, movement, and presentation of a myriad of menus, windows, and tables on 16 displays and 4 touch screens with 4 keyboards, 2 trackballs, 2 joysticks, and 8 levers. A novice RPA pilot said flying the aircraft from a computer console while interpreting morphing numbers and symbols was like "flying the matrix," and to keep from getting lost in the system, inexperienced pilots and sensor operators spent as much as a third of a mission to configure the workstation and automated tools to display information in a predictable manner[17]

Close attention to this experiential reality militates against the trope of drone operations as a sort of a video game. Many believe that push-button killing compromises the dignity of both drone operators, who no longer risk heroic death for a noble cause, and foreign populations, who become dehumanized targets of neocolonial repression.[18] Dave Grossman makes the argument that people find it easier to kill at a distance (via artillery or aerial bombing) because they do not have to overcome a visceral human aversion to close combat, they tend to dehumanize victims that they do not personally interact with, and they can view the battlefield as a rationalized abstraction to understand and affect more than just the most tactical level of war.[19] A United Nations special rapporteur extends Grossman's argument to drones: "Because operators are based thousands of miles away from the battlefield, and undertake operations entirely through computer screens and remote audiofeed, there is a risk of developing a 'Playstation' mentality to killing."[20]

These concerns are exaggerated. While the bodies of drone aircrew were eight thousand miles away from the fight, their eyes were only eighteen inches away from their sensor display screens.[21] Remote operators often experienced killing even more intimately than attack pilots or artillerymen working in theater. Drone aircrew followed their targets for hours or days to understand the pattern of life and discover windows of opportunity to kill militants without killing their wives and children, witnessed the deaths of targets and bystanders in high definition, and lingered overhead to observe the aftermath. An F-15 pilot who dropped bombs in Kosovo and later flew drones in Afghanistan said that the act of killing "was more difficult in Predator."[22] Reaper aircrew's dense multichannel data environment, often featuring direct interaction with troops on the ground, created a profound sense of social proximity to the action of the ground.

Aircrew, engineers, analysts, and managers throughout the system continually negotiated new ways to enable or restore connections in such a way that drone performance could be rendered more reliable and appropriate in any given operational situation. They conscientiously sought to recover some definition of the external problem from the inherent ambiguity of irregular warfare. Some of their enabling representations were precomputed weeks or even years in advance (the coding of commercial software and the publication of maps), and some were adjusted in real time (communication circuits and configuration settings). Some of this adjustment occurred in a drone ground control station, and some occurred at other intelligence and command nodes, like a DCGS unit or a CAOC. When the coordinating work that went into structuring and stabilizing control channels faded into the background for the users of the system, then operators gained an experiential immediacy in their interaction with remote targets. They were not passive technicians, but rather active participants in the flow of battle. When something went wrong, however, the sociotechnical implementation would come obstinately into view. Breakdowns prompted debugging, or formal investigations depending on their severity, to repair the technological mediation of experience. In the course of ongoing operations, and over time between them, practitioner attention regularly shifted back and forth between what information meant and how information worked. Modifications to the latter shaped the experience of the former.

## INFORMATION FRICTION

Breakdowns were inevitable in a complex system that personnel redesigned on the fly. Adaptation shaded into problematic practice as various parts of the system drifted into more tightly coupled relationships. Uniformed aircrew and General Atomics engineers were differentially focused on warfighting effectiveness and engineering efficiency, respectively. The relationship between these subcultures improved after 9/11 with a shared sense of urgency, but the ferment of exploratory innovation still undermined attempts at standardization. Ground stations all had idiosyncratic configurations. During Predator's early years the majority of the pilot's time was spent assessing and correcting workstation malfunctions. Major software failures likely occurred in half of all operations. The sensor plot on the map tracker could be off by as much as a mile, which increased the computational load on the sensor operator who had to correct it. Predator did not achieve initial operational capability status, which would enable transition of the program from a Big Safari experiment to a more normalized acquisition office, until March 2005. Yet it had flown in combat continuously since October 2001. As operators added in new communications circuits, they opened up not only new possibilities for situational awareness but also new portals for interference. Curt chat texts from anonymous observers caused confusion and were misinterpreted by operators, who became increasingly sensitive to criticism.

Signal delays and feedback discontinuities (caused by the transit of communications under the ocean and into space) created mysterious behavior that operators struggled to comprehend and accommodate. Nearly half of all Predators purchased by the air force were involved in a major flight mishap between 2001 and 2013, due to limitations in pilot situational awareness, aerodynamic peculiarities, a lack of mechanical redundancies, persistent hardware faults and software glitches, and fragile communications links. In one close call, a C-130 Hercules cargo plane was struck by an RQ-7 Shadow drone after its operator lost track of it. The badly damaged aircraft was able to make an emergency landing without loss of life, but had a larger Predator or Reaper been involved, the outcome would surely have been worse.[23]

As discussed in the previous chapter, unconventional warfare is by its nature ambiguously constrained. Afghanistan presented some additional challenges. Afghanistan's rugged rural geography complicated reliable representation. U.S. forces were stretched too thinly for persistent measurement, and logistical supply was a nontrivial task. Troops had difficulty understanding the social relationships across rugged valleys, and nation building was actively contested from across the Pakistani border. The same individuals might play various roles at different times: civilian official, tribal elder, cleric, criminal, narcotrafficker, soldier, warlord, and insurgent. An infusion of international development aid distorted and corrupted Afghan society, while political intrigue in Kabul insulated the military headquarters. The Taliban thus had plenty of "dead space" in which to maneuver as a result.[24]

Real-time footage sometimes reinforced illusions of certainty about target attribution. On February 21, 2010, a U.S. Special Forces commander in Uruzgan Province received intelligence over the radio that insurgents were massing for an attack on his position.[25] A Predator crew in Nevada identified a column of vehicles with "tactically maneuvering" adult males. They also noted via text chat that children appeared to be present, but later described them as "adolescents." DCGS screeners watching in Florida and in the special operations headquarters in Afghanistan also had access to the video and chat feeds from the Predator, but they did not pass on the concern about the children to the commander. Even though the vehicles were twelve miles away, leaving plenty of time to develop the situation, the ground commander requested aerial fire support. The strike killed twenty-three Afghani civilians and injured twelve others, including children and a woman. A senior officer said that the deaths might have been prevented "if we had just slowed things down and thought deliberately."[26]

In another incident in Sangin Province on April 6, 2011, a Predator Hellfire killed a U.S. Marine and a navy corpsman. The crew misidentified the U.S. troops as insurgents; the ground force commander did not understand exactly where all of his troops were located; and the Predator crew failed to report information that might potentially have clarified the commander's situational awareness.[27] This event bore all the hallmarks of what Charles Perrow calls a "normal accident."[28] A tightly coupled system featured a conjunction of process or equipment faults, misinterpretations that seemed plausible under the circumstances, and failures in the monitoring or fail-safe mechanisms designed to catch and correct such problems.

Manned aircraft, of course, also made mistakes, and usually with heavier munitions than the Hellfire carried by drones. On June 9, 2014, in Zabul Province, an airstrike by a B-1B bomber killed five U.S. troops and one Afghan soldier. They had maneuvered away from the main element in the midst of a firefight and did not appear to be wearing the infrared strobes that aircrew expected friendly troops to display to identify themselves.[29] Surveys of U.S. terminal air controllers on the ground suggest, nevertheless, that troops still prefer to receive close air support from manned aircraft rather than drones.[30] On October 3, 2015, an AC-130 gunship killed at least forty-two civilians and injured dozens more at a Médecins sans Frontières clinic in Kunduz as a result of a communication equipment malfunction that precluded the receipt of an email from headquarters with the coordinates of the clinic; confusion was exacerbated by nighttime combat conditions, an active Taliban fighting position across the street from the clinic, deviation by aircrew and supporting commands from established processes and rules of engagement, and unverified intelligence from Afghan troops.[31] These events highlight the fact that targeting errors are usually more a product of the sociotechnical information systems that guide strike platforms, manned or unmanned, rather than of the characteristics of the platforms themselves. Airstrikes by any platform that harmed Afghan civilians tended to decrease support for the NATO International Security Assistance Force (ISAF) and increase support for the Taliban in the locality of the strike.[32]

Flight mishaps, fratricides, or serious collateral damage usually made it obvious to participants *that* the system had failed, even if *how* it failed remained a mystery. Their attention shifted reflexively to the information system to begin debugging. In more insulated failures, by contrast, the existence of the problem itself was obscured. Practitioner intentionality remained focused on the world that the information system made available, even if its representations had fallen out of sync with reality. Major General Michael Flynn and his coauthors thus harshly criticized the U.S. intelligence effort in Afghanistan in a January 2010 report: "Having focused the overwhelming majority of its collection efforts and analytical brainpower on insurgent groups, the vast intelligence apparatus is unable to answer fundamental questions about the environment in which US and allied forces operate and the people they seek to persuade."[33]

As if to underscore this opinion, in September that same year, a U.S. SOF unit targeted what turned out to be a six-car election convoy in broad daylight. Manned fighters dropped bombs while a drone provided surveillance. The unit believed it was targeting Muhammad Amin, a Taliban shadow deputy governor. SOF analysts believed Amin was using the alias "Zabet Amanullah," based on the analysis of cell phone data. Zabet Amanullah, however, was a well-known personage in the region who happened to be campaigning for his nephew in parliamentary elections. Amanullah appears to have had several phone conversations with Amin and others who had Taliban affiliations, but this would not have been unusual behavior in the social world of rural Afghanistan, especially by someone looking to foster political support in an election season. The SOF unit "found" Amin, but they conflated him with Amanullah, whom they then "fixed" and "finished." Nine other civilians died in the strike; SOF erroneously assessed them to be Taliban operatives traveling with Amin. SOF tended to code dead men on the objective as "enemy killed in action" unless countervailing evidence emerged later, which they did not go out of their way to seek out. At a press conference in Kabul after the raid, Defense Secretary Robert Gates announced that the raid had killed a senior member of the "Islamic Movement of Uzbekistan," Muhammad Amin. A month later, the SOF unit continued to insist to the journalist Kate Clark that its technical intelligence showed that Muhammad Amin and Zabet Amanullah were the same person, even as they conceded that they had not investigated the background of either person in much depth. Yet Muhammad Amin, very much alive, gave an interview in March 2011 that corroborated the SOF unit's biography of him. Premature consensus about the reliability of technical analysis and the ambiguous social environment of Takhar Province undermined the referential integrity of targeting intelligence. The insulated SOF unit never discovered or affirmed its mistake.[34]

Practitioners used and debugged an ersatz control system amid the confusion of civil war. Mistakes were made, and some of the mistakes were never recognized by aggressive SOF units. Other breakdowns triggered official investigations and earnest effort by operators, administrators, and commanders to understand what had gone wrong so that they might impose additional safeguards to keep from repeating the same mistakes. We know what we know about the Uruzgan civilian casualty and Sangin fratricide incidents, after all, because of lengthy Pentagon investigations that resulted in reprimands and reforms.[35] As officials and commanders sought to improve military efficiency and reduce collateral damage, drone operations gradually took on a more managed character. Practitioners continued to make marginal improvements in tactics, techniques, and procedures, but within an increasingly bureaucratized context of common standards and processes.

## Management and Insulation in the United States

In Afghanistan, JSOC could apply the F3EA model pioneered in Iraq, which used raids from secure bases to generate intelligence. The complementary ISAF counterinsurgency campaign could provide additional human intelligence and, potentially, offset some of the counterproductive side effects of counterterrorism through civil affairs and information operations. Many drone strikes in Afghanistan, moreover, supported troops in contact with the enemy. In these events the enemy conveniently identified themselves by shooting. Drone strikes against Taliban commanders degraded enemy professionalism and morale, influenced enemy targeting decisions, and diminished the success of enemy attacks for a while.[36]

Al-Qaeda responded by relocating and regenerating beyond the institutionalized reach of ISAF. Senior leadership found sanctuary in northwestern Pakistan and encouraged franchises in other countries. The Bush administration, in turn, responded with an intensified covert-action campaign under CIA authority in Pakistan, which was justified by the presence of al-Qaeda and Taliban leadership in the Federally Administered Tribal Areas (FATA) and a discreet invitation from Pakistan president Pervez Musharraf.[37] The Obama administration intensified the CIA drone war in FATA, viewing it as a middle ground between doing nothing and launching a costly intervention like those in Afghanistan and Iraq. Both administrations relied on a liberal interpretation of the post-9/11 Authorization for Use of Military Force to justify lethal action in Pakistan, Yemen, Somalia, Libya, and elsewhere against groups that had rather tenuous connections with al-Qaeda.[38]

Counterterrorism campaigns in regions outside declared combat zones had to rely more on technical collection than human intelligence. Drones also tended to require longer transit times to range targets in areas bereft of friendly bases. Such campaigns thus never achieved the same pace or volume of operations that JSOC was able to in Iraq and Afghanistan. The legal challenges of the altered problem of targeted killing outside defined combat zones, moreover, spurred further institutionalization of the American solution. Legal tussles became a prominent feature of drone information practice at all levels. The heightened difficulty of the intelligence problem and reduced political tolerance for targeting error encouraged greater patience in target development. The crucible of this evolution was the intensification of U.S. drone operations in Yemen, a country with no U.S. troops at war, to hunt and kill Anwar al-Awlaki, a U.S. citizen who received no due process.

WHO IS A VALID TARGET?

A drone strike in Yemen on November 3, 2002, was both the first to be conducted outside a recognized combat zone and the first to unintentionally kill a U.S. citizen. The CIA targeted an alleged planner of the USS *Cole* bombing, but it also killed Kamal Derwish, who happened to be traveling in the same car as the intended target. Derwish had not been indicted with the other members of the "Lackawanna Six" terror cell in Buffalo, but he was wanted for questioning and suspected of recruiting activity. The CIA conducted no more Predator strikes in Yemen for years, even as the political situation deteriorated.[39]

In September 2008, Yemeni militants attacked the U.S. embassy in Sanaa. In January 2009, they announced the formation of al-Qaeda in the Arabian Peninsula (AQAP) in partnership with a group of Saudis. The Yemeni president promised full counterterrorism cooperation with the United States, and Obama authorized a navy cruise missile strike on a suspected AQAP camp in al-Majala in December 2009. The strike killed its target and a dozen other militants, but the cluster munitions from Tomahawk cruise missiles also killed forty-one civilians, including twenty-one children. As Anwar al-Awlaki wrote to a colleague, "The Americans just scored a big own goal." AQAP eagerly exploited the propaganda opportunity. According to an Obama aide, "The president wasn't happy with it, and so went through a very long process led by [counterterrorism adviser John O.] Brennan to tighten up how we take lethal action in Yemen."[40]

Al-Awlaki had been an imam in San Diego, not far from the General Atomics factory that built the drone that would kill him. Disillusioned by anti-Muslim sentiment in the wake of 9/11, al-Awlaki left the United States for Yemen in 2002 and later affiliated with AQAP. Al-Awlaki's inflammatory lectures, posted online in English, inspired Hasan Nidal to open fire at Fort Hood in November 2009, and Faisal Shahzad to attempt to bomb Times Square in May 2010. Al-Awlaki directly aided in a plot to bring down an airliner on Christmas 2009, just eight days after the al-Majala disaster, with an underwear bomb worn by Umar Farouk Abdulmutalab. In October 2010, an al-Awlaki plot nearly succeeded in detonating explosives in printer cartridges shipped aboard a cargo aircraft bound for Chicago, the president's adopted city. Obama said that the decision to target al-Awlaki was "an easy one" in principle, but justifying the deliberate killing of a U.S. citizen overseas based on secret intelligence without a trial was, nonetheless, a delicate legal matter uncountenanced since the U.S. Civil War.[41]

A memo from the Office of Legal Counsel argued that al-Awlaki could be considered a "continued and imminent" threat and thus targeted within the constraints of the Fourth and Fifth Amendments to the Constitution, an injunction against the overseas murder of U.S. citizens in U.S. Code Title 18, and the Authorization for Use of Military Force.[42] The decision to target al-Awlaki leaked to the press, which prompted his father, Nasser al-Awlaki, to sue the U.S. government on his behalf. The district judge dismissed the "extraordinary and unique" case on procedural grounds (because there was no evidence that Anwar supported his father's suit) and argued that targeting was ultimately a national security issue, not a judicial one.[43] The CIA and JSOC thus had permission to kill al-Awlaki if they could find him. Both organizations operated drones in Yemen, but under different authorities derived, respectively, from U.S. Code Title 50 governing intelligence activities and Title 10 governing military operations. Institutional practices built around this difference resulted in discrepant kill lists and oversight problems.[44]

JSOC was still adjusting its industrial counterterrorism model developed in Iraq for use beyond the reach of SOF teams, and its targeting was still prone to insulation. On May 24, 2010, JSOC attacked a pair of vehicles in Marib Province, killing two AQAP members but also, inadvertently, the popular deputy governor, Jabir al-Shabwani. Al-Shabwani had been meeting with the militants in an attempt to convince them to surrender to the government. JSOC found out about its mistake only when angry tribesmen blew up an oil pipeline to avenge al-Shabwani's death on behalf of his grieving father. The tribe turned against the Yemeni government, and some tribesmen became sympathetic to AQAP. The strike was a serious setback for attempts at nonviolent mediation.[45]

The CIA and JSOC often worked together, but not without friction. Their most famous collaboration resulted in the death of Osama bin Laden in Abbottabad, Pakistan, for which Obama relied on Navy SEALs from JSOC operating under CIA authority, rather than a CIA drone strike. Obama had feared that a drone strike would produce too many civilian casualties, and he wanted positive confirmation that bin Laden was dead. On May 5, 2011, just four days after the bin Laden raid, JSOC drones had an opportunity to target al-Awlaki. Alerted at the last minute, al-Awlaki managed to swap cars and escape. He eluded U.S. intelligence for a few months until the outbreak of civil war in Yemen fractured rebel groups, forced individuals to move, and generated chatter that the National Security Agency (NSA) could monitor. This time the CIA, rather than JSOC, took control. On September 30, 2011, a Reaper supported by two Predators destroyed trucks carrying al-Awlaki and several other militants. The strike also happened to kill another American named Samir Khan, who published al-Qaeda's internet magazine, *Inspire*. Khan had not been added to the U.S. lethal target list, and U.S. intelligence did not know that he was traveling with al-Awlaki that day. Two weeks later, JSOC targeted and missed an AQAP media emir but inadvertently killed al-Awlaki's teenage son and cousin. Both were U.S. citizens. In hindsight it appears that Abdulrahman al-Awlaki was seeking vengeance for his father's death by trying to join AQAP, but the presence of any Americans whatsoever came as a complete surprise to JSOC. In just two weeks the United States had managed to kill four of its own citizens without a trial, but it had targeted only one. Obama reportedly was furious that there had been so much care with the intelligence and legal analysis around Anwar al-Awlaki but nothing like it on the JSOC operation. Another round of regulatory reform followed.[46]

## SIGNATURE STRIKES

In the FATA campaign, the CIA made a distinction between "personality strikes" against known individuals and "signature strikes" against unknown militants. Both types of targets required analytical development, collateral damage evaluation, and approval at senior levels, but personality targeting was more time consuming. Signature strikes aimed to disrupt militant networks through a higher volume and tempo of attacks that targeted suspicious patterns of behavior and communications in areas controlled by insurgents. They necessarily accepted more uncertainty about the identity of targets. The CIA described a quarter of the fatalities during fourteen months of drone strikes in FATA in 2010–11 simply as "other militants" rather than specifically al-Qaeda. Military-aged males around suspicious targets were presumed guilty unless credible evidence emerged that they were not. This practice encouraged the overcounting of militants and the undercounting of civilians, which produced an overly optimistic perception of performance.[47]

Former CIA director Michael Hayden defended drone strikes on the basis of the danger of the al-Qaeda threat, their ability to discriminate militants, and their ultimate effectiveness. "Intelligence for signature strikes always had multiple threads and deep history," the general argued. "The data was near encyclopedic." The persistent surveillance and tactical patience of drones created an unprecedented ability to minimize mistakes. "Civilians have died," he conceded, "but in my firm opinion, the death toll from terrorist attacks would have been much higher if we had not taken action." Signature strikes were effective, furthermore, because they "drastically shrank the enemy's bench and made the leadership worry that they had no safe havens." Although the "program is not perfect," he concluded, "here is the bottom line. It works."[48]

The risks of signature strikes were underscored on January 15, 2015, in North Waziristan. For weeks CIA analysts had scrutinized the movements of four men and listened in on cell phone conversations for hundreds of hours. Pattern-of-life analysis led to a "high confidence" assessment that it was an al-Qaeda compound with no civilians present, and that the intelligence met the threshold of "near certainty" demanded by Obama. Yet imagery collected after the strike revealed locals near the target digging six fresh graves, not four, and the NSA intercepted conversations about "Westerners" having been killed. One victim turned out to be a seventy-three-year-old American contractor named Warren Weinstein who had been kidnapped in Pakistan in 2011. The other was an Italian aid worker named Giovanni Lo Porto who had been kidnapped in Pakistan in 2012.

One of the four militants killed (Ahmed Farouq) also turned out to be a U.S. citizen. Another signature strike in the same area days later happened to kill Adam Gadahn, a California-born al-Qaeda propagandist tracked by U.S. intelligence for over a decade but not known to be present at the time.[49]

The deaths of three more Americans brought the total number of U.S. citizens killed by their own government to nine. These included Kamal Derwish in Yemen in 2002, the four men associated with al-Awlaki in Yemen in 2010, and a North Carolina man named Jude Kenan Mohammad, who died in a 2011 signature strike in South Waziristan. Only Anwar al-Awlaki had been targeted by name. Five to seven of the others did indeed have an al-Qaeda affiliation, considering that al-Awlaki's aggrieved teenage son and cousin may have been trying to join AQAP, so some might consider them "bonus" collateral damage. Yet all were U.S. citizens who received no due process. In the immediate wake of this embarrassing breakdown, Washington launched an internal investigation and another round of process reform. In April, Obama publicly apologized for killing Weinstein and Lo Porto: "I take full responsibility for all our counterterrorism operations," he said. Drone strikes in general and signature strikes in particular would be subjected to much more rigorous scrutiny.[50]

## THE OBAMA ADMINISTRATION IMPROVES OVERSIGHT

After years of little official comment, Obama attempted to assuage concerns about the drone program in a May 2013 speech at the National Defense University. "This new technology," he said, "raises profound questions—about who is targeted, and why; about civilian casualties, and the risk of creating new enemies; about the legality of such strikes under U.S. and international law; about accountability and morality."[51] Ever the lawyer, Obama sought to reassure the nation that he had carefully considered these questions and grounded the answers in domestic and international law. The law of armed conflict is founded on clear distinctions between imminent and latent threats, self-defense and aggression, geographically defined combat zones and rear areas, and combatant and civilian personnel.[52] *Jus ad bellum* concerns the justification of going to war in self-defense, as a necessary last resort, and with consideration of the proportionality between the harm inflicted by a war and the harm prevented. *Jus in bello* concerns the legality of using force in a war already underway, requiring the discrimination of combatants from civilians and the weighing of the proportional risks of collateral damage relative to the military advantage of attack. While going to war and fighting a war were traditionally sequential problems, counterterrorism campaigns must deal with them in parallel. The protean nature of terrorist networks, the secrecy of terrorist plotting, and the potential consequences of a terrorist attack encourage preventative rather than simply preemptive rationales for the use of force. Yet this means that *ad bellum* necessity and proportionality have to be considered anew for each new terrorist franchise or even individual operatives, not just for declaring some categorical state of war against some actor in some region. The judgment of imminent threat also depends on assessment of the danger of foreign terrorism to the homeland, which a number of skeptical scholars have questioned.[53] The *in bello* aspects of counterterrorism are also tricky with regard to who can be considered a combatant or not, as exemplified in the debate over signature strikes.[54]

Obama's speech made the *ad bellum* case that "al Qaeda, the Taliban, and their associated forces . . . would kill as many Americans as they could if we did not stop them first. So this is a just war—a war waged proportionally, in last resort, and in self-defense." He also made the *in bello* argument that "before any strike is taken, there must be near-certainty that no civilians will be killed or injured—the highest standard we can set." The patience and precision of drones, along with the safety they afford to aircrew, made them attractive by the *in bello* criteria of distinction and proportionality.[55] Yet if this very advantage were to lead to a perpetual reliance on drones, as many critics worried, then the accumulated harms could increase the very civilian costs they were supposed to reduce. Compliance with *in bello* proportionality could, ironically, end up violating *ad bellum* proportionality.[56] A democratic polity that risked no casualties, furthermore, might not bother to hold its leaders accountable when they used force too hastily or over issues where vital national interests were not imperiled.[57] Obama attempted to mollify criticism by acknowledging the real dangers of insulation and proposing oversight to manage them:

> The very precision of drone strikes and the necessary secrecy often involved in such actions can end up shielding our government from the public scrutiny that a troop deployment invites. It can also lead a President and his team to view drone strikes as a cure-all for terrorism. And for this reason, I've insisted on strong oversight of all lethal action. . . . The last four years, my administration has worked vigorously to establish a framework that governs our use of force against terrorists—insisting upon clear guidelines, oversight and accountability that is now codified in Presidential Policy Guidance.

The Presidential Policy Guidance (PPG) was a novel type of policy that not only directed cabinet members to take certain actions but also spelled out the responsibilities of the president in the target review process. Drones enabled a degree of intervention so precise and so fraught with political risk that, remarkably enough, the president of the United States became personally involved in tactical targeting decisions. Obama micromanaged the use of force to a degree not seen since the Johnson administration. In an era where drones dropped their bombs at the margin of control and the margin of legitimacy, *jus ad bellum* had to be considered for tactical strikes and *jus in bello* could not be relegated to functionaries. The details of the PPG remained classified until they were released in redacted form in August 2016 through a Freedom of Information Act request filed by the American Civil Liberties Union.[58]

Target folders mapping out each individual's role in the evolving network and substantiating intelligence were maintained in a database called the "disposition matrix" at the National Counterterrorism Center. Input came from the NSA, the CIA, JSOC, the State Department, the FBI, and other agencies, which held regular meetings among midlevel officials to scrub the data. The analysis of any given individual's pattern of life could be developed for years before the quality of the intelligence and the accessibility of the target justified proceeding any further. Nominations for action were compiled into PowerPoint slides called "baseball cards" that summarized pertinent data about each individual. These were debated in regular meetings of the National Security Council Deputies Committee, chaired by counterterrorism adviser John Brennan. Brennan provided the first official acknowledgment and description of the program in a speech in April 2012, stressing a deliberative concern for accuracy and legitimacy: "We discuss. We debate. We disagree. We consider the advantages and disadvantages of taking action. We also carefully consider the costs of inaction." In assessing the merits of legal arguments, the quality of intelligence, and strategic assumptions about tactical actions, Obama pushed participants to tackle what Brennan characterized as "tough questions," such as, "Is capture really not feasible? Is this individual a significant threat to U.S. interests? Is this really the best option? Have we thought through the consequences, especially any unintended ones? Is this really going to help protect our country from further attacks? Is this going to save lives?"[59] After the committee nominated a target for "lethal action," Obama then personally approved (or rejected) every strike in Yemen or Somalia, as well as the riskier strikes in Pakistan, reportedly a third of the total.[60]

The case of Mohanad Mahmoud al-Farekh highlights the controls in place. Al-Farekh was a Texas-born al-Qaeda operative in Pakistan who attracted intensive attention from U.S. intelligence agencies in 2012. Drones monitored his movements and a secret warrant from the Federal Intelligence Surveillance Court authorized surveillance of his communications. The Pentagon nominated, and the CIA endorsed, al-Farekh for lethal action in 2013. However, Attorney General Eric Holder questioned whether the intelligence case supported the judgment that al-Farekh was an immediate threat or even very significant in the al-Qaeda hierarchy. Arguing that the targeted killing of a U.S. citizen should be a last resort, Holder instead recommended pursuing other options, such as capture.[61] Debate simmered in the Deputies Committee and congressional oversight committees for years under the more stringent rules Obama had put in place. Pakistani forces detained al-Farekh in spring 2015 and extradited him to the United States, where he was arraigned in a federal court in Brooklyn.[62] Alternatives to targeted killing thus did exist. Controversial *ad bellum* arguments of "imminent and continued" threat had to be carefully evaluated on a case-by-case basis, and the system was able to do so with the right leadership.

Personality and signature strikes alike were constrained by restrictive rules of engagement requiring positive identification of named targets and deliberate efforts to limit collateral damage. Pattern-of-life analysis helped not only to identify a target but also to identify windows of vulnerability to kill him while reducing the potential harm to family members and other civilians. Planners used a software-based methodology to model weapons effects against particular types of buildings and geographical features in order to select appropriate weapons, fuse settings, and delivery headings. The tool helped them to target combatants while shielding civilians. After taking all possible tactical measures to mitigate collateral damage, planners completed a Casualty Estimate Worksheet and submitted anything that exceeded a designated Non-combatant Casualty Cut-Off Value for a Sensitive Target Approval and Review process. Targets that passed review had to be endorsed by the secretary of defense or the president himself. To gain feedback from every action, the White House required each agency conducting a lethal direct action to produce an after-action report detailing the planning criteria, tactical details, combatants killed, collateral damage, and effects assessment. The Department of Defense maintained statistics on all aspects of drone operations—remote split ops afforded a large and legible trail of data—including the mix of intelligence sources used, timelines of the targeting process, and tactical outcomes. This feedback helped to identify and correct sources of false positives (hitting the wrong target) and false negatives (missing the right one), which enabled further regulatory reform.[63]

Critics of U.S. targeted killing policy argued that, as Philip Alston put it, there was "no meaningful domestic accountability for a burgeoning program of international killing. This in turn means that the United States cannot possibly satisfy its obligations under international law to ensure accountability for its use of lethal force."[64] Yet while coordination in the Obama administration was not publicly transparent, it was neither arbitrary nor uncontrolled. Decisions to undertake lethal strikes against particular individuals were the accumulated results of numerous well-constrained decisions. As Gregory McNeal argues, bureaucratic accountability was enacted through standard operating procedures, departmental regulations, formal performance evaluations, and career assignments.[65] Legal accountability was inherent in the specter of courts martial, criminal prosecution, or referral to international tribunal. Professional accountability operated through socialized concerns about informal reprimands and one's reputation with or shunning by peers. Political accountability was in place through congressional oversight committees, public interest groups, media reporting, and electoral approval. Public transparency might improve some, but not all, of these mechanisms, and could even weaken some of them.

Real and feared targeting errors, together with the political sensitivity of covert counterterrorism, led the Obama administration to build a substantial policy architecture to regulate all phases of targeted killing. Considerable procedural constraints and oversight were put into place to provide many of the checks and reviews that some feared did not exist. Critics could disagree about the goals of U.S. counterterrorism policy, but the process was substantially institutionalized.

Some critics felt that White House oversight had become too onerous, putting process ahead of effectiveness and undermining U.S. counterterrorism goals. Moreover, the very problem that Obama's PPG process was designed to manage—combating the al-Qaeda organization and its affiliates—began to change. ISIS (the Islamic State in Iraq and Syria, also known as Daesh) regenerated from the remnants of al-Qaeda in Iraq (AQI), the ruthlessly violent faction that broke with al-Qaeda central during the Iraq war, as discussed in the previous chapter. While Iraqi Awakening movements and relentless JSOC attacks had significantly weakened AQI, its administrative core remained intact. The outbreak of civil war in Syria provided a growth opportunity for ISIS, and affiliates sprung up in many other countries as well.[66] The U.S. military sometimes received temporary exemptions to operate outside "areas of active hostilities" (Afghanistan, Iraq, and Syria) where the PPG imposed more stringent guidelines for protecting civilians, most notably for operations against ISIS in Libya in 2016. Thus even a legalistic president like Obama could find ways to work around the constraints that he placed on himself.

## THE TRUMP ADMINISTRATION REDUCES OVERSIGHT

President Donald J. Trump shared neither Obama's legal sensibilities nor his work ethic. Trump's fiery rhetoric on the campaign trail and in office raised concerns that drones, and military force more generally, were about to be employed indiscriminately. CIA director Michael Pompeo pushed for and received more authority to expand the CIA drone campaign from Pakistan to Afghanistan, targeting not only al-Qaeda but also the Taliban and the Haqqani network. "When we've asked for more authorities, we've been given it," Pompeo said. "When we ask for more resources, we get it."[67] In March 2017, Trump approved a Pentagon request to exempt parts of Yemen and Somalia from Obama's PPG, targeting ISIS and not just al-Qaeda.[68] Trump ceded more authority to the military and CIA to take a "more aggressive approach," as a White House official explained, to enable them to conduct missions with "more speed and more efficiency." When things went wrong, however, Trump was reticent to take responsibility. After a raid in Yemen resulted in the death of Chief Petty Officer (SEAL) William Owens and over a dozen civilians, Trump deflected blame onto "the generals."[69]

The Trump administration reportedly replaced the Obama PPG with a new "Principles, Standards, and Procedures" (PSP) document, with elements of both continuity and change. The fact that the policy was revised rather than abolished altogether reflected an institutionalized commitment to oversight, in the "deep state" if not in the Trump White House. Importantly, the PSP retained the PPG's "near certainty" requirement that lethal strikes would not harm civilians, as well as the distinction between active war zones and external regions that required more caution. However, the PSP also lowered the targeting threshold. Individuals no longer had to pose a "continuing and imminent threat to U.S. persons." This meant that lower-ranking and support personnel could be targeted, and lethal actions could be conducted in support of U.S. allies where U.S. persons were not at risk. The requirement on the quality of intelligence that fixed the target's location was also reduced from "near certainty" to "reasonable certainty." Most significantly, the PSP discontinued the high-level White House coordination meetings that the PPG required, replacing them with an annual review of "country plans" while delegating operational decisions to the military and CIA.[70]

In short, the Trump administration both expanded the scope of lethal drone operations and reduced oversight over them. In the framework of this book, Trump's revision of Obama's managerial reforms cleared the way for more aggressive exploitation. This could potentially reduce bureaucratic insulation by empowering subordinate commanders to adapt to changing battlefields, thereby improving counterterrorism performance. Yet the increased play of uncoordinated adaptation also (re)opened the door to problematic practice. As discussed below, the civilian casualty rates of U.S. drone strikes have appeared to increase during the Trump presidency after a steady decline during the later Obama years.

## Ambiguous Performance

U.S. drone campaigns continue as of this writing, and much data remain classified. Nevertheless, two general observations can be made. First, the tactical performance of the drone targeting system has generally improved over time as targeting errors have come under better control. Second, consensus about the strategic effectiveness of targeted killing remains elusive.

TARGETING ACCURACY

Figure 6.1 depicts independent estimates of casualties caused by U.S. drone strikes in Pakistan, Yemen, and Somalia through 2018 from the

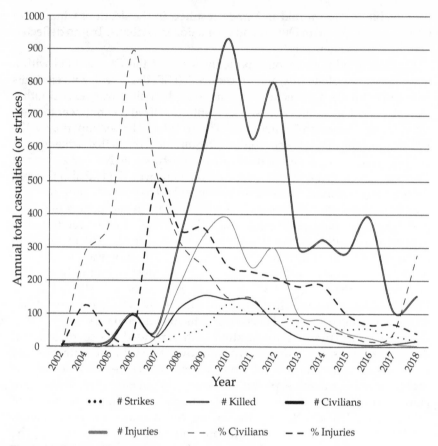

Figure 6.1. Estimated casualties due to U.S. drone strikes in Pakistan, Yemen, and Somalia, 2002–18

Bureau of Investigative Journalism.[71] The solid lines depict the total number of people killed each year, total civilians killed, and total injured each year. The dashed lines depict the ratio of civilian deaths and injuries to total deaths, which can be interpreted as a rough measure of targeting accuracy. Fewer civilians killed implies fewer false positives. Fewer injuries implies more decisive lethal operations. Both of these measures trend steadily downward from spikes in the early years of the drone program. The dotted line depicts the total number of strikes each year. Note that while the overall number of strikes is increasing during the height of the campaign in Pakistan, the civilian casualty and injury rates are decreasing. They continue to decrease even when the number of strikes starts to decline. Also note that the overall number of people killed decreases at a faster rate than the overall number of strikes. All of these trends are consistent with improvements in targeting accuracy and the increasingly managed character of drone targeting over time.

The exception to the decreasing trend is the spike in the civilian casualty rate in 2017–18. This is suggestive of more problematic practice under the Trump administration. Most of this spike is the result of dramatic increases in civilian casualty rates in Somalia (11 percent) and Yemen (39 percent) in 2018. In the full dataset, the rate of civilian casualties caused by drones and manned aircraft in these countries is the same (8 percent). If we omit the Trump years, however, then the ratio of civilian casualties due to drones (7 percent) versus manned aircraft (17 percent) is much lower. Drone targeting may have become relatively less accurate because of Trump's less restrictive rules of engagement.

One should be cautious about drawing conclusions from these data. The Bureau of Independent Journalism's estimates tend to be slightly higher than U.S. government estimates, which has prompted allegations of overcounting civilian deaths due to the exaggerations of sympathetic populations in insurgent-controlled areas. The discrepancy may also be due to a tendency in the U.S. government to code "military-aged males" as enemy combatants rather than civilian bystanders, on the dubious assumption that mere proximity to a known target indicates combatant status. If there is an upward bias in the bureau's data for whatever reason, however, then the gradual decline of civilian casualty percentages over time is all the more notable.[72]

AN AL-QAEDA ASSESSMENT

Al-Qaeda suffered steady attrition from relentless drone strikes, but it also adapted. In a letter found in the Abbottabad raid, Osama bin Laden expressed concern that drones were killing his most experienced commanders and urged them to take precautions:

A warning to the brothers: they should not meet on the road and move in their cars because many of them got targeted while they were meeting on the road. They also should not enter the market in their cars. . . . The Americans . . . can distinguish between houses frequented by men at a higher rate than usual. Also, the visiting person might be tracked without him knowing. This applies to locals too. Inform the brothers that this is the arrangement for every emir at this time. It is important to have the leadership in a far-away location. . . . When this experienced leadership dies, this would lead to the rise of lower leaders who are not as experienced as the former leaders and this would lead to the repeat of mistakes. Remind your deputies that all communication with others should be done through letters.[73]

A June 2011 report by a militant calling himself Abdullah bin Mohammed turned up in a safe house in Timbuktu in early 2013.[74] The report summarized lessons learned in Waziristan and discussed in a "Military Research Workshop" held in Yemen. Bin Mohammed offered prescient opinions on why the United States had resorted to drone strikes, followed by twenty-two tips on how to survive them and a strategy to work around them. In the triangular logic of counterinsurgency, the materially weaker insurgent hides within a supportive population below the legibility threshold of the materially stronger counterinsurgent and indefinitely imposes costs to erode its will to fight. The American innovation of drone warfare improved the threshold of legibility, in effect imposing a new solution on American terms. Al-Qaeda would have to counterinnovate to exploit this improvement.

Bin Mohammed begins by pointing out that the chief advantage of drones is not military—the Americans have plenty of "combat jets like the F16"—but political. A drone costs policymakers "nothing compared to the manned jets and it does not create public exasperation when it crashes." He notes that the United States is "in the 10th year of war," causing "[economic] exhaust[ion]" and "human losses" leading to "public pressure backed by the Congress" for "honorable and responsible withdrawal as a prime goal of the White House." In these circumstances, "the war of the drone appeared as a perfect solution" for the United States to wage "a comfortable war to prove to us their indifference to a long war." As al-Qaeda stiffened its resolve to fight, the United States resorted to the substitution of technology for labor in a desperate attempt to stay in the fight.

Bin Mohammed then discusses a range of denial and deception, electronic warfare, and defensive countermeasures to complicate drone operations. To defeat measurement, he recommends "not to use permanent headquarters"; "complete silence of all wireless contacts"; "stay in places unlit by the sun such as the shadows of the buildings or the trees"; "use smoke as cover by burning tires"; "deceive the drone by entering places of multiple entrances and exits"; and "formation of fake gatherings such as using dolls and statutes to be placed outside false ditches to mislead the enemy." To defeat coordination, he recommends hacking "the drone's waves and the frequencies" with "computer know-how" and "jamming . . . and confusing of electronic communication" by leaving mechanical equipment running. To defeat enforcement, he recommends "well-placed reconnaissance networks . . . to warn all the formations to halt any movement in the area" and the use of "skilled snipers to hunt the drone, especially the reconnaissance ones because they fly low." Meetings should be held in "underground shelters" or "forests and caves" to "avoid gathering in open areas," and if "a drone is after a car," operatives should "leave the car immediately and everyone should go in different directions because the planes are unable to get after everyone." As one U.S. Predator operator remarked, "These are not dumb techniques. It shows that they are acting pretty astutely."[75]

Finally, bin Mohammed returns to the strategic contest. He observes that it is no longer possible to impose political costs on the United States militarily, since it risks no casualties in battle. He then argues that al-Qaeda should find a new way to impose costs, by targeting civilians. He recommends that anybody suspected of providing intelligence should "be hanged in public places with a sign hanging from his neck identifying him as an 'American Spy'" to create a "decisive deterrent means against anyone who might dare to carry out this mission." Bin Mohammed expects intimidation to foment "public opinion against these [drone] attacks to instigate an alternative Arab and Islamic street." Further, he recommends, "We start kidnapping Western citizens in any spot in the world, whether in the Islamic Maghreb, Egypt, Iraq or any other easy kidnapping places and the only demand is the halt of attacks on civilians in Yemen which is a just and humanitarian demand that will create world support and a public opinion pressure in America as they are being hurt again." This strategy, he says, "is the golden solution that shortens the long distances" by turning the United States' respect for the law against itself.

If drones were attractive to the United States precisely because of their low political costs, bin Mohammed argues, then al-Qaeda should link U.S. drone warfare to higher political costs through murder and intimidation. Empirical studies of drone strikes in Pakistan offer some support for his logic. Strikes have been correlated with a short-term reduction in the incidence and lethality of terrorist attacks immediately following drone strikes in the same locale.[76] Yet drones are also associated with an increase in attacks against government targets and the displacement of militant activity into populated areas where drones are less likely to be used.[77] Once again, reforms that improved managed practice in drone operations appear to have encouraged militants to complicate the strategic problem.

## CONTESTED EFFECTS

Scholars are divided on the effects of drones in particular and targeted killing in general, even disagreeing about how to define it.[78] Drone surveillance and targeted strikes do appear to suppress insurgent violence and disrupt terrorist planning in the short run.[79] Leadership targeting appears to promote war termination and reduce violence under some circumstances.[80] Al-Qaeda leadership attrition, in conjunction with border controls and other counterterrorism measures, seems to have undermined the ability of the hierarchical organization to plan and communicate.[81] However, mature bureaucratic organizations like al-Qaeda evolve under constant selection pressure to improve resilience; they also tend to replace fallen martyrs with more desperate and ruthless leaders.[82] Audrey Kurth Cronin concludes that drones are effective at "killing operatives who aspire to attack the United States today or tomorrow. But they are also increasing the likelihood of attacks over the long term, by embittering locals and culti-vating a desire for vengeance."[83] Aqil Shah counters, "Interview and survey data from Pakistan, where, since 2004, the U.S. Central Intelligence Agency has launched more than 430 drone strikes, show little evidence that drone strikes have a significant impact on militant Islamist recruiting either locally or nationally."[84] Given the persistence of debate about counterterrorism effectiveness in the comfort of hindsight, it is reasonable to conclude that contemporary practitioners had a hazy understanding of their own effectiveness.

As drones made it easier to kill targets with more accuracy, furthermore, there was more equivocation over whom to kill, and why. Drones enhanced the possibility of precision, but they also fostered unreasonable expecta-tions of perfection. Senior leaders who were more sensitive to political risk, and public activists who were less tolerant of mistakes, began to take a keen interest in the drone targeting processes. The abundance of data collected, the tactical precision of strikes, residual uncertainty about target identity, and controversy about targeting error combined to incentivize leadership to pile on bureaucratic oversight. Operational personnel, in turn, had to negotiate not only the tactical and technical challenges of remote split oper-ations, but also the evolution of policy constraints far above their pay grade. The increasing prominence of lawyers in the conduct of war went hand in hand with the increasing importance of information work. Law, and with it public controversy about the law, penetrated into the inner workings of operational control. The normative dimension of control loomed especially prominently for operations beyond Afghanistan.[85]

## Killing without Dying

Drone warfare, far more than the advent of artillery or strategic bombing, has decoupled the ability to kill from the risk of dying. To an extreme matched or exceeded only in cyber operations, drone operators may take intellectual and moral risks, but they are themselves physically safe. In ear-lier historical eras, the logistics of control required commanders to be phys-ically present on the battlefield to see or direct anything at all. Physical courage and exposure to fire were necessary for effective measurement, coordination, and enforcement. Yet the industrial revolution in war began to separate the cognitive and physical aspects of operations into a division of labor between staff and field officers. Staff personnel worked with repre-sentational tools in distant headquarters. Line officers directed the use of force on the battlefield. The latter continued to merit awards for bravery, yet automation continued to increase the influence of the former.

The concept of "the warfighter" in Western military discourse has been changing subtly in recent decades. Increasingly it is used to refer to any professional who manages violence in uniform, regardless of whether he or she is also subject to it. This new identity remains contested. Military professionals who perform information work (i.e., intelligence, cryptology, network management, and psychological, cyber, drone, or space operations) have struggled to gain recognition as warfighters within a martial culture that valorizes courage under fire. At the same time, information work is increasingly vital in modern operations. The U.S. Department of Defense created a new Distinguished Warfare Medal in 2013 to honor the contribution of exceptional information work from afar. Yet because it appeared in the order of precedence above the Bronze Star, which can be awarded for combat valor, the new medal was widely ridiculed and was canceled later that year.[86]

Because a telescope on a hill cannot make out every detail that matters, some personnel still need to go out into the mud to detect subtleties of the combat situation. Forward air controllers who can identify concealed enemy positions and distinguish them from friendly units have often made a decisive difference in the application of airpower.[87] Some risk, moreover, still needs to be undertaken to make political commitments credible. The complete and perfect separation of killing and dying compromises the costly signal of resolve that a nation conveys to the enemy when citizens risk their lives in war. How can one know what an opponent is willing to die for if technology makes death unlikely? Honoring physical sacrifice in war, and valorizing courage, is not only a way of showing respect to warriors and building a military ethos. It also has some political utility.

Advanced industrialized countries have become adept at protecting their own territories and defending their vital interests. They have been less successful in peripheral interventions where they are less resolved to pay the costs of military occupation and economic development. Weaker but more resolved opponents can pursue subversive strategies to exploit stronger but less resolute powers. The improvement of military control systems, exemplified by drone campaigns on the other side of the planet, enables strong states to undertake more ambitious missions in more ambiguous environments. Yet those same states are more likely to experience friction and unsatisfying performance in those places—neither outright failure nor decisive victory. Drone counterterrorism attempts to extend control into areas where policymakers are less willing to pay the costs of traditional military intervention but still want to achieve some political goals. The targets of drones, in turn, are incentivized to shift their attacks to nonmilitary targets and find other ways to subvert the legitimacy of the war.

Debates about norms, ethics, and legitimacy in war are essentially debates about the intentionality of a military information system. Organizational control cycles attempt to maintain or move the state of the world into closer alignment with a goal state. But what goal, and for whom? What exactly is the system trying to achieve, and what is it trying to avoid? A system may be designed with many different values in mind, and it may rely on overlapping control systems designed to implement different intentions. Trade-offs across goals become sources of disagreement and inefficiency. A national security system aims to provide security from threats, but stakeholders differ on the degree to which vital interests include the security of the ruling regime, the integrity of territory, military power position, access to resources, protection of civil society, or something else. Counterterrorism relies on sensitive sources and methods for measurement and enforcement, and secrecy surrounds deliberations over individual targeting decisions. The protection of secrets is usually justified by the need to prevent an adversary from adapting to elude intelligence collection or kinetic strike, but the same secrecy can also insulate a system from criticism. Consensus about the rightness of intentions can flourish without being tested, while critics can be held at bay by the mystique of secrecy. The emergence of drone warfare as a primary tool of U.S. counterterrorism runs afoul of these issues and more. How imminent is the terrorist threat? What should a democracy give up to mitigate it? Is it moral to kill by remote control without risking death? Can the public trust their leaders with secret intelligence and secret deliberation to kill foreign terrorists—or their fellow citizens—on their own recognizance? These are political questions that technology cannot answer.

Increasing precision in the use of force has also created unrealistic expectations for perfection in targeting. Disagreement over the principles and objectives that guide targeting ensures that some expectations will not be met. The drafting, interpretation, and contestation of legal policy has thus emerged as an aspect of information practice that has become every bit as relevant for modern war as intelligence and planning. War has become more legalized in some respects but less legitimate in others.[88] Lawyers not only constrain military operations; they also enable them by helping commanders pursue favored objectives through the creative interpretation of existing legal frameworks, much as hackers use the rules of software to get machines to behave in ways that their architects do not intend. Drones exemplify this development insofar as lawyers, and policymakers who care about the rule of law, can virtually sit beside the operator who pulls the trigger. Yet unconventional operations in the gray zone between war and peace also expose a tension between the law of armed conflict and human rights norms. The former exonerates commanders from the tragic but inevitable loss of life that has to be expected in a just war; the latter aims to protect the inalienable right to life of combatants and civilians alike.[89] Militants who are not actively involved in combat might be legitimate targets under one framework but protected under the other. Different actors may selectively emphasize different legal principles to justify their preferred intentions or to undermine their rivals.[90]

In sum, the *why* of drone operations has become every bit as complicated as the *how*. The operational problem of finding, fixing, and finishing particular individuals on the opposite end of the earth is intrinsically difficult. In an information system this complex and distributed, practitioners can always find ways to better optimize their solution to the hunting problem (adaptive practice)—that is, improving the speed and acuity of targeting while protecting comrades and civilians. Yet in a tightly coupled system with many moving parts, tragic and embarrassing breakdowns must be expected now and again (problematic practice). Better oversight and deeper reforms might improve triangulation on individual militants identified though an increasingly legitimized process (managed practice). Yet this does not necessarily improve knowledge about local politics, let alone the factors that sustain or undermine militant strategies. The reduction of uncertainty in the tactical aspects of counterterrorism can leave a residual blind spot wherein strategic threats could continue to develop (insulated practice). Deliberate counteraction by enemy organizations and normative controversy among international audiences further muddy the strategic environment. The canonical cycling of the different ideal types of information practice has tended to build up an ever more complex information system over time. Yet more complexity and tactical precision do not translate into reliable strategic advantages and legitimate outcomes.

# Practical Implications of Information Practice

I tracked the network from the generals' plasma screens at Central
Command to the forward nodes on the battlefields in Iraq. What
I discovered was something entirely different from the shiny picture of
techno-supremacy touted by the proponents of the Rumsfeld doctrine.
I found an unsung corps of geeks improvising as they went, cobbling
together a remarkable system from a hodgepodge of military-built
networking technology, off-the-shelf gear, miles of Ethernet cable, and
commercial software. And during two weeks in the war zone, I never
heard anyone mention the revolution in military affairs.

Joshua Davis, "If We Run Out of Batteries,
This War Is Screwed"

Technological visions of future war err by forecasting success where trou-
bles are hidden or exaggerating threats where adversaries are weak. This
book offers an alternative perspective that focuses on the social context of
military information systems. It is not just the quality of technology that
matters in war, but the way in which practitioners use it. The quality of
information practice, in turn, depends on the relationship between operational problems and organizational solutions. Stable and defined military
tasks are well served by institutionalized or formalized systems; this interaction gives rise to performance-enhancing *managed practice*. More unstable
or ambiguous problems, by contrast, are better served by organic or
informal arrangements, which promote performance-improving *adaptive
practice*. Mismatches between solutions and problems, however, tend to
create information friction that undermines military performance. *Problematic practice* emerges when organic solutions to constrained problems make
accidents and targeting errors more likely. *Insulated practice* emerges when
institutionalized solutions to unconstrained problems reinforce myopia
and misperception. External and internal actors may act to alter operational
problems or organizational solutions, which changes the quality of practice.

This dynamic interaction tends to drive organizations through all four patterns in a canonical sequence—managed, insulated, adaptive, and problematic. As organizations cycle through them over time, the complexity of systems and practice tends to increase, but strategic advantage is never assured. Indeed, temporary success sets the conditions for unforeseen failure, and vice versa.

A richer understanding of the microfoundations of military power in the information age has important implications for defense strategy and policy. This book offers both negative correctives and positive suggestions. On the negative side, it cautions against the uncritical acceptance of the technology theory of victory. Scientific advances in artificial intelligence and quantum computing are already encouraging familiar hopes and fears. It is important to ask critical questions about how organizations will actually use computational technologies in the unpredictable circumstances of armed conflict. On the positive side, this book offers some hope that warfighters can find better ways to use their information technologies. A fundamental leadership challenge for organizations today is to develop an institutional support system that can both encourage user innovation and limit its liabilities. In this chapter I sketch out a synthesis of top-down management and bottom-up adaptation that I call *adaptive management*. I also discuss the strategic limits to adaptive management at the low and high ends of the conflict spectrum, which are exemplified by irregular warfare and nuclear warfighting, respectively. Cybersecurity, furthermore, is relevant across the spectrum because it uses information practice to protect or exploit information practice itself.

## The Variety of Information Practice

The problem with the technology theory of victory is not that it is wrong absolutely but that it relies on unstated strategic and organizational conditions. One reason for the enduring allure of ideas like the RMA is that advanced sensors, common networks, and precision munitions can and do deliver on the revolutionary promise, in some situations. Targets that once required hundreds of sorties to destroy can now be eliminated with a single satellite-guided bomb. Special operations forces (SOF) supported by airpower and precision fires can now generate influence in areas that once required tens of thousands of troops to control. Troop casualties can be minimized with standoff weapons that reduce the exposure of friendly personnel, or eliminated altogether in the case of drones flying on the other side of the planet. Strikes targeted more discriminately and proportionately, with greater potential for review and accountability, can improve adherence to the laws of war. A postmortem of any military operation, moreover, will inevitably reveal information-processing problems that might be fixed through the smart application of new computational technologies and processes. Creative personnel will keep finding ways to modify commercial information technologies to improve their mission performance. One reason that technology theory is perennially attractive, therefore, is that tactical improvement is always possible.

Yet the same information technologies that enable success can also fail in alarming ways. Precise weapons and imprecise intelligence combine to produce targeting errors that kill friendly forces or innocent bystanders. Expectations of perfection and pervasive media coverage amplify public outrage over targeting mistakes. Administrative oversight and digital accountability mechanisms that are supposed to control mistakes can inadvertently turn practitioner attention inward toward the maintenance of the bureaucracy. Policymakers who are no longer constrained by conscription or other political costs of war, meanwhile, become tempted to rely on military technologies in more ambiguous situations. The targets of remote military intervention then seek out new ways to hide and adapt that further increase the ambiguity of the strategic problem. The technology theory of victory falters as tactical success turns into strategic failure.

The popular imagery of information age warfare is full of streaming ones and zeros, autonomous weapons zipping through air and space, and rationalized command and control diagrams. When we look beyond these tropes, we invariably find personnel in office spaces struggling with glowing rectangles, and with each other. To understand military performance in the information age, therefore, it is important to understand how organizations actually use their computers. A long-term historical trend in the conduct of war is a growing emphasis on intellectual over physical labor. This transformation is both cause and consequence of military organizations' greater reliance on information technology of all kinds, from standardized forms and radio communications to digital apps and the internet. Warfighter experience has become more technologically mediated as sensors and weapons have become more sophisticated. I call this historical development the informational turn in military practice. One unintended consequence is that military personnel have become increasingly preoccupied with the frictions and breakdowns produced by the complexity of the information systems that enable them to know and influence the battlefield.

In the ideal information system, friction is minimized and efficiency is maximized. Many system architects aspire to build something like RAF Fighter Command—a rationalized and deliberately constructed information system that uses state-of-the-art technology to substitute information for mass. Because information is a relationship between internal representations and external realities, information systems work best when organizational and operational context is relatively stable. As I argued in chapter 3, the circumstances that enabled Britain's impressive performance in 1940, not least of which was the self-insulation of the

German adversary, were quite fortuitous. Yet even with strong permissive conditions, there was still some friction in the system. The realization of managed practice at Fighter Command also depended on a lot of low-level adjustment by scientists and operators to compensate for endless minor frictions. Fighter Command had some of the best radar scientists in the world, and they worked side by side with military operators. Because technical expertise was available in forward locations and personnel had a shared understanding of Fighter Command's mission, the runtime adjustment of rules and tools was more likely to enhance mission performance than to exacerbate coordination failures.

Champions of user innovation, who tend to be skeptical of ponderous bureaucratic management, would rather build something like FalconView— a flexible and responsive application designed by and for operators that is easy to change as warfighting requirements change. When the external information problem is ambiguous or unstable, personnel in an organically decentralized organization can take the initiative to adapt, repair, and improve their internal solution. Chapter 4 revealed that this case benefited from fortunate circumstances as well, to include a culture of tech-savvy pilots who were responsible for their own planning calculations and not (yet) beholden to a formal systems management bureaucracy. Moreover, the sustained performance of adaptive practice in this case actually depended on a symbiosis with managed practice. The core architecture of FalconView was managed by a small group of professional civilian engineers at Georgia Tech who ensured that there was a stable and flexible platform on which military users could build. FalconView also received top-cover from friendly supporters in the systems bureaucracy who protected and supported it while official programs were faltering.

The situation described in chapter 5 is perhaps more typical. Every unit is a garbage can full of ersatz Microsoft documents, abuzz with social interactions to make sense of it all. Inboxes are flooded with email, schedules are packed with meetings, and users struggle with the programs that are supposed to make them more productive. The level of technical expertise across units remains uneven, and many data processes and products are thus not as well constructed as they might be. Occasionally an alpha geek emerges in some unit with a preternatural aptitude for technology and enough entrepreneurial drive to cobble together a helpful solution for some operational problem. Then he or she runs afoul of network managers or simply moves on to the next assignment; the unsupported invention then fades away into the complicated milieu of working software and templates. Units continue to muddle through until someone blows up a Chinese embassy or bombs friendly troops, and then officers intervene to impose more onerous administrative controls.

Military units that design their own information systems can either help or harm the strategic effectiveness of a military intervention, but this depends critically on the goals and values of actors in the distributed system. Representations tend to become insulated from reality when there is a misalignment between actor preferences and the circumstances in which they find themselves. Chapter 5 showed how one SOF unit's bias for direct action in a situation where indirect action was arguably a better fit both distorted its information system and undermined its information expertise. The system's guiding intentionality was an institutionalized preference for direct action, but its sociotechnical implementation was an organic mélange of Microsoft expedients. Pervasive information friction turned the unit's attention inward as practitioners intervened to repair minor breakdowns along well-worn pathways. Problematic practice thus reinforced the epistemic insulation of the task force, resulting in a perverse converse of Fighter Command (where adaptation reinforced management). The enduring challenge for military leadership, system architects, and working operators alike, therefore, is to find ways to navigate the churn of exploitation and reform without getting stuck in problematic or insulated practice for too long.

Well-meaning system designs will inevitably run into problems in the field, systems will be reconfigured on the fly, and the turbulent cycle will repeat indefinitely. Chapter 6 traced the development of the U.S. military drone program as it lurched through all four patterns of information practice, evolving from a curious science experiment into a highly regulated apparatus for targeted killing. The coordination meetings held in the Obama White House were sometimes interpreted as micromanagement. Yet the traditional sequencing of *jus ad bellum* and *jus in bello* considerations (the justification for and conduct of war, respectively) was highly problematic in U.S. drone campaigns. When high-value targets emerged outside declared combat zones or turned out to be U.S. citizens, the president personally weighed the costs and ethics of individual targeting decisions. When the Trump administration rolled back oversight of the program, however, targeting accuracy suffered.

The dynamics of information practice matter for any organization that uses computers to understand and influence its world. War, curiously enough, is a useful laboratory for examining how knowledge is possible under conditions where it is least likely. Combatants try to build better information systems while their opponents try to break them. War thereby exposes many tensions between problems and solutions, people and machines, organizations and allies, and all of the different components in a command and control system. Yet these same tensions affect civilian information practice too. For example, the physician Atul Gawande describes a well-meaning effort to automate medical records with a $1.6 billion program known as Epic.[1] His Epic story features all four types of information practice. System architects planned to build an ambitious enterprise computer system to coordinate doctors, patients, nurses, pharmacists, lab technicians, and hospital administrators. Like the Air Force Mission Support System in chapter 4, Epic aimed to be something for everyone. Also like AFMSS, it became insulated by red tape and spiraling costs. Gawande laments, "I've come to feel that a system that promised to increase my mastery over my work has, instead, increased my work's mastery over me." Another physician found that her "In Basket" was "clogged to the point of dysfunction," so she had to rely on colleagues to alert her to anything important via "yet another in-box." Others adapted around some Epic failures by hiring medical scribes to interface with Epic screens so that doctors could pay more attention to patients. Yet while this expedient improved patient satisfaction, it also created new problems for doctors: "Their workload didn't lighten; it just shifted." Other physicians in a neurosurgery department started "removing useless functions and adding useful ones." As in the case of FalconView, these "innovations showed that there were ways for users to take at least some control of their technology—to become, with surprising ease, creators." Gawande concedes, "Granted, letting everyone tinker inside our medical-software systems would risk crashing them." He notes that Epic managers gradually started experimenting with an "App Orchard" of customized tools to allow users to extend the system's functionality, along with a deeper integration of physicians and technologists. Medical record management in a large distributed health care organization is an inherently difficult and dynamic problem. Therefore, much like complex military information systems, Epic management tended to drift into insulation, and adaptation tended to drift into problems. Over time, the Epic system reformed as more complex forms of management emerged that were better aligned with the complex problems of modern medicine.

## Maneuver Warfare for Information Systems

What is an organization to do? Forgoing technology altogether is not an option if competitors have strong incentives to seek out any advantage. Yet reliance on information systems is a Faustian bargain: the technologies designed to reduce uncertainty become new sources of it. The first rule, therefore, is for systems integrators and organizational leadership to accept the inevitability of friction and compromise. The pressing question then becomes how to turn friction into a source of traction that enables people to improve their systems.

### SOCIOTECHNICAL GENIUS

The Clausewitzian solution to fog and friction is "genius"—a bold, intuitive, creative, and above all pragmatic way of dealing with a situation as it unfolds. Clausewitz does acknowledge the importance of institutional-ization insofar as "routine . . . leads to *brisk, precise,* and *reliable* leadership, reducing natural friction and easing the working of the machine." Yet institutionalization has its limits: "The highest level that routine may reach in military action is of course determined not by rank but by the nature of each situation."[2] In the chaos and ambiguity of battle, according to Clausewitz, commanders must be able to perceive and implement solu-tions that are not specified by formal policy: "We must remind ourselves that it is simply not possible to construct a model for the art of war that can serve as a scaffolding on which the commander can rely for support at any time. Whenever he has to fall back on his innate talent, he will find himself outside the model and in conflict with it; no matter how versatile the code . . . *talent and genius operate outside the rules, and theory conflicts with this practice.*"[3]

If we indulge an anachronistic reading of the words "model" and "code," Clausewitz cautions against overreliance on institutionalized data systems. Adaptive practice is precisely this genius that operates "outside the rules." The big difference in the modern era is that the intellectual labor of war is no longer concentrated in the mind of a single commander (and was already beginning to diffuse across command staffs in Clausewitz's day). Military cognition is increasingly distributed across the entire organization, if not the planet. Genius, therefore, must be distributed too. Another differ-ence between today and the nineteenth century is that the expression of genius is no longer limited to the commander's appreciation of the stra-tegic situation and initiative in battle, if indeed it ever was. It can also be expressed in the creative reconfiguration of sociotechnical systems. The hackers in a military organization use "talent and genius" to reconfigure information systems as actual operations reveal new problems and oppor-tunities. If no war plan survives first contact with the enemy, then no archi-tectural design survives first contact with runtime conditions. Hackers thus engage in *runtime design.*

The administrative challenge today is the same as it was for Clausewitz: where do you find these mythical geniuses, and how do you ensure their availability in war? A Napoleon appears but rarely, and alpha geeks may be unicorns too. Clausewitz suggests two solutions. The first is for com-manders to acquire vicarious experience through the study of military his-tory. If genius depends on combat experience, which is both infrequent and dangerous, then officers would do well to learn from the experience of others. The second is for military organizations to develop institutions that can nurture a cadre of capable staff officers—for instance, through profes-sional military education, realistic training exercises, mentorship, staff rides, and lessons learned from history. While not every officer can become a genius thereby, a recruitment and training pipeline that rewards the desired qualities can at least up the odds that genius will be available when needed.[4] The Clausewitzian approach is still relevant in the twenty-first century, but it must be expanded to encompass the architecture of information systems as well as operational art. Personnel today would do well to gain experience in runtime design through cultivating technical mastery, prototyping data solutions, and engaging with other innovators. Furthermore, it is incumbent on military leadership to develop institutions that cultivate and enable user innovation. Exclusive reliance on formal procurement and command and control systems, "no matter how versatile the code," is a recipe for insulation and disaster.

Military organizations already have an intuitive understanding of concepts that could prove useful in this regard. A prominent leitmotif running through modern strategic thought is the complementarity of shared doctrine and independent maneuver.[5] To survive and thrive on complex battlefields, joint military operations have to strike a balance between top-down management and bottom-up adaptation. The doctrines of "combined arms warfare" in the army, "command by negation" in the navy, and "centralized control/decentralized execution" in the air force all stress the importance of combining unity of command with the initiative of subordinates in the field. The vaunted principle of "mission command" holds that commanders should tell their troops what to do, not how to do it. Mission orders specify the objective and rules of engagement, as contrasted with task orders that attempt to anticipate every contingency. Mission command enables headquarters to issue shorter communications more quickly, on the assumption that units in the field will adapt in the face of unforeseen circumstances. This accomplishment is possible, however, only because troops develop a common understanding of the commander's intent and common operational doctrine. To shoot, move, and communicate in harmony, decentralized units need to have a prior understanding of other units' tactics, techniques, and procedures. To gain proficiency with common doctrine, moreover, units must exercise it through prior training in realistic conditions. SOF take this philosophy to an extreme by granting junior commanders a lot of autonomy and discretion in how they accomplish their mission, but SOF must first be trained to a very high level of tactical proficiency. SOF can deploy in risky situations (i.e., behind enemy lines, in small numbers, with light weapons, at night) because their extensive preparation ensures that tactical improvisations tend to mitigate, rather than exacerbate, the risks. SOF are expensive to train and hard if not impossible to mass produce, but in the right circumstances they can be a potent force multiplier. To succeed in maneuver warfare, militaries must orient their institutions around manning, training, and equipping units to take the initiative in combat. Management thereby enables adaptation.

By contrast, the architecture of command and control systems is generally guided by something more like task orders—formal requirements documents that exhaustively specify how systems should be constructed and what functions they should perform. Defense contractors are legally bound to deliver on requirements, altering or adding functions only if program management offices explicitly change the requirements. Algorithms, moreover, are deterministic rule sets that do precisely and only what they are programmed to do. Computers have little capacity for creative invention in the face of unhandled exceptions, and they do not appreciate nuanced social dynamics that depart from scripted templates. It is up to people to adapt computing systems that cannot adapt themselves.[6] In any real organization, as we have seen, runtime design is prevalent because operational circumstances change faster than bureaucratic requirements. Yet system program managers are understandably reticent to let operational personnel take the reins with technical designs. Fears that amateur modifications will compromise the confidentiality, integrity, and availability of mission-critical data are often well founded. Yet well-meaning controls can introduce real risks of bureaucratic insulation. The leadership challenge, then, is to encourage and enable runtime design without courting the safety and security problems that it often generates. The point is not to eschew careful system management, which is critical, but to plan for extensibility in the face of unforeseeable circumstances. As President Dwight Eisenhower once said, "Plans are worthless, but planning is everything."[7] The same goes for system architecture.

## ADAPTIVE MANAGEMENT

The basic guideline for performance improvement is to use managed practice to enable adaptive practice. This approach aims to capture the advantages of both institutionalized and organic information solutions while avoiding their disadvantages. As a shorthand I call this *adaptive management*. Organizations that embrace this principle will employ complementary forms of practice for different aspects of the mission, or at different levels of the organization. The analogue in political economy is the use of regulatory and trade policy to enhance the efficiency and innovativeness of markets while limiting the harmful effects of negative externalities, unfair competition, and imperfect information. This all sounds fine in theory, but it is a complicated balancing act in practice. Actors often end up exploiting public institutions for private gain, or the market stumbles into catastrophes that regulators fail to foresee or control. Even more so in military affairs, as Clausewitz reminds us, "Everything in war is very simple, but the simplest thing is difficult."[8]

The four different patterns of information practice are ideal types that are usually mixed in real organizations. Chapter 5 analyzed a complex mixed case by distinguishing between the guiding intentionality and sociotechnical implementation of an information system. In that case, problematic practice reinforced insulation, which enacted the worst of both worlds. Table 7.1 uses this same analytical approach to highlight desiderata for enacting the best of both worlds. I further break out the social and technical dimensions of implementation. It would be contrary to the argument of this book, which stresses the contingent fit between external problems and internal solutions, to now suddenly suggest that there is, after all, a panacea for war in the information age. There is not. My aim is simply to help organizations improve their odds of adjusting internal solutions whenever external problems change, as they inevitably will.

*Common Vision and Values.* As used in this book, the concept of intentionality refers to the ideational aspects of an information system—the goals, values, beliefs, norms, preferences, and identities that guide information practice. For a military organization, intentionality encompasses an organization's strategic objectives, theory of victory, concept of operations, rules of engagement, and commander's intent. Aligning intentionality with the evolving nature of a strategic problem can be a difficult task, but also one that is widely recognized to be essential. Thus Clausewitz urges leaders to first understand the nature of any given political problem if they are to come up with an appropriate military solution:

> Wars must vary with the nature of their motives and of the situations which give rise to them. The first, the supreme, the most far-reaching act of judgment that the statesman and commander have to make is to establish by that test the kind of war on which they are embarking; neither mistaking it for, nor trying to turn it into, something that is alien to its nature.[9]

Many would argue that the alignment of ends and means is best achieved through ongoing conversation and negotiation among civilian and military leaders throughout the chain of command.[10]

All cybernetic systems are guided systems. Objectives or desires make it possible to compare the sensed state of the world with a goal state in order to narrow the gap between them (or keep it from opening). Intentionality plays a role not only in defining the overall goals of a sociotechnical

Table 7.1  Principles for adaptive management

|  |  | *Implementation* | |
| --- | --- | --- | --- |
| *Solution* | *Intentionality* | *Social* | *Technical* |
| Institutionalized | Common vision and values | Support system for runtime design | Extensible architecture |
| Organic | Hacker ethos for adaptive practice | Forward-deployed technical expertise | Small-scale experiments |

system but also in guiding its day-to-day operation. Beliefs and values shape the ways in which sensors detect and amplify battlefield features (measurement), data are represented in a distributed organization (coordination), and instructions are communicated (enforcement). The concept of institutionalized intentionality in table 7.1 refers not only to operational objectives and organizational identities, therefore, but also to a shared consensus about desirable system engineering qualities such as interoperability, reliability, security, and such. The socialization of shared values for system design across distributed units increases the chances that local modifications will help, or at least not interfere with, the rest of the network. Importantly, as in maneuver warfare, common vision and values do not simply emerge by fiat from the commander's brain. An ongoing conversation between command staffs and units in the field—reports, stories, interpretations, and questions that move up, down, and across the chain of command—can create a shared sense of mission and purpose encompassing both operations and architecture.

It is particularly important for leadership to figure out which systems and applications are more or less vulnerable to problematic or insulated practice. It is quite reasonable to restrict runtime design in cases where breakdowns could be catastrophic. Systems with extreme safety and security concerns, such as avionics and nuclear weapons, should be developed through deliberate systems integration and enterprise management techniques. The risks of insulation may be worth running if some types of problems are particularly unacceptable. Arguments about reliability and security should be interrogated critically, however, since bureaucrats will also make them to protect turf and acquire resources, which promotes insulation. It is also prudent, therefore, to empower runtime design in cases where bureaucratic myopia or sclerosis is a greater risk to mission performance. The chances of problematic breakdown are worth running if an organization can survive the learning process. As a general heuristic, when a computational input has a more indirect effect on, or is located farther upstream from, a sensitive operational outcome, then modifications to that input can be made with less risk of catastrophic breakdown. There are many command and control, intelligence, targeting, logistics, and administrative tasks that can be improved through the empowerment of user innovation communities, precisely because these tasks are just one of many inputs into the information practice that generates combat power. Again, adaptive practice must be guided by shared values regarding reliability, security, interoperability, and so on.

*Hacker Ethos for Adaptive Practice.* Intentionality also matters for organic adaptation, but here the emphasis is more on rapidly making sense of and shaping local circumstances. The word "hacker" often conjures up visions of malevolent nerds in hoodies, but the term originally referred to any type of creative problem solving. As Richard Stallman writes, "Hacking means exploring the limits of what is possible, in a spirit of playful cleverness."[11] Hackers discover and unlock unappreciated degrees of freedom to find a new way of solving a problem. There is an important social dimension in hacking too, as hackers aim to impress other hackers and not be embarrassed by making rookie mistakes. The hacker ethos thus encourages curiosity about possibilities where others see finality, as well as action on possibilities with a high degree of technical proficiency.

Hacking should be distinguished from mere improvisation that produces an inelegant kludge. Hacking may be expedient, but it also strives to be smart. Hackers value elegant solutions that use technology in ways that are playful or unexpected rather than crude or dangerous (except, of course, to the adversaries on the receiving end of improvised solutions). Hackers pride themselves on their technical acumen, much as SOF operators pride themselves on their tactical skills. The problem with the information system in the SOF unit described in chapter 5 was not simply its preference for direct action in a province recovering from war but also its lack of technical proficiency. The SOF culture of innovation ended up producing amateurish solutions even for the problems toward which it was most inclined. While the SOF model has many characteristics worth emulating, a different intentional orientation is needed in this context. The focus is on intellectual performance rather than tactical soldiering. Because information systems are more than software, furthermore, hackers should also nurture an ethnographic sensibility to better observe and articulate the ways in which technologies are used in an operational context. Finally, while military hackers may see their organization as a weapon that can be customized to become more lethal and effective, they should also be attuned to nonlethal alternatives that might be more clever solutions to a given strategic problem. Operational hackers will thus have to participate in, or at least be aware of, the ongoing conversation that generates and disseminates the common vision and values discussed above.

*Support System for Runtime Design.* A major leadership challenge in implementing adaptive management is the reorientation of system procurement and management bureaucracies toward the support of user innovation. All-volunteer forces in rich democracies like the United States tend to have a lot of tech-savvy personnel in their ranks. The organizational challenge is identifying and empowering them. Many talented hackers will self-identify through their own technical initiatives, but for them to have more than a fleeting impact on tactical performance, the organization must further enable and encourage them through an institutional support system.

Acquisition programs and network managers must assume that systems will break down and that runtime design is vital for repairing them. In the face of unorthodox user solutions, system managers should adopt the motto "If you can't beat 'em, join 'em!" The distinction between the use and design of information systems was already blurry in the industrial era, but it is increasingly complicated in an era of general-purpose computing devices and reprogrammable systems. Software applications offer many layers and opportunities for customization. As a result, many users will spontaneously adapt them to different degrees, which makes it imperative for organizations to guide this activity in productive directions. Writing Microsoft Office macros is a rudimentary form of programming. Creating PowerPoint templates for targets and operations is a form of knowledge engineering. Design and redesign can occur at many levels: hardware, protocol, application, data file and template, organizational policy, and enterprise integration. All of these activities can benefit from user innovation communities nurtured by supportive leadership and provided with resources to facilitate mindful experimentation.

To increase the quality of runtime design, personnel with technical acumen have to be recruited and retained in the force. Personnel who create useful prototypes should be recognized and rewarded with meaningful work, financial bonuses, and training opportunities. The more cross-training the better, so that technical experts understand mission requirements and operators understand technological possibilities. Realistic exercises should enable personnel to hack their systems with reduced risk of inflicting problems. Engineering expertise should be made available to them (e.g., through online forums or a pool of skilled consultants retained for the purpose) so that forward-deployed personnel can reach back for code samples and technical advice. Warfighters should be able to upload innovative solutions to forums similar to Github, yet on classified networks, where others can download them, modify them for their local needs, and upload new modifications for others. Hacker prototypes should be evaluated and, where appropriate, incorporated into more managed systems integration efforts, leaving the door open for further cycles of customization and evaluation. Workshops should be hosted to gather user innovators together to share stories and approaches and build trust with the systems management community. These suggestions are just the tip of the iceberg for an institutional support system for runtime design.[12]

*Forward-Deployed Technical Expertise.* Information about distant battlefields can be hard to obtain. Data are always embodied in physical media that must survive the journey back to remote locations. Even when they can be collected successfully, data are meaningful only for people or machines that know how to interpret them. Operationally relevant information thus tends to be localized or "sticky."[13] Therefore, operational hackers should be deployed as far forward as possible. Adaptive practice is a way of exploring the contours of an unconstrained problem to discover subtle constraints that can be incorporated into a new information solution. Expertise in both information technology and military operations have to be combined as close to the point of contact as possible (logically if not geographically) to improve the likelihood of timely discovery and productive exploitation.

Militaries have long incorporated combat engineers to build bridges and airstrips. Their exigent constructions might not conform to municipal building codes, but they get the job done under fire. Expeditionary hackers might provide a similar service for information systems. Hacking to improve rather than degrade, operational knowledge engineers could seek out information problems overlooked by formal requirements and prototype working solutions. The radar scientists of Fighter Command, for example, used their formidable technical expertise to improve both devices and work processes. Joint Special Operations Command embedded analysts from the CIA and NSA and interrogators from the FBI in its counterterrorism task forces. If the deployment of civilian technical experts is not feasible for whatever reason, then reserve forces may offer a source of polymaths with relevant information skills. However, most reserve programs would have to be overhauled to provide a deployable pool of high-quality consultants rather than just a stock of temporary workers. Mass-mobilization reserve systems developed for industrial age warfare are designed to provide adequate substitutes for active duty talent, but the reserves could be redesigned to provide high-value complements instead. A small number of forward-deployed, technically talented, mission-oriented hackers have the potential to outperform larger bodies of analysts and engineers coordinated through traditional formal requirements. Competitive analysis and design exercises could test this proposition.

*Extensible Architecture.* Some military systems have especially austere interoperability, safety, and security requirements. Safety-of-flight and weapons-release functions are not good candidates for runtime design for these reasons. Conversely, some applications are so localized or trivial (e.g., accounting for the officer's mess) that there is no reason to ever consider an institutionalized solution. The large and growing area in between these extremes can benefit from the adoption of a stable yet flexible architectural platform that can be extended to accommodate new functionality and emerging requirements. Extensible modular architectures are useful for supporting both management and adaptation. General-purpose applications like FalconView, Microsoft Office, and cloud-based software services can act as platforms for runtime design. By opening up application programming interfaces and software development kits, platforms enable third parties to extend or customize functionality without interfering with the platform itself. Platforms must be engineered to a high degree of reliability and security through managed practice, of course. Stable platforms are stable for all, while platform vulnerabilities affect everything built on them. Shared understandings and a hacker ethos can help end users to understand and mitigate risks—for instance, by prototyping alternative functionality in case the platform goes down and helping systems integrators repair the platform to keep it up.

Commercial technology that is flexible enough to support a human rights organization or a venture capital firm is also flexible enough to support a variety of military applications. Tech-savvy users accustomed to tinkering with their devices at home will look for ways to use them on deployment. Since government-designed systems are especially prone to insulation, systems architects should consider whether specialized military functions can be built on more robust commercial platforms. The use of commercial components and open standards may lower technical barriers to adaptation in forward units. Unfortunately, use of commercial components also exposes military systems to any software vulnerabilities in them, although it is not necessarily the case that software written by defense contractors will be any more reliable. On the contrary, commercial vendors of general-purpose software intended for a diverse set of customers have more opportunities (and incentives) to test their systems against a large set of use conditions. As I have stressed repeatedly, it is important that this technological potential is complemented by an institutional support system and a hacker ethos to guide runtime design in productive rather than destructive directions.

*Small-Scale Experiments.* Adaptive practice works best when it tackles localized problems that can be solved without interfering with other locales. Organic solutions work well for less constrained and loosely coupled problems. In those circumstances, local practitioners can exploit the slack and ambiguity in the system to experiment without adversely harming the larger organization. FalconView initially took off precisely because it was designed *not* for everyone, but rather just for mission planning supporting the F-16 fighter. Its creators endeavored to make a small number of features work really well. Later on, this provided the benefit of a stable geospatial platform that other users could build on. FalconView's wildly popular growth throughout the U.S. military was an opportunistic serendipity, not an initial design goal.

Forward-deployed hackers should tackle projects that have a modest scope. Low-hanging fruit can be harvested with a minimum of coordination. By starting small on local tactical or administrative problems, hackers can ensure that their prototypes will actually work to advance the mission. They should start small and debug constantly. Working prototypes can then become the foundation for something larger when they are fed back into an active user innovation community through the support system for runtime design. When forward units do identify novel approaches that work, then management programs and commercial manufacturers can take advantage of larger economies of scale and quality control to improve and disseminate user designs. A lot of innovative sporting equipment, scientific instruments, and household appliances have emerged through a similar process.[14] User-designed prototypes perform, in effect, a search function for emerging design requirements in a dynamic warfighting environment or market. Military organizations would be wise to encourage and leverage this search.

## Strategic Implications

Even if all of this advice is heeded, there are performative limits to adaptive management. Some strategic problems are simply harder or easier to solve, regardless of the type of information solution that a military adopts. Understanding the nature of the war, as Clausewitz insists, is the foremost challenge for the intentionality that guides information practice.

In the pages remaining, I offer tentative conjectures about several important contemporary problems. All of these are active debates that have already generated many perspectives and disagreements, so I can barely scratch the surface here. I raise these issues now simply to suggest areas in which information practice theory might be applied in new directions. Indeed, I would be surprised if more attention to the nuance of each case did not invite criticism and refinement of these speculative arguments. According to my own theory, solutions that appear promising when a problem is first considered will often turn out to be inappropriately insulated from the evolving structure of the actual phenomena. Complexity is an overarching theme of this book, and the jump from operational microfoundations to strategic macroimplications is nothing if not complex. I welcome empirical and theoretical counterarguments to facilitate the reformation of these conjectures.[15]

### CHINESE MILITARY POWER

"Informatization" is the technology theory of victory with Chinese characteristics. In the wake of the 1991 Gulf War, China concluded that the character of war had fundamentally changed, resulting in a new strategic guideline for "winning local wars under modern high-technology conditions." Chinese strategists highlighted in particular the vulnerability of U.S. digital networks, logistic communications, and space satellites, even as they sought to emulate the prowess of the networked superpower. In 2004 China further adjusted its strategic guideline to "winning local wars under conditions of informatization" and did so again in 2015 to "winning informatized local wars."[16] As stated in the 2013 *Science of Military Strategy*, the capstone doctrine of the People's Liberation Army (PLA), "Control of information is the foundation of seizing the initiative in battle. Without information supremacy, it is difficult to effectively organize fighting for control of air and control of sea." As a result, "the fast-movers dominate the slow-movers" and "unmanned, invisible, and inaudible" wars are emerging.[17] According to China's 2015 military strategy, "Integrated combat forces will be employed to prevail in system-vs-system operations featuring information dominance, precision strikes and joint operations."[18]

While U.S. conflicts since the end of the Cold War have pitted advanced networked forces against less capable or asymmetric adversaries, the U.S. RMA would certainly meet its match versus Chinese "antiaccess/area denial" (A2/AD)—a network of ships, submarines, fifth-generation aircraft, antiship missiles, antisatellite weapons, and cyberwarfare capabilities. Assessments of the operational potential of Chinese modernization suggest a growing and formidable threat, although analysts disagree about a number of technical implications.[19] Indeed, there are structural aspects of the A2/AD problem that benefit China. The United States must deploy across thousands of miles of ocean to intervene in an East Asian crisis, while China plays defense. China can base over-the-horizon radar, command and control nodes, and signals intercept stations on mainland territory, while the U.S. military will be more dependent on satellite and sea-based sensors and transmitters. While ground stations have the disadvantage of being fixed targets that the United States can easily strike, strikes on the mainland are also an escalatory threshold that the United States may be unwilling to cross in many conflict scenarios. China also has a significant quantitative advantage in terms of ships and missiles, and as military planners like to say, "Quantity has a quality of its own."

Unfortunately for PLA information practice, however, China hopes to contest a vast maritime area within the so-called second island chain. Any extension of the PLA's warning envelope beyond the chain requires monitoring an even larger area of operations. Undersea geography in East Asia, moreover, is more favorable to the U.S. Navy than the PLA Navy.[20] There is quite simply a lot of ocean on, over, and under which the U.S. military can maneuver as it presses its attack. Given the large and complex space(s) that the PLA must defend, it will have to integrate a vast amount of data from many different sensors to synchronize the employment of standoff precision weapons. The PLA thus becomes as dependent on data links and computers as the U.S. military. U.S. doctrine and military culture encourage commanders to attack from multiple axes at once to create decision dilemmas for the enemy. Given that an uncooperative adversary is one of the principal external causes of information friction, the United States can be expected to create a lot of friction for Chinese operations.

The PLA is also likely to suffer from a lot of internally generated friction in its pursuit of informatized warfare. The PLA recently underwent a major reorganization to consolidate cyber operations, electronic warfare, and space capabilities into a single Strategic Support Force. Yet the PLA remains a highly bureaucratized organization with rank-conscious officers presiding over silos of specialized capabilities. Joint military operations, military-intelligence coordination, civil-military integration, and cybersecurity all involve vexing management challenges. The U.S. military has struggled with these problems for decades, and has been constantly in combat since 2001. U.S. personnel have been able to experiment and struggle with new types of sensors, munitions, and time-critical-targeting processes. Granted, the unconventional warfare problems that the U.S. military has been trying to solve for the past two decades differ radically from future combat scenarios with China; this change in the external problem can be expected to put considerable stress on the U.S. military. At the same time, U.S. personnel have acquired some hard-earned experience with the integration of diverse capabilities in combat conditions. The U.S. military has been able to learn from many mistakes, targeting failures, civilian deaths, friendly-fire tragedies, and missed targeting opportunities. The PLA, by contrast, has not had the opportunity to experience information friction firsthand or to debug coordination problems in combat conditions. It is difficult for any nation to organize exercises that simulate an advanced adversary and reproduce a realistic intelligence environment. Chinese exercises are doubly challenged for domestic reasons. Big exercises are perceived as too big to fail, so maneuvers tend to be scripted in advance to demonstrate rather than test operational concepts. This raises nettlesome questions about how well PLA units will perform off script.[21]

The U.S. military is a professional all-volunteer force that encourages even its most junior personnel to exercise initiative and express opinions, even if they challenge military orthodoxy. The PLA, by contrast, remains an army-dominated force that is preoccupied with internal security threats and reliant on conscription. The PLA is the military arm of the Chinese Communist Party rather than the Chinese state. Political commissars are embedded with units throughout the chain of command to ensure party loyalty. Xi Jinping's anticorruption purges have eliminated some venal officers, which might improve military performance, but Xi has also eliminated political threats to his grip on power. Coup proofing is generally thought to be counterproductive for military effectiveness in modern war.[22]

This book stresses the importance of runtime design and repair. There is little doubt that the Chinese people are adept at innovating with whatever resources they have available, as any visit to a Shenzhen market will make clear.[23] The question is whether the PLA organization will actively encourage bottom-up adaptation, and if so, whether it will help or hinder information practice in wartime conditions. Uncoordinated amateurism or corruption (i.e., problematic practice) takes a toll on organizational performance. For example, a PLA cyber-exploitation unit based in

Shanghai was identified and described by U.S. cybersecurity researchers because of the sloppy tradecraft of Chinese cyber operators.[24] Friction at the low end, moreover, complements insulation at the high end. Even when China has successfully stolen technical data, the Chinese defense industry has struggled to reproduce advanced Western technologies, mainly owing to the difficulty of integrating complex sociotechnical weapon systems.[25]

To sum up, the A2/AD problem is only weakly defined for the PLA, and U.S. attacks on PLA information systems will heighten its instability. The PLA organization, meanwhile, suffers from bureaucratic rigidity that will inhibit adaptation and promote command insulation, even as local deviations from standards by corrupt or unprofessional units will make internal friction more likely. Chinese doctrine emphasizes rapid, preemptive, aggressive strikes, which could further encourage the PLA to overestimate its chances in combat against the U.S. military. A difficult war for China does not by any means imply an easy war for the United States. Any contest between two great-power militaries brims with dangers of inadvertent escalation and unanticipated protraction.[26] The United States thus has the opportunity to make many mistakes as well.

THE THIRD OFFSET

Pentagon strategists have proposed a "third offset strategy" to counter Chinese military power. The first offset, in this construct, used U.S. nuclear weapons to deter Soviet conventional invasion in the early Cold War. The second offset, in the late Cold War (i.e., the historical "offset strategy" that gave rise to the RMA in the first place), used advanced conventional weapons to counter Soviet preponderance. The third offset now aims to mobilize novel capabilities such as directed energy, hypersonic munitions, autonomous robotics, artificial intelligence, and quantum computing to counter Chinese A2/AD.[27]

In effect, the third offset doubles down on the RMA. Proponents emphasize that the third offset also requires new doctrinal concepts (e.g., "AirSea Battle," later rebranded "Joint Access and Maneuver in the Global Commons"). Yet the technological imperative to adopt new doctrine was also a common refrain in the RMA era. Strategists now suggest harnessing the innovative engine of Silicon Valley rather than the ponderous defense procurement bureaucracy. Yet the original RMA also looked to the civilian information economy for inspiration about improving organizational performance. A further similarity is that the problem of irregular warfare remains out of scope. The RMA and third offset are both intended to address the problem of major combat operations against a peer adversary. While the technological capabilities contemplated in the 1980s and 2010s differ considerably, both the RMA and the third offset aim to reduce the costs of war by building an even more sophisticated and automated reconnaissance-strike complex.

The third offset strategy envisions the U.S. military conducting integrated "cross-domain" or "multidomain" operations across the land, sea, air, space, and cyber domains.[28] Solving this massive coordination problem requires human beings to be able to cope with technological and social complexity. As Bernard Brodie wrote about the intellectual demands of mechanized sea power, "The tools of war become ever more complex and more deadly, but the net result of those changes on the personnel factor is to place ever greater demands on the spirit and intelligence of the men who plan and wage battles."[29] Brodie's point is even more apt for informatized warfare. American prowess in joint operations to date is not simply a function of technology. An aircraft carrier is not just a big ship with airplanes on top, but rather a floating organization that can integrate very different types of complicated weapon systems for many different missions. This "system of systems" is prone to constant breakdown and retrofit, but it works because even its most junior personnel are encouraged to identify and correct problems. A young sailor on the deck of a carrier can call for the cessation of flight operations if she perceives a hazard; she gains the insight and confidence to do so through constant training and conversation with more experienced veterans.[30] Moreover, the introduction of new weapons and mission requirements throughout the course of the carrier deployment cycle results in a constant churn of activity that can be accommodated only through proactive adjustment. Carrier warfare is an organizationally embodied capacity that has emerged through a long historical process of trial and error (i.e., the rules are "written in blood").

Proponents of the third offset often emphasize the importance of organizational capacity in the adoption of disruptive technologies. But as we have seen repeatedly, organizations may be more or less institutionalized or organic in their use of computers. Information practice can wring performance advantages out of older technologies or undermine the potential of newer technologies. Tapping the expertise of Silicon Valley for technology development will not automatically channel its hacker culture into the operational conduct of war. Runtime design is a critical yet undervalued asset of the U.S. military today. The viability of the third offset will turn on not just the sophistication of sensors, networks, and weapons (acquired at design time), but also the ability of adaptive management to knit them together into a coherent fighting force (implemented at runtime). Actual operations in which systems break, or realistic exercises and red-teaming in which information systems are allowed to break, give personnel vital experience fixing them. Adaptive practice is the only emollient for the inevitable friction of joint military operations, but this capacity is too often overlooked and unappreciated. Unfortunately, the viability of runtime design in U.S. C4ISR systems is at risk because of a growing emphasis on enterprise configuration management in pursuit of the elusive goals of interoperability and cybersecurity. The suggestions above about adaptive management can help the U.S. military to enhance rather than squander the advantages it currently has in runtime design. The United States should nurture its advantages in information practice while remaining mindful of its limitations.

IRREGULAR WARFARE

It is important to scope the potential of adaptive management for improving military performance. Much of this book has already discussed recent U.S. military interventions at the lower end of the conflict spectrum, so I will say comparatively less about irregular warfare here. Counterinsurgency is an ambitious strategy of reform that attempts to improve public infrastructure and military protection to encourage local civilians to inform on insurgents. A shadowy adversary thereby becomes legible against a background of newly sympathetic civilians. Counterterrorism pursues a more traditional strategy of control that endeavors to enhance intelligence acuity and operational precision. Geographical, urban, demographic, and transactional constraints (e.g., on bank accounts and phone calls) provide a degree of structure that modern counterterrorists can leverage to triangulate their quarry. Underground networks, however, can counter with exploitation strategies that bypass, insulate, or subvert the counterinsurgent's bureaucratic strengths. Irregular combatants thus remain disadvantaged in power but advantaged in information.

While U.S. personnel are both safer and more lethal than ever before in history, they have also been employed in more ambiguous conflicts where these strengths become liabilities. Bottom-up, decentralized, adaptive practice is essential for military effectiveness in the relatively unconstrained problem of irregular warfare, but even the most adaptive organization is likely to enter the realm of diminishing returns in pursuit of a determined irregular adversary. Most unconventional actors have strong incentives to erode the constraints on U.S. information problems. The U.S. military, meanwhile, has weaker incentives to reform its more insulated solutions. This situation recalls Aesop's parable about the hare running faster for its life than the hound runs for its dinner. As a general tendency, the dynamic nature of the unconventional problem interacts with recalcitrant military subcultures to produce counterproductive insulation. Improvement in intelligence sensors and analysis has not yet eliminated dead space in the sociotechnical structure of civil society where insurgents, terrorists, and other miscreants can hide. The cumulative costs from the chronic frictions that they inflict, and that insular forces inflict on themselves, may eventually outweigh the stakes of the conflict. The counterterrorist dream of an intelligence panopticon, moreover, could turn into an

Orwellian nightmare. The nature and desirability of "victory" in counter-insurgency ultimately depend more on political goals and values than on technical factors.

The basic political question that must be answered in irregular warfare is whether the burn rate of counterinsurgency or counterterrorism can or should be sustained indefinitely. The aggregate costs of targeting errors and risks of escalation, which technology can reduce but never eliminate, can gradually exceed the value of achieving modest political goals. What is the least harmful of several bad options in this situation? Aggressive rollback of the enemy's underground organization risks entrapment in an expensive quagmire. Retrenchment risks ceding advantages to a dangerous foe that may recover and strike back with greater ferocity. Maintaining a low level of engagement, or containment, requires ongoing investment and a willingness to sustain losses for a long period of time. Yet this approach can also buy time for an insurgency to make its own mistakes, exhaust its potential, and catalyze effective local resistance. An opportunity like the Anbar Awakening cannot be created at will, but it can be encouraged when it does happen. Policymakers bent on intervention in irregular scenarios ought to bide their time, emphasize information gathering, intervene only where local alliances are feasible, and avoid counterproductive overreactions. Containment does not guarantee that adversaries can be outlasted, especially if the enemy is resolved enough to absorb great costs and resilient enough to find new ways to inflict them. Civilian populations become especially attractive targets for weak aggressors as the invulnerability of strong militaries makes it harder to test their resolve, which heightens the tragedy of containment. Success in irregular warfare depends on persistence and adaptation, and this entails a willingness to make mistakes and pay costs indefinitely. Nations that are unwilling to pay these political costs should avoid getting involved in irregular contests.

## NUCLEAR COUNTERFORCE

At the other end of the conflict spectrum, adaptive management is limited by the risk of catastrophic blunder. Information practice works by making and repairing mistakes, but some mistakes are simply too massive to make. Yet whenever organic adaptation is suppressed, institutional insulation becomes a serious liability. The risks mount as combat interactions begin to change the external constraints to which internal institutions are anchored. In irregular warfare, an insulated military organization can survive its misadventures. Insulated practice in nuclear war, however, courts existential disaster.

While most strategists agree that nuclear weapons are useful for deterrence, there is a long-running debate about whether they are further useful for warfighting. Proponents of "counterforce" and "damage limitation" strategies argue that targeting enemy nuclear forces can improve deterrence by denial and, should deterrence fail, reduce the level of destruction that the enemy arsenal can inflict.[31] Keir Lieber and Daryl Press argue that a "new era of counterforce" is dawning that undermines traditional verities. Changes in technology, they write, "are eroding the foundation of nuclear deterrence. Rooted in the computer revolution, these advances are making nuclear forces around the world far more vulnerable than ever before."[32] Thus the survivability of secure second-strike arsenals can no longer be taken for granted. Furthermore, technology empowers the United States to consider preemptive attacks with lower risk of catastrophic retaliation against nations with small or immature deterrents (i.e., North Korea and Iran).

The conventional wisdom about "the nuclear revolution" is itself a sort of technological determinism. The capability to annihilate cities is thought to make great-power war unthinkable, transforming contests of strength into contests of resolve. Lieber and Press counter with the alternative determinism of the RMA. They list five factors in "the accuracy revolution" that make "low-casualty nuclear counterforce strikes" feasible: (1) "sensor platforms have become more diverse," (2) "sensors are collecting a widening array of signals for analysis using a growing list of techniques," (3) "remote sensing platforms increasingly provide persistent observation," (4) "steady improvement in sensor resolution," and (5) "huge increase in data transmission speed."[33] Without disputing these developments, and indeed because of them, the resulting explosion of intelligence data will create difficult coordination and data fusion problems. Human beings will still have to provide essential interpretive services in the system of distributed cognition that controls any nuclear exchange. Lieber and Press include no discussion of the problems of multisource analysis and intelligence-operations fusion that must be solved to knit all of their technologies together. Their argument implicitly assumes that the information systems that constrain and enable counterforce targeting will be reliable.[34]

There is no doubt that the U.S. Air Force and Special Operations Command (SOCOM) in particular have made great strides in adapting reconnaissance-strike networks to solve the problem of time-critical targeting. The United States has fielded the most discriminating counterterrorism targeting system in history. As we have seen, however, it still sometimes kills the wrong people and misses the right ones. Even after significant improvements in target development and validation processes, incredibly accurate weapons have accidentally bombed medical clinics, wedding parties, election convoys, and U.S. citizens. High-value terrorist targets have continued to elude the intelligence dragnet as operatives find ways to disguise their signatures and deceive the hunters. Analytical and legal checks and balances have been piled on, and tactical decision-making has drifted up the chain of command. Yet the residual possibility for error and insulation has not been eliminated.

To be sure, there are aspects of the nuclear counterforce problem that make it much easier than hunting human targets. Enemy missiles, transporters, and submarines are all big metal machines. There are real constraints on where they can be concealed or relocated, which considerably simplifies the search problem. Bases, underground facilities, road networks, and deployment patterns can be identified and targeted in advance, and overhead collection and unattended ground sensors can be focused on areas most likely to be traversed by targets. Yet ambiguities remain.[35] "Wartime reserve mode" protocols (i.e., technical settings not used in peacetime but activated in war) can suddenly cause intelligence sources to "go dark" in combat. Decoys can simulate real targets, and real targets can be disguised as decoys. It may be possible to discriminate some types of decoys from targets due to their technical characteristics, but the additional filtering protocol necessarily increases the computational burden of analysis. The adversary might also use unconventional means to deliver nuclear weapons—for instance, by smuggling them in on commercial trucks or fishing vessels. Meanwhile, the organizational and psychological processes of command and control in the enemy organization remain as inscrutable as ever for remote sensors. The United States was able to track Soviet submarines in the Cold War to support a potential counterforce mission, yet this depended on an intelligence process that was subject to friction. As one U.S. intelligence analyst who tracked Soviet submarines recalls, "One tends to get caught up in the 'truth' of the technological representation. . . . Occasionally, a unit described in reports as having 'probably returned to port' had only fallen off the map after losing magnetism."[36] Digital representations would simply suffer from different types of friction.

While false positives and false negatives in counterterrorism create local tragedies for civilian and military families, targeting errors in nuclear war can incinerate metropolitan cities. Because the consequences of errors in irregular warfare are smaller, states can sustain repeated errors over the course of a long campaign as they work through cycles of exploitation and reform. Many counterforce scenarios, by contrast, are conceived as fast-paced preemptive campaigns. Slower and more methodical campaigns can be imagined, but it is almost inconceivable that anyone would ever justify a slow-rolling nuclear war just to allow the targeting system to make mistakes, understand them, and improvise corrections. On the contrary, the social and technical stresses of a nuclear campaign can be expected to degrade rather than improve command and control. Improvisational learning is anathema to nuclear targeting. To take just one example, Ron Rosenbaum tells a disconcerting story about user innovation in a Minuteman missile silo. Missileers apparently devised a way to defeat the two-person control requirement of turning two different launch keys simultaneously, motivated by a professional concern that "one of the guys turns peacenik and refuses to twist his key." Their hack, as Rosenbaum writes, was "called the 'spoon-and-string' trick. Here is how it works: First shoot or immobilize the other crewman. Then tie a long string to the handle of a spoon. Insert your launch key in your slot and the other launch key into the other guy's slot, and tie the other end of the string to the other launch key's top. Go back to your launch console and then twist your key at the same time that you yank the spoon and string forcefully enough to twist the other key in sync with yours." Fortunately, in this case, problematic practice prompted reforms to restore managed practice. As Rosenbaum concludes, "Nuclear engineers have reconfigured the launch control consoles so that it is no longer possible for the spoon-and-string trick to work."[37]

There are good reasons why nuclear targeting processes are highly institutionalized. Yet for the same reasons, it would be a mistake to assume that U.S. Strategic Command (STRATCOM) has internalized the lessons learned by SOCOM and the air force about time-critical targeting. Indeed, despite all the technological upheaval of the information revolution, nuclear targeting protocols have changed little. As recently as 2012, as one government report revealed, "the process for developing nuclear targeting and employment guidance . . . remained virtually unchanged since 1991, according to [Department of Defense] officials."[38] The Single Integrated Operational Plan (SIOP) was originally planned against fixed targets in the Soviet Union, China, and other Communist countries.[39] Modification and implementation of the SIOP was a very methodical and deliberate process. The forbearer of STRATCOM, Strategic Air Command, developed a "zero defect" culture to implement the stringent "always-never" criteria of nuclear command and control: deterrence requires that weapons are always available to the president, while reassurance requires that they are never used without presidential authorization. Even so, numerous glitches and mishaps in nuclear command and control have been documented.[40] Changes to the SIOP in the 1970s introduced more flexibility with limited, selective, and regional nuclear options, but these were all still preplanned options.[41] The U.S. military tried throughout the latter half of the Cold War to find ways to make nuclear warfighting manageable and limitable. Yet planners always fell back on large-scale preplanned options, in part because they were painfully aware of the targeting difficulties of nuclear warfighting.

The 2018 U.S. Nuclear Posture Review similarly envisions "tailored strategies and flexible capabilities" that can signal differently to different adversaries and conduct limited nuclear warfighting if deterrence fails.[42] Yet the difficulty of the counterforce targeting problem and the risks of getting it wrong will still tend to incentivize reliance on massive preemptive strikes (overkill). STRATCOM planners have never fought a nuclear war for which they have planned, thank goodness. The Chinese PLA, as discussed above, is in a similar situation. Insulation is thus a risk for both organizations, but the potential costs are far more extreme in a nuclear scenario. The risks are doubly great if a myopic PLA starts a conventional war that China cannot win, which then leads Chinese leadership to gamble with nuclear war to coerce the United States to back down, which in turn prompts a massive counterforce effort from a risk-averse STRATCOM.[43] Insulation in nuclear targeting practice thus produces two distinct risks. The first is that overconfidence in the representation of target identities, characteristics, and locations makes limited nuclear war more attractive for the United States, when in fact accuracy is an illusion that sparks a bigger war than intended. The second risk is that underconfidence in targeting—fear that accuracy is indeed an illusion—leads U.S. nuclear planners to fall back on a traditional preference for massive nuclear attack.

The "new era of counterforce" is an argument based on technological possibility, and it is wise to revisit invalid operational assumptions about the nuclear revolution. At the same time, the reality of information practice suggests that counterforce will remain a very difficult problem to solve. Nuclear command and control relies on a strongly institutionalized information system that resists change, for good reasons. Yet nuclear warfighting will have to cope with a very dynamic problem. Despite all the improvements in the technology of counterforce, there remains a certain illogic to U.S. nuclear strategy.[44]

## CYBERSECURITY

Nuclear war, thankfully, remains an unlikely event. Conflict in cyberspace, by contrast, has become a global epidemic. Cyber operations should be understood as a second-order form of information practice that protects and exploits the first-order information practice that enables organizational performance. An important difference between cybersecurity and the cases examined in this book is the scope and scale of the relevant information system. Cyberspace is the most complex control system ever built, enabling firms and governments to rapidly and reliably coordinate transactions across the entire planet. Yet any potential for more reliable and trustworthy computing creates more potential to abuse trust. Cybersecurity is a complex contest of deception and counterdeception conducted through global information systems. It marries the classic logic of intelligence (espionage and subversion) to the unprecedented scope and scale of cyberspace. This is bad news from a counterintelligence perspective, since digital spies can now steal more information, and in more ways, than human spies could ever manage in the past. Yet there is also some surprising good news: reliance on deception creates incentives to limit the severity of cyber conflict.

A full treatment of this important topic is beyond the scope of this book. Here I will simply sketch out five implications of information practice theory. First is that the "cyberwar" variant of the technology theory of victory takes the substitution of information for mass to an extreme. Without firing a shot or exposing themselves to risk, in this narrative, malicious hackers steal intellectual property, corrupt vital data, alter public opinion, and paralyze critical infrastructure. Global connectivity and cheap malware seem to give weak actors advantages against strong states that are overdependent on networked computers.[45] China's *Science of Military Strategy* succinctly sums up the conventional wisdom:

> Offensive weapons in cyberspace can be developed cheaply and quickly. The risk of being punished after a cyber attack is relatively small. It is inevitable that complex, sophisticated networks will have weaknesses and loopholes. Meanwhile, cyber defenders only succeed in fixing weaknesses that have been attacked, not those that remain inactive and unidentified. In general, cyberspace is easy to attack and difficult to defend, which makes it asymmetrical.[46]

There is no shortage of threat reporting to substantiate fears. Yet prominent cyber operations to date have been more ambiguous than decisive, more persistent than rapid, more restrained than disruptive, and more like intelligence than war.[47] Once again, attention to the social and political context of information technology calls into question deterministic claims about radical disruption.

The second implication is that cyberspace is not a space at all.[48] The viability of a global information infrastructure relies on the voluntary adoption of common standards. Software is quite literally nothing but code, protocol, program, registry, file, folder, procedure, and so on. Information technology is bureaucracy by other means. Cyberspace, therefore, is better understood a collection of sociotechnical institutions that provide mutual benefits to the actors who participate in them. This large-scale assemblage of digital infrastructures and governance processes has been shaped by many actors over the course of many decades. Most of them share an interest in having a stable information-processing environment to transact business. Hardly an experiment in digital anarchy, the internet is a triumph of liberal institutionalization at a global scale.[49] In the terms of this book, the "internal solution" used by cyber operators becomes, in part, its own "external problem" insofar as attackers and targets share the same infrastructure. Attackers and defenders may be in different political organizations, but they participate in the same information system. Cybersecurity thus requires a relaxation of the internal-external distinction I have used throughout this book. The more relevant distinction is between local operators and remote operations, which still gives rise to difficult coordination challenges that can be analyzed with information practice theory.

Third, the cooperative constitution of cyberspace hardly means the end of conflict. On the contrary, the increasing potential for trustworthy and stable transactions at a planetary scale also creates unprecedented potential for deception. Deception—broadly construed to include concealment, encryption, espionage, subversion, sabotage, disinformation, and manipulation—exploits trust through simulation and dissimulation. As the managed practice of states and firms shades into insulation for whatever reason, malicious hackers exploit their networks and steal their data. Yet hackers can achieve this only by working within a system of common rules and tools. Cyber operators do not impose their will through kinetic force; rather, they attempt to instruct deterministic machines to do things that system engineers and users do not intend. Similarly, a clever lawyer exploits loopholes in the law to gain advantages unanticipated by lawmakers, rather than breaking laws outright. In this respect, hackers are more like spies or criminals in a foreign society than soldiers on a battlefield. Traditional military combat is a duel between control systems that attempt to materially overpower the adversary and impose their favored scheme of control. Subversives, by contrast, operate within an adversary's control system to undermine it from the inside. They must adopt the norms and practices of the enemy to get close enough to steal secrets or disrupt operations. Cybersecurity, and intelligence more generally, is more of a contest of wits than a contest of strength. With the advent of ubiquitous global information infrastructure, intelligence-counterintelligence contests become more sophisticated than ever before.

Fourth, it follows that the intelligence exploitation of common information infrastructure produces incentives for restraint.[50] Intelligence agents (including cyber operators) must rely on stealth and stratagem to learn about or influence a target, but to do so they must maintain access to and participate in the shared institutions of the host society. Common protocols are the condition for the possibility for deception. Paradoxically, actors in cyberspace have to collaborate to compete. Whenever cooperation is a condition for the possibility of conflict, attackers have incentives to moderate the severity of attacks in order to protect operational security and preserve future opportunities for attack. The complexity and abstraction of cyber networks create information friction for attackers, who must take care to avoid making mistakes that will compromise their operations. Even in wartime, cyber operators must pull their punches if they want to maintain access to valuable targets and to protect sources and methods that are hard to make and easy to lose. Compromise can undo months or years of secret preparation in a few hours if the defender is adept at patching or reconfiguring the network. The attacker's intelligence problem becomes even more difficult if the defender is employing active deception and network monitoring as part of its counterintelligence solution.[51]

Finally, the same incentives for restraint that limit the "vertical" escalation of conflict severity also promote the "horizontal" proliferation of cyber conflict. Cold War strategists described a stability-instability paradox wherein nuclear deterrence discouraged war between the superpowers but encouraged conventional posturing and proxy wars in peripheral theaters. Cyber conflict is constrained by a new variation on this paradox, bounded not only by the risks of escalation to war, as in the nuclear case, but also by the benefits of interdependence that make cyber exploitation possible in the first place.[52] The attacker's desires to avoid compromise, preclude retaliation, and continue transacting in cyberspace combine to incentivize restraint. While cyber operations blur the distinctions between peace and war, cyber actors also have incentives to avoid becoming too warlike. One important but unlikely exception is produced by the "cross-domain" combination of cyber operations and nuclear deterrence, in which the deception inherent in the former undermines the transparency required by the latter.[53] Most cybersecurity interactions, however, take place beneath the threshold of armed conflict, which has prompted interest in strategies of "persistent engagement."[54]

To improve cybersecurity, organizations must engage in adaptive management on a grand scale. The buildout of global trust systems enables the decentralized abuse of trust. Threat actors have (strong) incentives to search for and exploit unguarded degrees of freedom in the self-insulated information systems of their targets. Network defenders have (weaker) incentives to improve the sophistication and stability of the systems that detect and discourage threat actors. The state of the art in cybersecurity thus develops increasingly complex patterns of information practice through cycles of disruptive exploitation and generative reform. The result is more sophisticated networks of trust (managed practice), together with more complicated patterns of deception (adaptive practice) by attackers and defenders alike. Ambiguity is inevitable in information systems because the complexity of control contests keeps increasing through cycles of exploitation and reform.

## The Paradox of Control

We live in an era when drone pilots in Nevada can kill insurgents in Yemen, computer code from Maryland can wreck centrifuges in Iran, and stealth bombers from Missouri can blow up the Chinese embassy in Serbia. The United States has weapons that are fast and precise, but its recent wars have been slow and messy. Complex technologies mediate human experience of the battlefield to an unprecedented degree, but all of that complexity can prevent human beings from actually seeing the world.

As weapons, organizations, and operations become more sophisticated, the coordination of digital representations with distant realities becomes simultaneously more difficult and more vital for military success. Mere representations are translated into consequential outcomes through complex distributed data processes that are difficult, if not impossible, to understand in detail. Any organization's understanding of its own information systems in action, therefore, becomes a limiting factor for its operational performance. The infrastructure of control, moreover, becomes a limiting factor for effective strategy. Debates about the potential applications and limits of modern military power make implicit assumptions about where information practice will be more or less likely to enhance operational performance. Organizations that learn how to manage informational complexity (or are mindful of their limitations when they cannot) are more likely to prevail on (or judiciously avoid) the complex battlefields of the future. As the military application of force becomes more reliable, finally, political questions about why and where to use force become more complicated. The growing importance of technology in military affairs is increasing the importance of nontechnological aspects of military performance.

Information technology has become vital in peace and war at the same time that the distinction between them is eroding. The same systems that improve socioeconomic coordination at a planetary scale also expand the play of friction. It may be the case that chronic exploitation, subversion, and deception are the price to pay for reduced exposure to the horrors of war. Even this possibility is cold comfort for practitioners who have to make decisions about defending organizations and using force. It is also true that our dependence on advanced technology for almost everything that we do also creates new possibilities for catastrophe. When computers are used for every vital function, the disruption of computation can be a matter of life and death. Information practice—the frictions of control, the imperfections of knowledge, the tussles of organizational life, and the tireless effort of practitioners to keep information systems running—is all that keeps some of the most terrible possibilities from being actualized. This is an unsettling state of affairs. It is the nature of the information age that we are doomed to struggle with our own success.

The essence of information technology is control. The word "cyberspace" and the word "government" both share the same Greek root meaning to steer or pilot a boat. Bureaucratic files and digital networks enhance control by improving the organizational capacity for measurement, coordination, and enforcement. Yet because organizations must live in a world full of other organizations, and because some organizations have incompatible intentions, more potential for control also means more competition. One actor can use information technology to improve its ability to assert control while another can use the same technology to improve its ability to resist. As competitors come to rely on the same global information systems for control, moreover, the digital means of control themselves become contested. The more we are willing to trust our computers, the more vulnerable we become to deception. The more dependent we become on control technology, ironically, the more we feel like we are losing control. The paradox of control is that a degree of ignorance is the price of knowledge.

# Methodology

Before it is possible to explain variation in phenomena, it is first necessary to see that those phenomena exist. Information practice is vital for command and control, but it has been hiding in plain sight. My case studies have been highly granular, in part, to reveal something important that needs to be explained. As John Gerring notes, "Case studies enjoy a natural advantage in research of an exploratory nature," even as discussion of "social science methodology has focused almost exclusively" on testing.[1] In Karl Popper's terms, this is a project of conjecture rather than refutation. For ease of exposition I have encapsulated the theoretical framework in chapter 2, but the logic of discovery was very different from the logic of presentation. Too often the former is swept under the rug in an attempt to conform to the scientific norms of an academic discipline. The resulting scientific representations hide the details of their construction.

Consider the following: this book develops a theory to explain the conditions under which information practice creates knowledge (or confusion) to improve (or degrade) control. I employ ethnographic and historical methods to test the theory by leveraging variation on the key explanatory factors—the degree of constraint in the external problem and the internal solution—both across and within cases. By tracing measurement, coordination, and enforcement processes within different organizations in different wartime conditions, I show how the external problem and the internal solution interact to produce different patterns of information practice. The cases roughly confirm the logic of the theory, albeit with nuances aplenty.

This is all true enough. Yet the same cases were also important as generative studies. To understand the origin of theoretical concepts, which is the taproot of the scientific enterprise, some confessional acknowledgment of the path of discovery is helpful. Kenneth Waltz writes that the origin of concepts is something of a mystery: "At some point a brilliant intuition flashes, a creative idea emerges. One cannot say how the intuition comes and how the idea is born."[2] In fact, it is possible to say something about the origin of ideas.

This book's theoretical framework can be applied reflexively to explain the very process that created it. Science and scholarship, after all, are forms of information practice that connect representations—theories, figures, models, books, articles—to an intellectual domain of human concern. Conversely, we can think of military information practice as a form of pragmatic science that tries to come up with actionable models of an operational domain. As Waltz observes, "A theory is a picture, mentally formed, of a bounded realm or domain of activity. A theory is a depiction of the organization of a domain and the connections among its parts."[3] Systematic, formalized, and deductive theory can be likened to an institutionalized solution. Empirical, intuitive, and inductive research can be likened to organic adaptation. The former is most appropriate when the target domain is well defined ("a bounded realm"), but the latter is better when phenomena are ambiguous ("pictures, mentally formed" are still hazy). Mismatches create a lot of friction, and scientists have to cope with it. As Steven Shapin points out, "The making, maintaining, and modification of scientific knowledge is a local and a mundane affair."[4] Shapin echoes Clausewitz's insight that war is "an area littered by endless minor obstacles" rather than "great, momentous questions, which are settled in solitary deliberation."[5] A cycle of intellectual exploitation and reform in scientific practice generates four familiar patterns. Table A.1 provides new labels, but the logical framework is the same as that described in chapter 2.

This method is generally known as *abduction*, which is distinct from logical deduction or empirical induction.[6] Abduction works by extending extant theory into conjectures about a new domain. Empirical exploration and experimentation then seek to find areas where theory works well or breaks down. The discovery and interpretation of challenges along the way, in turn, prompt reformulation of the theory. The hermeneutic cycle of scientific inquiry thus stabilizes a new (subjective) explanation that better fits with the (objective) structure of the world, which itself is clarified through the same process.[7] As with information practice generally, researchers must be mindful of their own motivation and situation in selecting and engaging with any given material. An earnest commitment to understanding the phenomena of concern, moreover, should guide the scientific endeavor, lest

**Table A.1  The method of abduction as information practice**

|  | *Defined domain* | | *Ambiguous phenomena* |
|---|---|---|---|
| Logical deduction | Theory | ◄ Abduction ⟩ | Conjecture |
| Empirical induction | Challenges | | Experimentation |

science get hijacked by more parochial concerns with empire building, careerism, or dogmatic consistency.

Ethnographic methods are especially well suited to abduction, which is also known as grounded theory development.[8] The problem of "choosing on the dependent variable" is less important for case selection in this type of research since its goal is to identify and characterize the microlevel processes that create meaning for a community. Close engagement with repeated patterns of social interaction over an extended period of time, together with attention to the structural factors that shape and are shaped by those patterns, provides some confidence that the results are generalizable beyond the study community. I focus in particular on the mundane, tacit, taken-for-granted practice that animates working information systems.[9] An ethnographic sensibility is particularly useful for understanding military information practice because data that get archived for eventual declassification can be difficult to interpret without understanding the pragmatic context of inscription. Ethnography is a way of bringing a calibrated instrument—a researcher sensitized to the relevant phenomena by prior engagement with the relevant scholarship—into a real-world laboratory to observe uncontrolled social practice as it unfolds, then returning to articulate more general concepts.[10]

My initial intuitions developed through reading academic scholarship while reflecting on my own military experience. I read widely across the disciplines of security studies, political science, cognitive science, and the sociology of technology and science. The diversity of conceptual assumptions and rhetorical styles in these intellectual communities posed a major challenge for synthesis. This tension ultimately was generative, but only after a lot of wrangling with empirical material, and no little frustration. Of the cases examined in this book, the first that I engaged with was Falcon-View. The application and I both entered the U.S. Navy at about the same time in the late 1990s. During my active duty years, I gained some experience writing software that interfaced with FalconView, which later facilitated my interaction with the mission planning community. Our modest projects (Quiver, A3) also failed where FalconView succeeded. Indeed, FalconView is the most sustained and successful case of bottom-up innovation in military software that I have been able to identify. I recognized that the conditions for FalconView's success were unique in many ways, even as the impulse for users to innovate in the military was not.

I left active duty in 2003 but remained in the Naval Reserve. I thus had the opportunity to mobilize to Iraq while I was in graduate school. I knew that a protracted ground war would put interesting stress on theories about military innovation (i.e., the RMA). I did not yet appreciate how significantly the culture of Naval Special Warfare would determine the quality of representational practice in the unit to which I deployed in 2007–8. My active duty role imposed major constraints on field research, to be sure.

Official responsibilities and classified work spaces precluded the use of standard ethnographic methods like on-site interviews or video recording. I was open with military colleagues about my civilian life as an academic researcher, but as a mobilized reservist I was duty bound to be a participant first and an observer second. Fortunately, there was a natural synergy between these roles given my substantive interest in information systems. As discussed in the main text, military personnel often become preoccupied with the technologies that mediate their knowledge as they anticipate and respond to breakdowns. Conversations about data management and organizational effectiveness thus occurred naturally in the absence of any research agenda because they facilitated the ongoing debugging of our information systems.[11] Over the course of the deployment I filled up several little green notebooks, in which comments about staff work and systems issues were admixed with countless operational minutiae, and I would have done so in the absence of any research agenda.

After redeployment I mined my notebooks and recorded additional recollections as autoethnographic data.[12] I focused on recurring patterns of human-computer interaction at the SOTF rather than the content of specific operations or methods, which remain classified as of this writing. A subsequent classification review of my findings resulted in no redaction. Fortunately for this research, informal practices tend not to be formally classified. The everyday interactions between people and machines often go unnoted in explicit rules and prohibitions, precisely because they are part of the unarticulated milieu of practice, or what Pierre Bourdieu calls "habitus."[13] The ethnographic data prompted me to reformulate the theoretical framework to take organizational culture more seriously. I also had to grapple with mixed modes of organization since the SOTF was simultaneously unified in its intentionality and fragmented in its implementation.

Once I developed more refined concepts about the relationship between organizational representations and operational realities, I wanted to examine some outside cases in order to test, extend, and generalize the concepts. The Battle of Britain (chapter 3) was an obvious choice for many reasons: simpler technology, intrinsic historical importance, a different warfighting domain, a successful outcome, and abundant archival data. Furthermore, a new appreciation of the decades-long development of the British information system militated against prominent arguments, by my dissertation adviser among others, about civilian intervention at the eleventh hour.[14] While variation on so many factors at once would rightly be seen as anathema in a research design for theory testing via controlled comparison, the additional complexity can actually be useful for theory building. Rich detail, thickly described, can highlight unconsidered explanatory factors and underscore surprising similarities. The Battle of Britain case thus both broadened and improved my confidence in the insights that I developed in the U.S. cases.

I also wanted to examine an outside case from the digital eta. The RMA thesis is about networked computers, not electromagnetic tabulators, after all. I decided to examine U.S. drone campaigns (chapter 6) because the bodily location of drone pilots, situated remotely from the battlefield, together with their close mental proximity and the attendant controversy over drone strikes, exemplified the problem of technologically mediated experience, which had become central to my argument. The case also offered opportunities for comparison with, and chronological extension of, the other U.S. cases. Drones were used for a similar counterterrorism process as the special operations unit examined in chapter 5, and they were used by some of the same organizations that used FalconView in chapter 4. Most fortuitously, the appearance of an outstanding ethnographic study on MQ-9 Reaper aircrew by Timothy Cullen provided me with comparable fine-grained detail on information practice.[15]

All four cases focused on the tactical or operational level of war. At this level of granularity I already had to account for tremendous heterogeneity in causal factors: different services, cultures, missions, tasks, weapons, infrastructures, protocols, and so on. To accommodate all of the particularity and idiosyncrasy within and across cases, I was driven to articulate the theory at a very high level of abstraction. I surmised that the fundamental tension in political economy between markets and governments also mediated the basic epistemological relationship between subject and object in distributed cognition. Therefore, the performance of any cybernetic feedback loop that enabled knowledge and control in an organization was also an institutional collective action problem. I combined these insights into the very simple framework presented in this book. While I was very sensitive to the fact that my synthesis rode roughshod through four very different traditions (security studies, political economy, cognitive science, sociology of technology), I felt that a parsimonious approximation could nonetheless reveal something important about the organizational context of information technology. These coarse but time-tested distinctions (inside-outside, sensing-acting, market-government, success-failure) seemed to carve nature at its joints. Given the degree of abstraction in my framework, moreover, the external and internal constraints that determine the quality of information practice necessarily require considerable interpretation in each case. Fortunately, the very abstraction of the theory also lends itself naturally to generalization. The concepts in this book should thus be relevant for many problems beyond those considered in it, whether beyond the military realm or even, reflexively, for science itself.

One outside problem is particularly important in the context of this book, and notable for its absence. At the same time that I was working on this project, I was also publishing on cybersecurity in international relations. Cybersecurity can be conceived as a second-order problem of protecting and exploiting information practice. To foreground the phenomena of information practice itself, I eventually scoped the book to just military combat operations. I ultimately decided not to include any cyber research in this book, both because it focused on broader themes and because I did not have access to details of practitioner interaction comparable to the other cases. I leave the detailed argument for another time, but suffice it to say that my ideas about cybersecurity have both shaped and been shaped by my ideas about information practice.[16]

To summarize, my theoretical intuitions about representation emerged through personal experience in war and engagement with a diverse range of scholarship. An initial articulation of concepts guided further research in the field and in the library on cases of particular salience. Empirical exploration in turn prompted further refinement of concepts, which prompted the revisiting of cases and selection of new cases, and so on repeatedly. The entire manuscript then went through several major revisions during a lengthy review process, which pushed me to further refine and clarify the core theoretical concepts, their application to the cases, and their broader implications. My journey was full of unexpected opportunities, contingencies, and difficulties, but the endpoint was a more generalizable theory. The main text clears away the messy historical scaffolding, which I have revealed in this appendix. In it I attempt to deductively argue the theory on its own merits and perform a set of empirical tests through process tracing. We thus end up where we began.

My emphasis on the sociotechnical construction of knowledge by a community of practitioners has a clear affinity with the constructivist paradigm of international relations theory.[17] Yet I also argue that military power is sensitive to strategic competition and geopolitical structure, which is more aligned with a realist worldview.[18] I also stress the political-economic trade-offs in the design of information systems, which draws on liberal insights about the relative strengths of markets and institutions.[19] My framework assumes that structural and organizational factors interact to produce military power, which is perhaps closest to the (neo)classical realist paradigm.[20] Ultimately I am less interested in paradigmatic allegiance than in making sense of the phenomena of information practice as they appear in the real world.[21] Pragmatic criteria should be used to evaluate a pragmatic theory of knowledge. The ultimate measure of my success or failure is whether or not warfighters recognize something of their own experience in the ideas offered here.

# Notes

## Introduction

1. Steven Lee Myers, "Chinese Embassy Bombing: A Wide Net of Blame," *New York Times*, April 17, 2000.

2. See Peter Hays Gries, "Tears of Rage: Chinese Nationalism and the Belgrade Embassy Bombing," *China Journal*, no. 46 (2001): 25–43; Susan L. Shirk, *China: Fragile Superpower* (New York: Oxford University Press, 2007), 212–33; Kevin Pollpeter, "Chinese Writings on Cyberwarfare and Coercion," in *China and Cybersecurity: Espionage, Strategy, and Politics in the Digital Domain*, ed. Jon R. Lindsay, Tai Ming Cheung, and Derek S. Reveron (New York: Oxford University Press, 2015), 149. China almost certainly would have "informatized" its military even if the bombing had never occurred. As a rising great power, China has structural incentives to act on its understanding of the evolving nature of military power. Nevertheless, the bombing did help to underscore the urgency of the threats and opportunities posed by the U.S. RMA.

3. See *Bombing of the Chinese Embassy, Before the House Permanent Select Committee on Intelligence*, 106th Cong. (July 22, 1999) (statement of George Tenet, Director of Central Intelligence); Myers, "Chinese Embassy Bombing." Conspiracy theories—e.g., John Sweeney, Jens Holsoe, and Ed Vulliamy, "Nato Bombed Chinese Deliberately," *Guardian*, October 16, 1999—have never been substantiated.

4. Myers, "Chinese Embassy Bombing."

5. See the appendix on methodology.

6. Carl von Clausewitz, *On War*, trans. Michael Howard and Peter Paret (Princeton, NJ: Princeton University Press, 1976), 102. Emphasis in the original.

## 1. The Technology Theory of Victory

1. Colin S. Gray, "Nuclear Strategy: The Case for a Theory of Victory," *International Security* 4, no. 1 (1979): 54–87; William C. Martel, *Victory in War: Foundations of Modern Military Policy* (New York: Cambridge University Press, 2007).

2. On airpower, see Tami Davis Biddle, *Rhetoric and Reality in Air Warfare: The Evolution of British and American Ideas about Strategic Bombing, 1914–1945* (Princeton, NJ: Princeton

University Press, 2002); Phil M. Haun, *Coercion, Survival, and War: Why Weak States Resist the United States* (Stanford, CA: Stanford University Press, 2015). On counterinsurgency, see Austin Long, *The Soul of Armies: Counterinsurgency Doctrine and Military Culture in the US and UK*, Cornell Studies in Security Affairs (Ithaca, NY: Cornell University Press, 2016); Jacqueline L. Hazelton, "The 'Hearts and Minds' Fallacy: Violence, Coercion, and Success in Counterinsurgency Warfare," *International Security* 42, no. 1 (July 1, 2017): 80–113.

3. Alfred von Schlieffen, "Der Krieg in Der Gegenwart," *Deutsche Revue* 34, no. 1 (1909): 18. I am grateful to Lennart Maschmeyer for translation assistance.

4. Timothy T. Lupfer, *Dynamics of Doctrine: The Changes in German Tactical Doctrine during the First World War*, Leavenworth Papers (Fort Leavenworth, KS: U.S. Army Command and General Staff College, July 1981); Peter Simkins, "'Building Blocks': Aspects of Command and Control at Brigade Level in the BEF's Offensive Operations, 1916–1918," in *Command and Control on the Western Front: The British Army's Experience, 1914–1918*, ed. G. D. Sheffield and Daniel Todman (Stroud, UK: Spellmount, 2004), 141–72.

5. General William C. Westmoreland, speech to the Association of the U.S. Army, Sheraton Park Hotel, Washington, DC, October 14, 1969, reprinted in Paul Dickson, *The Electronic Battlefield* (Bloomington: Indiana University Press, 1976), 218.

6. Westmoreland, speech, 220–21.

7. Westmoreland, speech, 221.

8. Westmoreland, speech, 221.

9. On the Cold War roots of the RMA, see Andrew F. Krepinevich Jr., ed., *The Military-Technical Revolution: A Preliminary Assessment* (Washington, DC: Center for Strategic and Budgetary Assessments, 2002); Dima Adamsky, *The Culture of Military Innovation: The Impact of Cultural Factors on the Revolution in Military Affairs in Russia, the US, and Israel* (Stanford, CA: Stanford University Press, 2010); Andrew F. Krepinevich and Barry D. Watts, *The Last Warrior: Andrew Marshall and the Shaping of Modern American Defense Strategy* (New York: Basic Books, 2015).

10. Assessments of the RMA debate include Tim Benbow, *The Magic Bullet? Understanding the Revolution in Military Affairs* (London: Brassey's, 2004); Sean Lawson, *Nonlinear Science and Warfare: Chaos, Complexity and the U.S. Military in the Information Age* (New York: Routledge, 2014); Andrew Futter and Jeffrey Collins, eds., *Reassessing the Revolution in Military Affairs: Transformation, Evolution and Lessons Learnt* (New York: Palgrave Macmillan, 2015).

11. E.g., Williamson Murray and Allan Reed Millett, *Military Innovation in the Interwar Period* (New York: Cambridge University Press, 1996); Stephen Peter Rosen, *Winning the Next War: Innovation and the Modern Military* (Ithaca, NY: Cornell University Press, 1991).

12. Andrew W. Marshall, foreword to Krepinevich, *Military-Technical Revolution*, ii.

13. On Boyd's life and legacy, see Robert Coram, *Boyd: The Fighter Pilot Who Changed the Art of War* (New York: Little, Brown, 2002).

14. Arthur K. Cebrowski and John J. Garstka, "Network-Centric Warfare: Its Origin and Future," *U.S. Naval Institute Proceedings* 124, no. 1 (1998): 28–35.

15. Office of the Chairman of the Joint Chiefs of Staff, *Joint Vision 2010* (Washington, DC: U.S. Department of Defense, 1996), 17.

16. Donald Rumsfeld, "Defense Transformation" (speech delivered at the National Defense University, Fort McNair, Washington, DC, January 31, 2002).

17. William A. Owens and Edward Offley, *Lifting the Fog of War* (New York: Farrar, Straus and Giroux, 2000), 15.

18. Owens and Offley, *Lifting the Fog of War*, 15.

19. John Ferris, "Netcentric Warfare, C4ISR and Information Operations: Towards a Revolution in Military Intelligence?," *Intelligence and National Security* 19, no. 2 (2004): 199–225; Milan N. Vego, "Operational Command and Control in the Information Age," *Joint Forces Quarterly*, no. 35 (2004): 100–107; Keith L. Shimko, *The Iraq Wars and America's Military Revolution* (New York: Cambridge University Press, 2010).

20. Harvey M. Sapolsky, Benjamin H. Friedman, and Brendan R. Green, eds., *US Military Innovation since the Cold War: Creation without Destruction* (New York: Routledge, 2009); Peter J. Dombrowski and Eugene Gholz, *Buying Military Transformation: Technological Innovation and the Defense Industry* (New York: Columbia University Press, 2006).

21. James N. Mattis, "USJFCOM Commander's Guidance for Effects-Based Operations," *Parameters*, Autumn 2008, 24, 19.

22. See chapter 5.

23. David Deptula and Mike Francisco, "Air Force ISR Operations: Hunting versus Gathering," *Air & Space Power Journal* 24, no. 4 (2010): 16.

24. Keir A. Lieber and Daryl G. Press, "The New Era of Counterforce: Technological Change and the Future of Nuclear Deterrence," *International Security* 41, no. 4 (April 1, 2017): 9–49.

25. For the argument that cyberwarfare is transformational, see, inter alia, Mike McConnell, "Cyberwar Is the New Atomic Age," *New Perspectives Quarterly* 26, no. 3 (2009): 72–77; Richard A. Clarke and Robert K. Knake, *Cyber War: The Next Threat to National Security and What to Do about It* (New York: Ecco, 2010); Lucas Kello, "The Meaning of the Cyber Revolution: Perils to Theory and Statecraft," *International Security* 38, no. 2 (2013): 7–40. Skeptical perspectives include Jon R. Lindsay, "Stuxnet and the Limits of Cyber Warfare," *Security Studies* 22, no. 3 (2013): 365–404; Erik Gartzke, "The Myth of Cyberwar: Bringing War in Cyberspace Back Down to Earth," *International Security* 38, no. 2 (2013): 41–73; Brandon Valeriano and Ryan C. Maness, *Cyber War versus Cyber Realities: Cyber Conflict in the International System* (New York: Oxford University Press, 2015). See chapter 7 for further discussion of cybersecurity.

26. State Council Information Office, *China's Military Strategy* (Beijing: State Council Information Office of the People's Republic of China, May 2015), China Daily, http://www.china daily.com.cn/china/2015-05/26/content_20820628.htm.

27. Jacqueline Newmyer, "The Revolution in Military Affairs with Chinese Characteristics," *Journal of Strategic Studies* 33, no. 4 (2010): 483–504; Kevin Pollpeter, "Controlling the Information Domain: Space, Cyber, and Electronic Warfare," in *Strategic Asia 2012–13: China's Military Challenge*, ed. Ashley J. Tellis and Travis Tanner (Seattle: National Bureau of Asian Research, 2012), 163–94.

28. Bob Work, "The Third U.S. Offset Strategy and Its Implications for Partners and Allies" (speech delivered at the Willard Hotel, Washington, DC, January 28, 2015), U.S. Department of Defense, https://www.defense.gov/News/Speeches/Speech-View/Article/606641/the-third-us-offset-strategy-and-its-implications-for-partners-and-allies/. See chapter 7 for further discussion of Chinese modernization and the third offset.

29. Li Chunyuan and Zang Jiwei, "Intelligentization Will Revolutionize Modern Warfare" (in Chinese), *PLA Daily*, December 14, 2017. I am grateful to Jinhui Jiao for translation assistance.

30. Westmoreland, speech, 221.

31. *Summary of the 2018 Department of Defense Artificial Intelligence Strategy: Harnessing AI to Advance Our Security and Prosperity* (Washington, DC: Department of Defense, February 12, 2019), 4, U.S. Department of Defense, https://media.defense.gov/2019/Feb/12/2002088963/-1/-1/1/SUMMARY-OF-DOD-AI-STRATEGY.PDF.

32. H. R. McMaster, "Thinking Clearly about War and the Future of Warfare—The US Army Operating Concept," *Military Balance* (blog), October 23, 2014, https://www.iiss.org/en/mili tarybalanceblog/blogsections/2014-3bea/october-831b/thinking-clearly-about-war-and-the-future-of-warfare-6183. On the fraught history of military futurology, see Lawrence Freedman, *The Future of War: A History* (New York: PublicAffairs, 2017).

33. Reviews include Adam Grissom, "The Future of Military Innovation Studies," *Journal of Strategic Studies* 29, no. 5 (October 1, 2006): 905–34; Stuart Griffin, "Military Innovation Studies: Multidisciplinary or Lacking Discipline?," *Journal of Strategic Studies* 40, no. 1–2 (January 2, 2017): 196–224. Historical perspectives include Murray and Millett, *Military Innovation*; MacGregor Knox and Williamson Murray, *The Dynamics of Military Revolution, 1300–2050* (New York: Cambridge University Press, 2001); Emily O. Goldman, ed., *Information and Revolutions in Military Affairs* (New York: Routledge, 2013).

34. Langdon Winner, *Autonomous Technology: Technics-out-of-Control as a Theme in Political Thought* (Cambridge, MA: MIT Press, 1977); Merritt Roe Smith and Leo Marx, eds., *Does Technology Drive History? The Dilemma of Technological Determinism* (Cambridge, MA: MIT Press, 1994).

35. Carl von Clausewitz, *On War*, trans. Michael Howard and Peter Paret (Princeton, NJ: Princeton University Press, 1976), 89. Applications of Clausewitz to information-age war

include Alan Beyerchen, "Clausewitz, Nonlinearity, and the Unpredictability of War," *International Security* 17, no. 3 (1992): 59–90; Barry D. Watts, *Clausewitzian Friction and Future War*, rev. ed., McNair Paper (Washington, DC: Institute for National Strategic Studies, National Defense University, 2004).

36. Practitioner perspectives include *DOD Authorization—Information Warfare, Before the Procurement Subcommittee and Research and Development Subcommittee of the House National Security Committee*, 105th Cong. (March 20, 1997) (statement of Lieutenant General Paul K. Van Riper, U.S. Marine Corps Commanding General); John A. Gentry, "Doomed to Fail: America's Blind Faith in Military Technology," *Parameters* 32, no. 2 (2002): 88–103; H. R. McMaster, *Crack in the Foundation: Defense Transformation and the Underlying Assumption of Dominant Knowledge in Future War*, Student Issue Paper (Carlisle Barracks, PA: U.S. Army War College Center for Strategic Leadership, 2003). Academic critics include Stephen Biddle, "The Past as Prologue: Assessing Theories of Future Warfare," *Security Studies* 8, no. 1 (1998): 1–74; Richard J. Harknett, "The Risks of a Networked Military," *Orbis* 44, no. 1 (2000): 127–43; Jacquelyn Schneider, *Digitally-Enabled Warfare: The Capability-Vulnerability Paradox* (Washington, DC: Center for a New American Security, August 29, 2016).

37. Charles Perrow, *Normal Accidents: Living with High-Risk Technologies*, 2nd ed. (Princeton, NJ: Princeton University Press, 1999). For application to military systems, see Scott D. Sagan, *The Limits of Safety: Organizations, Accidents, and Nuclear Weapons* (Princeton, NJ: Princeton University Press, 1995); Bart Van Bezooijen and Eric-Hans Kramer, "Mission Command in the Information Age: A Normal Accidents Perspective on Networked Military Operations," *Journal of Strategic Studies* 38, no. 4 (June 7, 2015): 445–66.

38. Scott A. Snook, *Friendly Fire: The Accidental Shootdown of U.S. Black Hawks over Northern Iraq* (Princeton, NJ: Princeton University Press, 2000).

39. Gene I. Rochlin, *Trapped in the Net: The Unanticipated Consequences of Computerization* (Princeton, NJ: Princeton University Press, 1997); Matt Spencer, "Brittleness and Bureaucracy: Software as a Material for Science," *Perspectives on Science* 23, no. 4 (2015): 466–84.

40. Chris C. Demchak, *Military Organizations, Complex Machines: Modernization in the U.S. Armed Services* (Ithaca, NY: Cornell University Press, 1991).

41. Fred Luthans and Todd I. Stewart, "A General Contingency Theory of Management," *Academy of Management Review* 2, no. 2 (1977): 181–95; Lex Donaldson, "The Normal Science of Structural Contingency Theory," in *Studying Organization: Theory and Method*, ed. Stewart R. Clegg and Cynthia Hardy (Thousand Oaks, CA: Sage Publications, 1996), 51–70.

42. Rosen, *Winning the Next War*; Eliot A. Cohen, "A Revolution in Warfare," *Foreign Affairs* 75, no. 2 (1996): 37–54; Murray and Millett, *Military Innovation*; Watts, *Clausewitzian Friction*; Thomas G. Mahnken, *Technology and the American Way of War since 1945* (New York: Columbia University Press, 2010).

43. Michael C. Horowitz, *The Diffusion of Military Power: Causes and Consequences for International Politics* (Princeton, NJ: Princeton University Press, 2010).

44. Owen R. Cote Jr., *The Third Battle: Innovation in the U.S. Navy's Silent Cold War Struggle with Soviet Submarines*, Newport Paper (Newport, RI: Naval War College, 2003); Norman Friedman, *Network-Centric Warfare: How Navies Learned to Fight Smarter through Three World Wars* (Annapolis, MD: Naval Institute Press, 2009).

45. Stephen D. Biddle, *Military Power: Explaining Victory and Defeat in Modern Battle* (Princeton, NJ: Princeton University Press, 2004).

46. Ryan Grauer, *Commanding Military Power* (New York: Cambridge University Press, 2016).

47. On the notion of mission command, see Martin van Creveld, *Command in War* (Cambridge, MA: Harvard University Press, 1985); Michael A. Palmer, *Command at Sea: Naval Command and Control since the Sixteenth Century* (Cambridge, MA: Harvard University Press, 2005); Meir Finkel, *On Flexibility: Recovery from Technological and Doctrinal Surprise on the Battlefield*, Stanford Security Studies (Stanford, CA: Stanford University Press, 2011); Caitlin Talmadge, *The Dictator's Army: Battlefield Effectiveness in Authoritarian Regimes* (Ithaca, NY: Cornell University Press, 2015).

48. I. B. Holley Jr., *Ideas and Weapons* (New Haven, CT: Yale University Press, 1957); David E. Johnson, *Fast Tanks and Heavy Bombers: Innovation in the U.S. Army, 1917–1945* (Ithaca, NY: Cornell University Press, 1998); Biddle, *Rhetoric and Reality*.

49. Carl H. Builder, *The Masks of War: American Military Styles in Strategy and Analysis* (Baltimore: Johns Hopkins University Press, 1989); Elizabeth Kier, *Imagining War: French and British Military Doctrine between the Wars* (Princeton, NJ: Princeton University Press, 1997); Isabel V. Hull, *Absolute Destruction: Military Culture and the Practices of War in Imperial Germany* (Ithaca, NY: Cornell University Press, 2006); Long, *Soul of Armies*.

50. Graham T. Allison and Morton H. Halperin, "Bureaucratic Politics: A Paradigm and Some Policy Implications," *World Politics* 24, no. S1 (April 1972): 40–79; Jack L. Snyder, *The Ideology of the Offensive: Military Decision Making and the Disasters of 1914* (Ithaca, NY: Cornell University Press, 1984); Barry Posen, *The Sources of Military Doctrine: France, Britain, and Germany between the World Wars* (Ithaca, NY: Cornell University Press, 1984); Stephen Van Evera, "The Cult of the Offensive and the Origins of the First World War," *International Security* 9, no. 1 (July 1, 1984): 58–107.

51. Robert W. Komer, *Bureaucracy Does Its Thing: Institutional Constraints on U.S.-GVN Performance in Vietnam* (Santa Monica, CA: RAND Corporation, 1972).

52. Deborah D. Avant, "The Institutional Sources of Military Doctrine: Hegemons in Peripheral Wars," *International Studies Quarterly* 37, no. 4 (December 1, 1993): 409–30; Amy B. Zegart, *Flawed by Design: The Evolution of the CIA, JCS, and NSC* (Stanford, CA: Stanford University Press, 2000); Risa Brooks, *Shaping Strategy: The Civil-Military Politics of Strategic Assessment* (Princeton, NJ: Princeton University Press, 2008).

53. Colin F. Jackson, "Defeat in Victory: Organizational Learning Dysfunction in Counterinsurgency" (PhD diss., Massachusetts Institute of Technology, 2008); Long, *Soul of Armies*.

54. Richard Betts, "The Downside of the Cutting Edge," *National Interest*, no. 45 (1996): 80–83; McMaster, *Crack in the Foundation*; Frederick Kagan, *Finding the Target: The Transformation of American Military Policy* (New York: Encounter Books, 2006). Interestingly, this objection is anticipated by the Office of Net Assessment in Krepinevich, *Military-Technical Revolution*, 31–32.

55. Diane Vaughan, *The Challenger Launch Decision: Risky Technology, Culture, and Deviance at NASA* (Chicago: University of Chicago Press, 1996), 62.

56. Lynn Eden, *Whole World on Fire: Organizations, Knowledge, and Nuclear Weapons Devastation* (Ithaca, NY: Cornell University Press, 2004).

57. van Creveld, *Command in War*, chaps. 5, 7; Eliot A. Cohen and John Gooch, *Military Misfortunes: The Anatomy of Failure in War* (New York: Free Press, 1990); Vego, "Operational Command and Control."

58. Eliot A. Cohen, "Change and Transformation in Military Affairs," *Journal of Strategic Studies* 27, no. 3 (2004): 395–407; Grissom, "Future of Military Innovation Studies."

59. Herbert Kaufman, *The Forest Ranger: A Study in Administrative Behavior* (Washington, DC: Resources for the Future, 1967); Michael Lipsky, *Street-Level Bureaucracy: The Dilemmas of the Individual in Public Service* (New York: Russell Sage Foundation, 1980).

60. Kimberly Marten Zisk, *Engaging the Enemy: Organization Theory and Soviet Military Innovation, 1955–1991* (Princeton, NJ: Princeton University Press, 1993); Adamsky, *Culture of Military Innovation*; Robert T. Foley, "A Case Study in Horizontal Military Innovation: The German Army, 1916–1918," *Journal of Strategic Studies* 35, no. 6 (2013): 799–827.

61. Theo Farrell, "Improving in War: Military Adaptation and the British in Helmand Province, Afghanistan, 2006–2009," *Journal of Strategic Studies* 33, no. 4 (August 1, 2010): 567–94; Jon R. Lindsay, "'War upon the Map': User Innovation in American Military Software," *Technology and Culture* 51, no. 3 (2010): 619–51; Williamson Murray, *Military Adaptation in War: With Fear of Change* (New York: Cambridge University Press, 2011); James A. Russell, *Innovation, Transformation, and War: Counterinsurgency Operations in Anbar and Ninewa Provinces, Iraq, 2005–2007* (Stanford, CA: Stanford University Press, 2011); Nina A. Kollars, "Military Innovation's Dialectic: Gun Trucks and Rapid Acquisition," *Security Studies* 23, no. 4 (October 2, 2014): 787–813; Nina A. Kollars, "War's Horizon: Soldier-Led Adaptation in Iraq and Vietnam," *Journal of Strategic Studies* 38, no. 4 (June 7, 2015): 529–53.

62. Thematic treatments include James Jay Carafano, *GI Ingenuity: Improvisation, Technology, and Winning World War II* (Mechanicsburg, PA: Stackpole Books, 2006); Center of Military History, *Improvisations during the Russian Campaign* (Washington, DC: U.S. Army, 1986); John H. Hay, *Tactical and Materiel Innovations*, Vietnam Studies (Washington, DC: Department of the Army, 1974).

63. Peter Maas, *The Terrible Hours: The Greatest Submarine Rescue in History* (New York: HarperCollins, 1999), 98–103.

64. Michael D. Doubler, *Closing with the Enemy: How GIs Fought the War in Europe, 1944–1945* (Lawrence: University Press of Kansas, 1995), 43–48.

65. Lars Heide, "Monitoring People: Dynamics and Hazards of Record Management in France, 1935–1944," *Technology and Culture* 45, no. 1 (2004): 80–101.

66. Rick Atkinson, "Left of Boom: The Fight against Roadside Bombs," *Washington Post*, four parts, September 30–October 3, 2007.

67. James C. Scott, *Seeing like a State: How Certain Schemes to Improve the Human Condition Have Failed* (New Haven, CT: Yale University Press, 1998); James C. Scott, *Weapons of the Weak: Everyday Forms of Peasant Resistance* (New Haven, CT: Yale University Press, 2008).

68. STS is also known as science and technology studies. Overviews of the STS perspective include Thomas Parke Hughes, *Human-Built World: How to Think about Technology and Culture* (Chicago: University of Chicago Press, 2004); David E. Nye, *Technology Matters: Questions to Live With* (Cambridge, MA: MIT Press, 2006); Lucy A. Suchman, *Human-Machine Reconfigurations: Plans and Situated Actions*, 2nd ed. (New York: Cambridge University Press, 2007).

69. Ruth Schwartz Cowan and Deborah G. Douglas, "The Consumption Junction: A Proposal for Research Strategies in the Sociology of Technology," in *The Social Construction of Technological Systems*, ed. Wiebe E. Bijker, Thomas P. Hughes, and Trevor Pinch (Cambridge, MA: MIT Press, 1987), 261–80; Lucy A. Suchman, "Working Relations of Technology Production and Use," in *The Social Shaping of Technology*, ed. Donald A. MacKenzie and Judy Wajcman, 2nd ed. (Buckingham, UK: Open University Press, 1999), 258–68; Nelly Oudshoorn and Trevor J. Pinch, eds., *How Users Matter: The Co-Construction of Users and Technology* (Cambridge, MA: MIT Press, 2003).

70. Ronald Kline and Trevor Pinch, "Users as Agents of Technological Change: The Social Construction of the Automobile in the Rural United States," *Technology and Culture* 37, no. 4 (1996): 775–77.

71. David N. Lucsko, *The Business of Speed: The Hot Rod Industry in America, 1915–1990* (Baltimore: Johns Hopkins University Press, 2008), 9.

72. An economic theory of user innovation is developed and tested by Eric von Hippel, *The Sources of Innovation* (New York: Oxford University Press, 1988); Eric von Hippel, *Democratizing Innovation* (Cambridge, MA: MIT Press, 2005).

73. Vernon Felton, "Matter: Bag to the Future," *Bike Magazine*, December 2013, http://www.bikemag.com/ephemera/matter-bag-to-the-future/.

74. Claudio Ciborra, *The Labyrinths of Information: Challenging the Wisdom of Systems* (New York: Oxford University Press, 2002), 3. See also Wanda J. Orlikowski, "Improvising Organizational Transformation over Time: A Situated Change Perspective," *Information Systems Research* 7, no. 1 (1996): 63–93.

75. Edwin Hutchins, *Cognition in the Wild* (Cambridge, MA: MIT Press, 1995), 172.

76. The critical security studies field has produced a rich literature on the aesthetics and rhetoric of information age war: inter alia, Paul Virilio, *War and Cinema: The Logistics of Perception*, trans. Patrick Camiller (London: Verso, 1989); James Der Derian, *Virtuous War: Mapping the Military-Industrial-Media-Entertainment Network* (New York: Routledge, 2009); Hugh Gusterson, *Drone: Remote Control Warfare* (Cambridge, MA: MIT Press, 2016).

77. Some have even tried to add "target acquisition" (C4ISTAR) and "combat systems and information operations" (C5ISR/IO). Concatenation in this domain has gone well beyond the point of diminishing returns. In this book I use "command and control" as a shorthand.

78. On the transition to symbolically mediated perception in tactical naval operations, see Timothy S. Wolters, *Information at Sea: Shipboard Command and Control in the U.S. Navy, from Mobile Bay to Okinawa* (Baltimore: Johns Hopkins University Press, 2013).

79. James Donald Hittle, *The Military Staff: Its History and Development*, 3rd ed. (Harrisburg, PA: Stackpole, 1961); van Creveld, *Command in War*, chaps. 2–4.

80. Dallas D. Irvine, "The Origin of Capital Staffs," *Journal of Modern History* 10, no. 2 (June 1, 1938): 173–74.

81. John J. McGrath, *The Other End of the Spear: The Tooth-to-Tail Ratio (T3R) in Modern Military Operations*, The Long War Series Occasional Paper (Fort Leavenworth, KS: Combat Studies Institute Press, 2007), 103.

82. U.S. Department of Defense, *Selected Manpower Statistics, Fiscal Year 2005* (Washington, DC: Defense Manpower Data Center, 2005), 49–61.

83. Daniel R. Lake, *The Pursuit of Technological Superiority and the Shrinking American Military* (New York: Palgrave Macmillan, 2019).

84. James R. Beniger, *The Control Revolution: Technological and Economic Origins of the Information Society* (Cambridge, MA: Harvard University Press, 1986); JoAnne Yates, *Control through Communication: The Rise of System in American Management* (Baltimore: Johns Hopkins University Press, 1989); Daniel R. Headrick, *The Invisible Weapon: Telecommunications and International Politics, 1851–1945* (New York: Oxford University Press, 1991); Alfred D. Chandler Jr. and James W. Cortada, eds., *A Nation Transformed by Information: How Information Has Shaped the United States from Colonial Times to the Present* (New York: Oxford University Press, 2000); David Paull Nickles, *Under the Wire: How the Telegraph Changed Diplomacy* (Cambridge, MA: Harvard University Press, 2003); James W. Cortada, *The Digital Hand*, 3 vols. (New York: Oxford University Press, 2008).

85. Russell F. Weigley, *The American Way of War: A History of United States Military Strategy and Policy* (Bloomington: Indiana University Press, 1973), xxii; Alex Roland, "Technology, Ground Warfare, and Strategy: The Paradox of the American Experience," *Journal of Military History* 55, no. 4 (1994): 447–67; Jonathan D. Caverley, "The Myth of Military Myopia: Democracy, Small Wars, and Vietnam," *International Security* 34, no. 3 (January 1, 2010): 119–57.

86. Morris Janowitz, "Changing Patterns of Organizational Authority: The Military Establishment," *Administrative Science Quarterly* 3, no. 4 (1959): 473.

87. John Ferris and Michael I. Handel, "Clausewitz, Intelligence, Uncertainty and the Art of Command in Military Operations," *Intelligence and National Security* 10, no. 1 (January 1, 1995): 1–58; Anthony King, *Command: The Twenty-First-Century General* (New York: Cambridge University Press, 2019).

88. Harvey M. Sapolsky, Eugene Gholz, and Caitlin Talmadge, *US Defense Politics: The Origins of Security Policy*, 2nd ed. (New York: Routledge, 2014), chap. 5.

## 2. A Framework for Understanding Information Practice

1. My understanding of representation and practice draws on Edwin Hutchins, *Cognition in the Wild* (Cambridge, MA: MIT Press, 1995); Bruno Latour, *Science in Action: How to Follow Scientists and Engineers through Society* (Cambridge, MA: Harvard University Press, 1987); Wanda J. Orlikowski, "Using Technology and Constituting Structures: A Practice Lens for Studying Technology in Organizations," *Organization Science* 11, no. 4 (2000): 404–28; Geoffrey C. Bowker and Susan Leigh Star, *Sorting Things Out: Classification and Its Consequences* (Cambridge, MA: MIT Press, 1999); John Seely Brown and Paul Duguid, *The Social Life of Information* (Cambridge, MA: Harvard Business Press, 2000); Lucy A. Suchman, *Human-Machine Reconfigurations: Plans and Situated Actions*, 2nd ed. (New York: Cambridge University Press, 2007).

2. Winn quoted in Donald McLachlan, *Room 39: A Study in Naval Intelligence* (New York: Atheneum, 1968), 115; On Winn's system, see Patrick Beesly, *Very Special Intelligence: The Story of the Admiralty's Operational Intelligence Centre, 1939–1945* (London: Greenhill Books, 2000), 113–14.

3. Gene I. Rochlin, *Trapped in the Net: The Unanticipated Consequences of Computerization* (Princeton, NJ: Princeton University Press, 1997), 156–68.

4. Beesly, *Very Special Intelligence*, 3–4; Michael A. Palmer, *Command at Sea: Naval Command and Control since the Sixteenth Century* (Cambridge, MA: Harvard University Press, 2005), 242–50.

5. Carl von Clausewitz, *On War*, trans. Michael Howard and Peter Paret (Princeton, NJ: Princeton University Press, 1976), 101.

6. Clausewitz, *On War*, 104.

7. Clausewitz, *On War*, 119.

8. Paul N. Edwards, *A Vast Machine: Computer Models, Climate Data, and the Politics of Global Warming* (Cambridge, MA: MIT Press, 2010), 97. See also Paul N. Edwards, Matthew S. Mayernik, Archer Batcheller, Geoffrey Bowker, and Christine Borgman, "Science Friction: Data, Metadata, and Collaboration," *Social Studies of Science* 4, no. 6 (2011): 667–90.

9. Oliver E. Williamson, "The Economics of Organization: The Transaction Cost Approach," *American Journal of Sociology* 87, no. 3 (1981): 552.

10. Peter Paret, "The Genesis of *On War*," in *On War*, by Carl von Clausewitz (Princeton, NJ: Princeton University Press, 1976), 16.

11. Bruno Latour, "Drawing Things Together," in *Representation in Scientific Practice*, ed. Michael Lynch and Steve Woolgar (Cambridge, MA: MIT Press, 1990), 25.

12. Quoted in Harvey M. Sapolsky, *The Polaris System Development: Bureaucratic and Programmatic Success in Government* (Cambridge, MA: Harvard University Press, 1972), 124.

13. James C. Scott, *Seeing like a State: How Certain Schemes to Improve the Human Condition Have Failed* (New Haven, CT: Yale University Press, 1998), 53–83.

14. Latour, "Drawing Things Together," 56. Emphasis in the original.

15. Anna Lowenhaupt Tsing, *Friction: An Ethnography of Global Connection* (Princeton, NJ: Princeton University Press, 2005).

16. On the general phenomenology of breakdown, see Martin Heidegger, *Being and Time*, trans. John Macquarrie and Edward Robinson (New York: Harper and Row, 1962), 95–102; Hubert L. Dreyfus, *Being-in-the-World: A Commentary on Heidegger's "Being and Time, Division I"* (Cambridge, MA: MIT Press, 1991), 69–83.

17. Hutchins, *Cognition in the Wild*.

18. Kourken Michaelian and John Sutton, "Distributed Cognition and Memory Research: History and Current Directions," *Review of Philosophy and Psychology* 4, no. 1 (2013): 1–24.

19. Kim Sterelny, *Thought in a Hostile World: The Evolution of Human Cognition* (Malden, MA: Blackwell, 2003), 155.

20. David Kirsh and Paul Maglio, "On Distinguishing Epistemic from Pragmatic Action," *Cognitive Science* 18, no. 4 (1994): 513–49.

21. James J. Gibson, "The Theory of Affordances," in *Perceiving, Acting, and Knowing: Toward an Ecological Philosophy*, ed. Robert Shaw and John Bransford (Hillsdale, NJ: Lawrence Erlbaum Association, 1977), 67–82; Ian Hutchby, "Technologies, Texts and Affordances," *Sociology* 35, no. 2 (2001): 441–56; Emanuel Adler and Vincent Pouliot, "International Practices," *International Theory* 3, no. 1 (February 2011): 1–36.

22. Peter Galison, *Image and Logic: A Material Culture of Microphysics* (Chicago: University of Chicago Press, 1997); Davis Baird, *Thing Knowledge: A Philosophy of Scientific Instruments* (Berkeley: University of California Press, 2004).

23. See, inter alia, Francisco J. Varela, Evan Thompson, and Eleanor Rosch, *The Embodied Mind: Cognitive Science and Human Experience* (Cambridge, MA: MIT Press, 1991); Andy Clark and David Chalmers, "The Extended Mind," *Analysis* 58, no. 1 (January 1, 1998): 7–19; Kim Sterelny, "Externalism, Epistemic Artefacts and the Extended Mind," in *The Externalist Challenge: New Studies on Cognition and Intentionality*, ed. Richard Schantz (New York: De Gruyter, 2004), 239–54; Alva Noë, *Out of Our Heads: Why You Are Not Your Brain, and Other Lessons from the Biology of Consciousness* (New York: Farrar, Straus and Giroux, 2009).

24. The British system is examined in detail in chapter 3. The logic of figures 2.3 and 2.4 is adapted from Don Ihde, *Technology and the Lifeworld: From Garden to Earth* (Bloomington: Indiana University Press, 1990), 47. The inset drawing of a track marker in figures 2.3, 2.4, and 2.5 appears in Air Ministry, *Air Defence Pamphlet Number Four: Telling and Plotting*, April 1942 (AIR 10/3760).

25. On the role of connection and disconnection in semantic reference, see Brian Cantwell Smith, *On the Origin of Objects* (Cambridge, MA: MIT Press, 1996), 202–9.

26. Karl E. Weick and Karlene H. Roberts, "Collective Mind in Organizations: Heedful Interrelating on Flight Decks," *Administrative Science Quarterly* 38, no. 3 (1993): 357–81.

27. David A. Mindell, *Between Human and Machine: Feedback, Control, and Computing before Cybernetics* (Baltimore: Johns Hopkins University Press, 2002), 22–23.

28. Douglass C. North, *Institutions, Institutional Change, and Economic Performance* (New York: Cambridge University Press, 1990), 3.

29. Langdon Winner, "Do Artifacts Have Politics?," *Daedalus* 109, no. 1 (1980): 121–36.

30. Bruno Latour, "Where Are the Missing Masses? The Sociology of a Few Mundane Artifacts," in *Shaping Technology/Building Society: Studies in Sociotechnical Change*, ed. John Law and Wiebe E. Bijker (Cambridge, MA: MIT Press, 1992), 225–58.

31. John Law, "Technology and Heterogeneous Engineering: The Case of Portuguese Expansion," in *The Social Construction of Technological Systems: New Directions in the Sociology and History of Technology*, anniversary edition, ed. Wiebe E. Bijker, Thomas Parke Hughes, and Trevor Pinch (Cambridge, MA: MIT Press, 2012), 105–28.

32. On the military-industrial origins of cybernetics, see Mindell, *Between Human and Machine*. On its military legacy, see Paul N. Edwards, *The Closed World: Computers and the Politics of Discourse in Cold War America* (Cambridge, MA: MIT Press, 1996). On its broader impact on science and society, see Jean Pierre Dupuy, *The Mechanization of the Mind: On the Origins of Cognitive Science* (Princeton, NJ: Princeton University Press, 2000); Ronald R. Kline, *The Cybernetics Moment: Or Why We Call Our Age the Information Age* (Baltimore: Johns Hopkins University Press, 2015). On its relevance for cybersecurity, see Thomas Rid, *Rise of the Machines: A Cybernetic History* (New York: W. W. Norton, 2016).

33. James G. March and Herbert A. Simon, *Organizations* (New York: John Wiley and Sons, 1958); John Steinbruner, *The Cybernetic Theory of Decision: New Dimensions of Political Analysis* (Princeton, NJ: Princeton University Press, 1974).

34. Karl Wolfgang Deutsch, *Nerves of Government: Models of Political Communication* (New York: Free Press, 1963); James R. Beniger, *The Control Revolution: Technological and Economic Origins of the Information Society* (Cambridge, MA: Harvard University Press, 1986).

35. An excellent discussion of the social and operational trade-offs that are inherent in the design of any military control technology is found in Thomas P. Coakley, *Command and Control for War and Peace* (Washington, DC: National Defense University Press, 1992).

36. Brentano discussed in Dupuy, *Mechanization of the Mind*, 99–102.

37. My distinction between internal and external factors does some violence to the richness of informational phenomena. Information may well be about things located beyond the boundaries of an organization. Yet data can also refer to abstract, nonexistent, past, or future entities. Information can and certainly does reside outside organizations, which often have fuzzy boundaries in any case. Reference can also be deictic or reflexive (i.e., information about its own context). The pragmatic and phenomenological traditions of philosophy from which I draw make a point of emphasizing, in contrast to the Cartesian vision of an internal subject isolated from the external objective world, that human beings are always already out in the world. See, inter alia, Dreyfus, *Being-in-the-World*; Richard Rorty, *Philosophy and the Mirror of Nature*, 30th anniversary ed. (Princeton, NJ: Princeton University Press, 2009); Andy Clark, *Supersizing the Mind: Embodiment, Action, and Cognitive Extension* (New York: Oxford University Press, 2008); Dan Zahavi, *Phenomenology: The Basics* (New York: Routledge, 2018). Objects emerge for subjects only on the background of pragmatic engagement with meaningful projects, social relationships, and embodied situations. This is why I stress the importance of distributed cognition or extended mind. The reason that I use the internal-external distinction here is that it is a convenient way to open up the problem of intentional disconnection and pragmatic reconnection in a distributed organization, which is the fundamental problem that any information practice must solve. I focus on internal representations of external realities for explanatory convenience, but the concepts should be understood to apply more generally.

38. William James, *Principles of Psychology* (New York: Henry Holt, 1890), 488.

39. "Cascade of inscriptions" is discussed in Latour, "Drawing Things Together," 40–42; "circulating reference" is discussed in Bruno Latour, *Pandora's Hope: Essays on the Reality of Science Studies* (Cambridge, MA: Harvard University Press, 1999), 24–79.

40. Latour, *Science in Action*, 215–57.

41. Jon Tetsuro Sumida, *Decoding Clausewitz: A New Approach to "On War"* (Lawrence: University Press of Kansas, 2008), 64–77.

42. Gregory Bateson, *Steps to an Ecology of Mind* (Chicago: University of Chicago Press, 2000), 315.

43. Smith, *On the Origin of Objects*, 27–76.

44. William James, *The Writings of William James: A Comprehensive Edition*, ed. John J. McDermott (Chicago: University of Chicago Press, 1977), 433.

45. Clausewitz, *On War*, 117.

46. Offensive and defensive strategies can target each of the four basic control functions to disrupt adversarial control or protect friendly control from adversarial action. This suggests eight basic moves in the control contest. Persuasion and dissuasion target enemy intentionality. Deception and concealment target enemy measurement. Disruption and protection target enemy coordination. Interdiction and evasion target enemy enforcement.

47. W. V. O. Quine, *Word and Object* (Cambridge, MA: MIT Press, 1960), 73–112.

48. Guido Vetere and Maurizio Lenzerini, "Models for Semantic Interoperability in Service-Oriented Architectures," *IBM Systems Journal* 44, no. 4 (2005): 887–90.

49. Michael I. Handel, "Clausewitz in the Age of Technology," *Journal of Strategic Studies* 9, no. 2–3 (1986): 51–92.

50. To describe this literature as vast would be an understatement. On the general trade-offs, see, inter alia, Oliver E. Williamson, *The Economic Institutions of Capitalism: Firms, Markets, Relational Contracting* (New York: Free Press, 1985); Charles Wolf Jr., *Markets or Governments: Choosing between Imperfect Alternatives*, 2nd ed. (Cambridge, MA: MIT Press, 1993); Robert Gilpin, *Global Political Economy: Understanding the International Economic Order* (Princeton, NJ: Princeton University Press, 2001).

51. R. H. Coase, "The Problem of Social Cost," *Journal of Law and Economics* 3 (1960): 1–44; Joseph E. Stiglitz, "Information and the Change in the Paradigm in Economics," *American Economic Review* 92, no. 3 (2002): 460–501; Charles P. Kindleberger and Robert Z. Aliber, *Manias, Panics, and Crashes: A History of Financial Crises*, 6th ed. (New York: Palgrave Macmillan, 2011).

52. Robert O. Keohane, "The Demand for International Regimes," *International Organization* 36, no. 2 (1982): 325–55.

53. Joseph E. Stiglitz, "The Private Uses of Public Interests: Incentives and Institutions," *Journal of Economic Perspectives* 12, no. 2 (1998): 3–22; Terry M. Moe, "Political Institutions: The Neglected Side of the Story," *Journal of Law, Economics, & Organization* 6, special issue (1990): 213–53; Paul Milgrom and John Roberts, "An Economic Approach to Influence Activities in Organizations," *American Journal of Sociology* 94 (1988): S154–79.

54. Carl H. Builder, *The Masks of War: American Military Styles in Strategy and Analysis* (Baltimore: Johns Hopkins University Press, 1989); Harvey M. Sapolsky, Eugene Gholz, and Caitlin Talmadge, *US Defense Politics: The Origins of Security Policy*, 2nd ed. (New York: Routledge, 2014), 96–118.

55. See also Alexander Cooley, *Logics of Hierarchy: The Organization of Empires, States, and Military Occupations* (Ithaca, NY: Cornell University Press, 2005).

56. Wendy Nelson Espeland and Mitchell L. Stevens, "Commensuration as a Social Process," *Annual Review of Sociology* 24 (1998): 313–43; Bowker and Star, *Sorting Things Out*; Latour, "Drawing Things Together."

57. Owen R. Cote Jr., "The Politics of Innovative Military Doctrine: The U.S. Navy and Fleet Ballistic Missiles" (PhD diss., Massachusetts Institute of Technology, 1996), 41–94; Sapolsky, Gholz, and Talmadge, *US Defense Politics*, chap. 3.

58. Lex Donaldson, "The Normal Science of Structural Contingency Theory," in *Studying Organization: Theory and Method*, ed. Stewart R. Clegg and Cynthia Hardy (Thousand Oaks, CA: Sage, 1996), 51.

59. Charles Perrow, "A Framework for the Comparative Analysis of Organizations," *American Sociological Review* 32, no. 2 (1967): 194–208; Charles Perrow, *Normal Accidents: Living with High-Risk Technologies*, 2nd ed. (Princeton, NJ: Princeton University Press, 1999).

60. Bart Van Bezooijen and Eric-Hans Kramer, "Mission Command in the Information Age: A Normal Accidents Perspective on Networked Military Operations," *Journal of Strategic Studies* 38, no. 4 (June 7, 2015): 445–66.

61. Scott, *Seeing like a State*, 54–56.

62. Thomas P. Hughes, *Rescuing Prometheus: Four Monumental Projects That Changed the Modern World* (New York: Vintage, 1998), 60–61; Andrea Prencipe, Andrew Davies, and Michael Hobday, eds., *The Business of Systems Integration* (New York: Oxford University Press, 2003).

63. Guido Vetere and Maurizio Lenzerini, "Models for Semantic Interoperability in Service-Oriented Architectures," *IBM Systems Journal* 44, no. 4 (2005): 887–90.

64. Nicholas S. Argyres, "The Impact of Information Technology on Coordination: Evidence From the B-2 'Stealth' Bomber," *Organization Science* 10, no. 2 (1999): 162–80.

65. Quoted in Daniel Todman, "The Grand Lamasery Revisited: General Headquarters on the Western Front, 1914–1918," in *Command and Control on the Western Front: The British Army's Experience, 1914–1918*, ed. G. D. Sheffield and Daniel Todman (Stroud, UK: Spellmount, 2004), 39.

66. M. D. Feld, "Information and Authority: The Structure of Military Organization," *American Sociological Review* 24, no. 1 (1959): 21.

67. The contrast between chateau headquarters and trench warfare has encouraged a historiography of "lions led by donkeys," exemplified by the satire *Blackadder Goes Forth*. Historians have recently cast more sympathetic light on the administrative performance of the giant armies, yet, according to Daniel Todman, "the continuing expansion of GHQ, combined with its well-established location, tended to isolate it from its front line units . . . and the result was a breakdown of trust between the Staff and those they directed and supplied." Todman, "Grand Lamasery Revisited," 58–59.

68. Paul N. Edwards, *The Closed World: Computers and the Politics of Discourse in Cold War America* (Cambridge, MA: MIT Press, 1996).

69. First Marine Division, *Operation Iraqi Freedom (OIF): Lessons Learned*, May 2003, 11–12.

70. For an example see Barry D. Watts, "Unreported History and Unit Effectiveness," *Journal of Strategic Studies* 12, no. 1 (1989): 88–98.

71. Peter Andreas and Kelly M. Greenhill, eds., *Sex, Drugs, and Body Counts: The Politics of Numbers in Global Crime and Conflict* (Ithaca, NY: Cornell University Press, 2010).

72. Joshua Rovner, *Fixing the Facts: National Security and the Politics of Intelligence* (Ithaca, NY: Cornell University Press, 2011).

73. Peter J. Dombrowski and Eugene Gholz, *Buying Military Transformation: Technological Innovation and the Defense Industry* (New York: Columbia University Press, 2006); Gordon Adams, *The Politics of Defense Contracting: The Iron Triangle* (New York: Council on Economic Priorities, 1981).

74. Defense Science Board, *Report of the Defense Science Board Task Force on Defense Software*, November 2000, ES1–2, 3. Similarly, see U.S. Government Accountability Office, *Stronger Management Practices Are Needed to Improve DOD's Software-Intensive Weapon Acquisitions*, GAO-04-393, March 2004; David C. Gompert, Charles L. Barry, and Alf A. Andreassen, *Extending the User's Reach: Responsive Networking for Integrated Military Operations*, National Defense University, Center for Technology and National Security Policy, Defense and Technology Paper No. 24, 2006.

75. Frederick P. Brooks Jr., *The Mythical Man-Month: Essays on Software Engineering* (Reading, MA: Addison-Wesley Longman, 1975), 25.

76. Henry Chesbrough, "Towards a Dynamics of Modularity: A Cyclical Model of Technical Advance," in Prencipe, Davies, and Hobday, *Business of Systems Integration*, 174–200.

77. Scott, *Seeing like a State*, 310–11.

78. Paul Attewell, "Information Technology and the Productivity Paradox," in *Organizational Linkages: Understanding the Productivity Paradox*, National Research Council (Washington, DC: National Academy Press, 1994), 36.

79. Rochlin, *Trapped in the Net*, 213. See also Thomas K. Landauer, *The Trouble with Computers: Usefulness, Usability, and Productivity* (Cambridge, MA: MIT Press, 1996).

80. Matt Spencer, "Brittleness and Bureaucracy: Software as a Material for Science," *Perspectives on Science* 23, no. 4 (2015): 466–84.

81. Harry Collins, *Artificial Experts: Social Knowledge and Intelligent Machines* (Cambridge, MA: MIT Press, 1990); Diana Forsythe, *Studying Those Who Study Us: An Anthropologist in the*

*World of Artificial Intelligence* (Stanford, CA: Stanford University Press, 2001); Meredith Broussard, *Artificial Unintelligence: How Computers Misunderstand the World* (Cambridge, MA: MIT Press, 2018); Brian Cantwell Smith, *The Promise of Artificial Intelligence: Reckoning and Judgment* (Cambridge, MA: MIT Press, 2019).

82. Weick and Roberts, "Collective Mind in Organizations."

83. Friedrich A. Hayek, "The Use of Knowledge in Society," *American Economic Review* 35, no. 4 (1945): 519.

84. On the economics of sharing software and other public goods, see Josh Lerner and Jean Tirole, "Some Simple Economics of Open Source," *Journal of Industrial Economics* 50, no. 2 (2002): 197–234; Yochai Benkler, *The Wealth of Networks: How Social Production Transforms Markets and Freedom* (New Haven, CT: Yale University Press, 2007). Note that Benkler distinguishes networks as a novel alternative to markets and hierarchies based on a narrow reading of markets as requiring priced commodities; however, his arguments for the features of decentralized networked peer production turn on local knowledge and innovation and are essentially Hayekian.

85. Eric S. Raymond, *The Cathedral and the Bazaar*, rev. ed. (Cambridge, MA: O'Reilly, 2001), 30.

86. On the theory of lead user innovation, see Eric von Hippel, *The Sources of Innovation* (New York: Oxford University Press, 1988); Eric von Hippel, *Democratizing Innovation* (Cambridge, MA: MIT Press, 2005); Eric von Hippel, *Free Innovation* (Cambridge, MA: MIT Press, 2017).

87. David L. Boslaugh, *When Computers Went to Sea: The Digitization of the United States Navy* (Los Alamitos, CA: IEEE Computer Society Press, 2003).

88. Norman Friedman, *Seapower and Space: From the Dawn of the Missile Age to Net-Centric Warfare* (Annapolis, MD: Naval Institute Press, 2000), 217–22, 354n.

89. James G. March, "Exploration and Exploitation in Organizational Learning," *Organization Science* 2, no. 1 (1991): 71–87.

90. Mancur Olson, *The Logic of Collective Action: Public Goods and the Theory of Groups*, 2nd ed. (Cambridge, MA: Harvard University Press, 1971), 5–52.

91. Elinor Ostrom and Charlotte Hess, "A Framework for Analyzing the Knowledge Commons," in *Understanding Knowledge as a Commons: From Theory to Practice*, ed. Charlotte Hess and Elinor Ostrom (Cambridge, MA: MIT Press, 2007), 41–81.

92. Widespread convergence on the WGS-84 datum in modern military cartography has largely eliminated this problem, but it is still a risk when using older maps.

93. George A. Akerlof, "The Market for 'Lemons': Quality Uncertainty and the Market Mechanism," *Quarterly Journal of Economics* 84, no. 3 (August 1, 1970): 488–500.

94. Donald MacKenzie, "The Credit Crisis as a Problem in the Sociology of Knowledge," *American Journal of Sociology* 116, no. 6 (May 1, 2011): 1778–841. On the role that finance theory and exchange automation played in precipitating the financial crises of 1987 and 1998, see also Donald MacKenzie, *An Engine, Not a Camera: How Financial Models Shape Markets* (Cambridge, MA: MIT Press, 2006).

95. Robert Jervis, "Reports, Politics, and Intelligence Failures: The Case of Iraq," *Journal of Strategic Studies* 29, no. 1 (2006): 3–52.

96. Robert Jervis, *System Effects: Complexity in Political and Social Life* (Princeton, NJ: Princeton University Press, 1998).

97. It is regrettable that I use this term in the exact opposite sense from the one used by March, "Exploration and Exploitation." I am trying to be consistent with the military usage, which stresses "exploitation" as the pursuit of local opportunities, to include the subversion of enemy networks. This creative search process is more akin to what March calls "exploration."

98. Scott A. Snook, *Friendly Fire: The Accidental Shootdown of U.S. Black Hawks over Northern Iraq* (Princeton, NJ: Princeton University Press, 2000), 179–201. A similar mechanism is used to explain the coevolution of human cognition and culture in Michael Tomasello, *The Cultural Origins of Human Cognition* (Cambridge, MA: Harvard University Press, 1999), 1–12.

99. On path-dependent economic development, see W. Brian Arthur, *Increasing Returns and Path Dependence in the Economy* (Ann Arbor: University of Michigan Press, 1994); North, *Institutions*, 92–106.

100. Langdon Winner, *Autonomous Technology: Technics-Out-of-Control as a Theme in Political Thought* (Cambridge, MA: MIT Press, 1977), 282–94; Beniger, *Control Revolution*, 426–37.

## 3. Strategic and Organizational Conditions for Success

1. Quoted in M. Kirby and R. Capey, "The Air Defence of Great Britain, 1920–1940: An Operational Research Perspective," *Journal of the Operational Research Society* 48, no. 6 (1997): 564.

2. Thomas P. Coakley, *Command and Control for War and Peace* (Washington, DC: National Defense University Press, 1992), 29.

3. Air Chief Marshal Sir Michael Knight notes that "so much has been researched, written, documented and filmed from all possible standpoints, that there can, in truth, have been only the odd gap, the occasional reminiscence, the last 'recollection in tranquility' to complete the picture." Quoted in Henry Probert and Sebastian Cox, *The Battle Re-Thought: A Symposium on the Battle of Britain* (Shrewsbury, UK: Airlife, 1990), 86. The sociotechnical history of radar in particular has only recently been examined, most notably by David Zimmerman, *Britain's Shield: Radar and the Defeat of the Luftwaffe* (Stroud, UK: Sutton, 2001).

4. C. P. Snow, *Science and Government: Appendix* (Cambridge, MA: Harvard University Press, 1962), 7.

5. On RAF neglect of defense, see Williamson Murray, "British and German Air Doctrine between the Wars," *Air University Review*, March–April 1980; Malcolm Smith, *British Air Strategy between the Wars* (New York: Oxford University Press, 1984). On civilian intervention, see Barry Posen, *The Sources of Military Doctrine: France, Britain, and Germany between the World Wars* (Ithaca, NY: Cornell University Press, 1984), 141–78.

6. Stephen Peter Rosen, *Winning the Next War: Innovation and the Modern Military* (Ithaca, NY: Cornell University Press, 1991), 13–18.

7. Probert and Cox, *Battle Re-Thought*, 56.

8. John Ferris, "Fighter Defence before Fighter Command: The Rise of Strategic Air Defence in Great Britain, 1917–1934," *Journal of Military History* 63, no. 4 (1999): 884; Zimmerman, *Britain's Shield*, 111. On LADA, see Neil Young, "British Home Air Defence Planning in the 1920s," *Journal of Strategic Studies* 11, no. 4 (1988): 492–508; Zimmerman, *Britain's Shield*, 2–3. On the acoustic mirrors, see Zimmerman, *Britain's Shield*, 9–27; for technical detail on the mirrors and the notion of "tactile perception," see Raviv Ganchrow, "Perspectives on Sound-Space: The Story of Acoustic Defense," *Leonardo Music Journal*, December 2009, 71–75.

9. Air Marshal Sir Robert Brooke-Popham, quoted in Ferris, "Fighter Defence," 882.

10. P. R. C. Groves, quoted in Brett Holman, *The Next War in the Air: Britain's Fear of the Bomber, 1908–1941* (London: Ashgate, 2014), 58.

11. Brett Holman, author of *Next War in the Air*, estimates the time for bombers to fly 50 miles up the Thames Estuary from their first detection by visual observers to London, and the time for fighters to climb to intercept them, in "The Widening Margin," *Airminded* (blog), May 27, 2008, http://airminded.org/2008/05/27/the-widening-margin. The relative time for intercept decreases from fifteen minutes at ten thousand feet in the 1910s to five minutes at twenty-five thousand feet in the 1940s. Improvements in early warning out to an additional 25 miles from the coast with acoustic mirrors (very optimistic) could increase the gap from five to ten minutes, and out to 120 miles with radar could increase it to forty minutes, more than enough time to position interceptors at altitude.

12. Young, "British Home Air Defence Planning," 495–96, 502; Zimmerman, *Britain's Shield*, 41, 60; Derek Wood and Derek D. Dempster, *The Narrow Margin: The Battle of Britain and the Rise of Air Power, 1930–1940*, 3rd ed. (Washington, DC: Smithsonian Institution Press, 1990),

44, 108; Ferris, "Fighter Defence," 857–63; David Zimmerman, "Information and the Air Defence Revolution, 1917–40," *Journal of Strategic Studies* 27, no. 2 (2004): 370–94.

13. Quoted in Zimmerman, *Britain's Shield*, 79.

14. Zimmerman, *Britain's Shield*, 172.

15. "Radar" is an acronym for "Radio Detection and Ranging." To disguise its true function, the British chose the ambiguous acronym RDF from 1935 to 1943. "Range and Direction Finding," the British term for radar, could be plausibly conflated with "Radio Direction Finding," which was the widely known technique used to vector fighters.

16. £200,000 in 1935 is the equivalent of $18.5 million in 2019. Zimmerman, "Information," 370; Zimmerman, *Britain's Shield*, 45–48, 89–90.

17. Wood and Dempster, *Narrow Margin*, 85, 110–12; Zimmerman, *Britain's Shield*, 109–15, 161; Kirby and Capey, "Air Defence of Great Britain."

18. John Law, "Technology and Heterogeneous Engineering: The Case of Portuguese Expansion," in *The Social Construction of Technological Systems: New Directions in the Sociology and History of Technology*, anniversary edition, ed. Wiebe E. Bijker, Thomas Parke Hughes, and Trevor Pinch (Cambridge, MA: MIT Press, 2012), 105–28.

19. Kirby and Capey, "Air Defence of Great Britain," 560–61; Snow, *Science and Government*; P. M. S. Blackett, "Tizard and the Science of War," *Nature* 185, no. 4714 (1960): 647–53; Zimmerman, *Britain's Shield*, 6–19, 37–77, 136–51, 230–32. For more details on this peculiar relationship, see Thomas Wilson, *Churchill and the Prof* (London: Cassell, 1995).

20. This counterfactual is analyzed by Robert Stanhope-Palmer, *Tank Trap 1940, or No Battle in Britain* (North Devon, UK: Arthur Stockwell, 1976). For the complementary argument that "because of an overwhelmingly powerful British Fleet and [the] decimation of the German Kriegsmarine around Norway, the German planners came to believe that even with air superiority, invasion was not a viable operation," see Anthony J. Cumming, "The Air Marshal versus the Admiral: Air Marshal Sir Hugh Dowding and Admiral of the Fleet Sir Charles Morton Forbes in the Pantheon," *History* 94, no. 2 (314) (2009): 203–28.

21. Zimmerman, *Britain's Shield*, 56. On British fear and enthusiasm for strategic bombing, see Tami Davis Biddle, *Rhetoric and Reality in Air Warfare: The Evolution of British and American Ideas about Strategic Bombing, 1914–1945* (Princeton, NJ: Princeton University Press, 2002), 69–127; Holman, *Next War in the Air*.

22. Quoted in Young, "British Home Air Defence Planning," 493.

23. The RAF principle of 1924 that "the bombing squadrons should be as numerous as possible and the fighters as few as popular opinion and the necessity for defending vital objectives permit" should be taken to mean that the RAF understood that it needed to provide for defense first. See Ferris, "Fighter Defence," 852.

24. Ferris, "Fighter Defence," 883; Young, "British Home Air Defence Planning," 494–98; Wood and Dempster, *Narrow Margin*, 57.

25. Paul Bracken, "Unintended Consequences of Strategic Gaming," *Simulation Gaming* 8, no. 3 (1977): 310–12; Ferris, "Fighter Defence," 848; Kirby and Capey, "Air Defence of Great Britain," 565.

26. Zimmerman, *Britain's Shield*, 134.

27. Quoted in Zimmerman, *Britain's Shield*, 134.

28. Smith, *British Air Strategy*, 198–226.

29. Hugh C. T. Dowding, "The Battle of Britain [Republication of Despatch Submitted to the Secretary of State for Air, 20 August 1941]," *Supplement to the London Gazette*, September 11, 1946, 4545; Richard Overy, *The Battle of Britain: The Myth and the Reality* (New York: W. W. Norton, 2000), 8; Kirby and Capey, "Air Defence of Great Britain," 563–64.

30. Stephen Bungay, *The Most Dangerous Enemy: A History of the Battle of Britain* (London: Aurum, 2000), 353–59; Robin Prior, *When Britain Saved the West: The Story of 1940* (New Haven, CT: Yale University Press, 2015), 281–87.

31. Carl von Clausewitz, *On War*, trans. Michael Howard and Peter Paret (Princeton, NJ: Princeton University Press, 1976), 358. Emphasis in the original.

32. Clausewitz, *On War*, 357.

33. Clausewitz, *On War*, 358.

34. The British surprisingly did not have a dedicated sea rescue service, but the Germans did, so this advantage was partially reversed over the Channel. Eleventh Group commander Keith Park thus instructed his fighters to attack the enemy only within "gliding distance of the coast." Prior, *When Britain Saved the West*, 203.

35. Horst Boog, "The Luftwaffe and the Battle of Britain," in Probert and Cox, *Battle Re-Thought*, 18–32; Bungay, *Most Dangerous Enemy*, 35–54, 120–27, 259–61; Prior, *When Britain Saved the West*, 155–65; Zimmerman, *Britain's Shield*, 175–76.

36. Bungay, *Most Dangerous Enemy*, 186–94.

37. Chief of Luftwaffe Intelligence, Colonel Josef "Beppo" Schmidt, to Oberkommando der Luftwaffe, July 16, 1940, quoted in Wood and Dempster, *Narrow Margin*, 67.

38. Sebastian Cox, "A Comparative Analysis of RAF and Luftwaffe Intelligence in the Battle of Britain, 1940," *Intelligence and National Security* 5, no. 2 (1990): 425–43; Zimmerman, *Britain's Shield*, 204–7; Wood and Dempster, *Narrow Margin*, 68; Probert and Cox, *Battle Re-Thought*, 69.

39. David Pritchard, *The Radar War: Germany's Pioneering Achievement, 1904–45* (Wellingborough, UK: Patrick Stephens, 1989), 206.

40. There is wide agreement that this was a major failure in Luftwaffe strategy. See Wood and Dempster, *Narrow Margin*, 116; Probert and Cox, *Battle Re-Thought*, 58; Boog, "Luftwaffe," 22; Cox, "Comparative Analysis."

41. Prior, *When Britain Saved the West*, 188; Derek Wood, "The Dowding System," in Probert and Cox, *Battle Re-Thought*, 9; Wood and Dempster, *Narrow Margin*, 70; Boog, "Luftwaffe," 23–26; Bungay, *Most Dangerous Enemy*, 44; Overy, *Battle of Britain*, 125–26.

42. Bungay, *Most Dangerous Enemy*, 203–6, 377–78.

43. Jon Agar, *The Government Machine: A Revolutionary History of the Computer* (Cambridge, MA: MIT Press, 2003), 209–16.

44. Air Ministry, *Air Defence Pamphlet Number One: An Outline of Air Defence Organization*, February 1942 (AIR 10/3757).

45. My account of RDF draws on Air Ministry, *Air Defence Pamphlet Number Two: Radiolocation Systems of Raid Reporting*, April 1942 (AIR 10/3758); Zimmerman, *Britain's Shield*; Alan Beyerchen, "From Radio to Radar: Interwar Military Adaptation to Technological Change in Germany, the United Kingdom, and the United States," in *Military Innovation in the Interwar Period*, ed. Williamson Murray and Allan R. Millett (New York: Cambridge University Press, 1996), 265–99; B. T. Neale, "CH: The First Operational Radar," *GEC Journal of Research* 3, no. 2 (1985): 73–83; Dick Barrett, the Radar Pages, accessed October 15, 2019, http://www.radarpages.co.uk; Wood and Dempster, *Narrow Margin*, 88–93.

46. According to B. T. Neale, "Signals at extreme ranges, well below 'noise' level, were detected and tracked. The mechanism by which this was achieved is still not fully understood but believed to be due to an unconscious form of pattern recognition within the noise structure." Neale, "CH," 76.

47. Ilkka Tuomi, "Data Is More Than Knowledge: Implications of the Reversed Knowledge Hierarchy for Knowledge Management and Organizational Memory," *Journal of Management Information Systems* 16, no. 3 (December 1, 1999): 103–17.

48. Wood and Dempster, *Narrow Margin*, 96–98.

49. F. W. Winterbotham, *The Ultra Secret* (New York: Harper and Row, 1974), 61–62, claims that Ultra was critical in the battle, but this claim is disputed; see Edward Thomas, "The Intelligence Aspect," in Probert and Cox, *Battle Re-Thought*, 42–46. The official British history of intelligence in the war states that "for all his major decisions [Dowding] depended on his own strategic judgment, with no direct assistance from the Enigma," quoted in Wood and Dempster, *Narrow Margin*, 77. For an account of the workings of station X, see Christopher Grey and Andrew Sturdy, "A Chaos That Worked: Organizing Bletchley Park," *Public Policy and Administration* 25, no. 1 (2010): 47–66.

50. Aileen Clayton, *The Enemy Is Listening* (London: Hutchinson, 1980), provides a first-person account of the RAF Y service.

51. Cox, "Comparative Analysis"; Wood and Dempster, *Narrow Margin*, 73–77.

52. Zimmerman, *Britain's Shield*, 158; Wood and Dempster, *Narrow Margin*, 108–13.

53. Wood and Dempster, *Narrow Margin*, 110, 116.

54. The RAF considered women to have "higher power of sustained concentration on a limited field of observation devoid of 'entertainment value' . . . finesse in relatively delicate setting of light moving parts . . . general conscientiousness . . . [and] lower average tendency to magnify individual importance by partial disclosure of secrets," and could thus "release men for other duties," according to Zimmerman, *Britain's Shield*, 167. On the gendered aspect of early computing generally, see Jennifer S. Light, "When Computers Were Women," *Technology and Culture* 40, no. 3 (1999): 455–83.

55. Neale, "CH," 81; Wood, "Dowding System"; Wood and Dempster, *Narrow Margin*, 89, 118.

56. Air Ministry, *Air Defence Pamphlet Number Four: Telling and Plotting*, April 1942 (AIR 10/3760).

57. Beyerchen, "From Radio to Radar," 285–86; Zimmerman, *Britain's Shield*, 175, 180.

58. Dowding, "Battle of Britain," 4546.

59. Zimmerman, *Britain's Shield*, 160–68, 183–90; Wood and Dempster, *Narrow Margin*, 118, 258; Neale, "CH," 80.

60. Zimmerman, *Britain's Shield*, 190–91, 202, 209; Zimmerman, "Information," 389–91.

61. Zimmerman, *Britain's Shield*, 173, 180, 185, 201; Beyerchen, "From Radio to Radar," 282.

62. Wood, "Dowding System," 5, notes that they weren't supposed to know about radar but it was obvious they did from the remarks in their logbooks.

63. Wood and Dempster, *Narrow Margin*, 89, 98, 105.

64. Wood and Dempster, *Narrow Margin*, 105; Zimmerman, *Britain's Shield*, 173. A contrarian argument that "a high proportion of the German aircraft intercepted between July and October 1940 were detected by the unsophisticated methods of the [Royal Observer Corps]" is made by Anthony J. Cumming, "Did Radar Win the Battle of Britain?," *Historian* 69, no. 4 (2007): 700. Cumming correctly points out that there was a lot of friction in the radar reporting system, but he does not demonstrate that the observer system fared any better in this regard; indeed, there is reason to believe that the effects of friction there were even more severe.

65. Wood and Dempster, *Narrow Margin*, 118; Dowding, "Battle of Britain," 4547; Agar, *Government Machine*, 214.

66. Neale, "CH," 81, notes that "many ingenious devices, including optical converters, and calculators, too numerous to describe here, were introduced in the latter stages of the war which made the Chain Home system extremely efficient and reliable."

67. Wood and Dempster, *Narrow Margin*, 83, 116; Kirby and Capey, "Air Defence of Great Britain"; Zimmerman, *Britain's Shield*, 50–51, 159, 179.

68. Air Ministry, *Air Defence Pamphlet Number Five: The Operational Control of Fighter Aircraft*, April 1942 (AIR 10/3761); Wood and Dempster, *Narrow Margin*, 119.

69. Air Ministry, *Air Defence Pamphlet Number Four*, 3.

70. Wood and Dempster, *Narrow Margin*, 112; Zimmerman, *Britain's Shield*, 116.

71. Wood and Dempster, *Narrow Margin*, 113, 121.

72. Wood and Dempster, *Narrow Margin*, 75, 79, 121; Zimmerman, *Britain's Shield*, 183; Zimmerman, "Information," 387; Beyerchen, "From Radio to Radar," 285.

73. Dowding, "Battle of Britain," 4546–47.

74. Bungay, *Most Dangerous Enemy*, 239–48; Dowding, "Battle of Britain," 4548; Wood and Dempster, *Narrow Margin*, 226.

75. Wood and Dempster, *Narrow Margin*, 206; Dowding, "Battle of Britain," 4558.

76. Dowding, "Battle of Britain"; Boog, "Luftwaffe," 31.

77. Bungay, *Most Dangerous Enemy*, 97; Prior, *When Britain Saved the West*, 192–93, 196–97, 228, 231. The total of 752 consisted of 469 Hurricanes and 283 Spitfires. Britain also had about 100 Defiant and Blenheim fighters, but these aircraft were so outclassed as to be irrelevant. Disputing the ineffectiveness of the Me-110, cf. Christer Bergstrom, *The Battle of Britain: An Epic Conflict Revisited* (Oxford: Casemate UK, 2015), 59–61.

78. Bungay, *Most Dangerous Enemy*, 371.

79. Prior, *When Britain Saved the West*, 228, 234; Bungay, *Most Dangerous Enemy*, 370; Overy, *Battle of Britain*, 127.

80. Prior, *When Britain Saved the West*, 231, 237.

81. Zimmerman, *Britain's Shield*, 211–24; Dowding, "Battle of Britain," 4560.

82. Dowding, "Battle of Britain," 4560.

83. Details about Chain Home, such as station site selection, writing design specs, manufacturer oversight, and electronics configuration, were all closely held secrets, which exacerbated the calibration problems discussed above. See Zimmerman, *Britain's Shield*, 126.

84. Zimmerman, *Britain's Shield*, 224. On the controversies surrounding Dowding's role in the Night Blitz and his dismissal, see John Ray, *The Battle of Britain, New Perspectives: Behind the Scenes of the Great Air War* (London: Arms and Armour, 1994).

85. Bungay, *Most Dangerous Enemy*, 374; Overy, *Battle of Britain*, 109.

86. Paul Kennedy, *Engineers of Victory: The Problem Solvers Who Turned the Tide in the Second World War* (New York: Random House, 2013), 5–74.

87. Dowding, "Battle of Britain," 4546.

## 4. User Innovation and System Management

1. Michael W. Kometer, *Command in Air War: Centralized versus Decentralized Control of Combat Airpower* (Maxwell, AL: Air University Press, 2007), 155–57. See also Alexander S. Cochran, *Gulf War Air Power Survey*, vol. 1, *Planning and Command and Control* (Washington, DC: Government Printing Office, 1993).

2. Kometer, *Command in Air War*, 161–67.

3. Jumper and Fogleman, quoted in John A. Tirpak, "Find, Fix, Track, Target, Engage, Assess," *Air Force Magazine*, July 2000.

4. Arthur K. Cebrowski and John J. Garstka, "Network-Centric Warfare: Its Origin and Future," *U.S. Naval Institute Proceedings* 124, no. 1 (1998): 28–35.

5. Thomas L. Kelly and John P. Andreasen, "Joint Fires: A BCD Perspective in Operation Iraqi Freedom," *Field Artillery*, December 2003, 20–25.

6. Kometer, *Command in Air War*, 176–80.

7. Jake Thorn, interview with the author, October 13, 2005.

8. Martin Campbell-Kelly, *From Airline Reservations to Sonic the Hedgehog: A History of the Software Industry* (Cambridge, MA: MIT Press, 2003), 202–21.

9. Mike Bartgis, interview with the author, October 28, 2005.

10. Jake Thorn, "Mission Planning History" (slide presentation, 2005); Thorn, interview.

11. Thorn, interview.

12. An economic theory of user innovation is presented by Eric von Hippel, *Democratizing Innovation* (Cambridge, MA: MIT Press, 2005).

13. Mark A. Gillott, *Breaking the Mission Planning Bottleneck: A New Paradigm*, Air University Air Command and Staff College Research Report (Montgomery, AL: Maxwell Air Force Base, April 1998), 4–6; Thorn, interview; Thorn, "Mission Planning History"; Mike Bartgis, "AFMSS MPS and PFPS News," *ANG/DOOM Newsletter*, May 2000, 6.

14. John Pyles, interview with the author, October 28, 2005. After 1986 the system was known as "MSS II," but I will continue to refer to it here simply as MSS.

15. Pyles, interview; John Pyles, email correspondence with the author, October 26, 2005; Gillott, *Breaking the Mission Planning Bottleneck*, 7; Jake Thorn, military biography (typescript provided to the author, 2005).

16. Gillott, *Breaking the Mission Planning Bottleneck*, 4–6.

17. John Pyles, email correspondence, October 26, 2005.

18. Thomas A. Keaney and Eliot A. Cohen, *Gulf War Air Power Survey Summary* (Washington, DC: Government Printing Office, 1993), 177–78; Jane Glaser, "When Reflex Is a Matter of Life and Death," typescript provided to the author, appearing in revised form as Bill Gates, "When Reflex Is a Matter of Life and Death," in *Business @ the Speed of Thought: Succeeding in the Digital Economy* (New York: Warner Books, 1999), 372–86; Gillott, *Breaking the Mission Planning Bottleneck*, 8–12.

19. Pyles, email correspondence, October 26, 2005; Robert Sandford, interview with the author, October 12, 2005.

20. Sandford, interview; Campbell-Kelly, *From Airline Reservations to Sonic the Hedgehog*, 250–51.

21. Pyles, interview; Pyles, email correspondence, October 26, 2005; John Pyles, email correspondence with the author, December 7, 2005. The fresh-out-of-college naiveté expressed itself in other ways too. Someone on the FalconView team included a copy of the video game *Doom* on the CD with version 1.0. *Doom* remained on the CD for three years until a squadron intelligence officer called GTRI and told them he was having trouble getting FalconView on his machines because security managers told him *Doom* wasn't accredited.

22. Pyles, interview; Sandford, interview.

23. Sandford, interview; Frederick P. Brooks Jr., *The Mythical Man-Month: Essays on Software Engineering* (Reading, MA: Addison-Wesley Longman, 1975).

24. Bartgis, interview.

25. Joe Webster, interview with the author, October 25, 2005. Unix purists, who preferred the command line interface, disparaged the Microsoft genre as the "WIMP" interface, short for "windows, icons, menus, and pointers."

26. Quoted in Gillott, *Breaking the Mission Planning Bottleneck*, 15.

27. Webster, interview; Pyles, interview; Sandford, interview.

28. Bartgis, "AFMSS MPS and PFPS News," 6–7; Sandford, interview.

29. Bartgis, "AFMSS MPS and PFPS News," 7; Sandford, interview.

30. Linda D. Kozaryn, "Air Force Releases Brown Crash Investigation Report," *American Forces Information Services*, June 13, 1996, http://www.defenselink.mil/news/Jun1996/n06131996_9606132.html; Chris Bailey, "FalconView: Mission Planning Tools at GTRI" (slide presentation, 2005).

31. This was still only a small portion of the overall $32.6 million AFMSS budget during the same time, FY1999–2000, which covered legacy MPS development and JMPS development. *Department of Defense RDT&E Budget FY2001–2002*, PE#0208006F.

32. Chris Bailey, interview with the author, October 14, 2005; Thorn, interview; Gates, "When Reflex Is a Matter of Life and Death"; Department of Defense, "Air Reserve Software Program Wins Worldwide Recognition," press release, June 3, 1997, http://www.defenselink.mil/releases/1997/b060397_bt285-97.html.

33. Pyles, interview; Webster, interview.

34. Chris Bailey, *Department of Defense Usage of FalconView*, GTRI White Paper, 2005.

35. Hank Davison, interviews with the author, October 18, 2005, and April 3 and 6, 2006; John Bennett, interview with the author, October 19, 2005.

36. Bennett, interview; Davison, interviews.

37. Bartgis, "AFMSS MPS and PFPS News," 1, 7; Paul Hastert, interview with the author, October 8, 2005; Mike Bartgis, email correspondence with the author, October 28, 2005.

38. Shawn Fleming, "Using FalconView for Situational Awareness and Post Mission Reconstruction" (presentation at the Mission Planning User Conference Moving Map Working Group, 1999); Davison, interviews; Hastert, interview; Bartgis, interview; Pamela Bowers, "CrossTalk Honors the 2002 Top 5 Quality Software Projects Finalists," *CrossTalk: The Journal of Defense Software Engineering*, July 2003, http://www.stsc.hill.af.mil/crosstalk/2003/07/top5finalists.html.

39. William A. Hastings, interview with the author, October 28, 2005.

40. Bailey, *Department of Defense Usage of FalconView*; Bailey, interview.

41. Eric Lipton, "3-D Maps from Commercial Satellites Guide G.I.'s in Iraq's Deadliest Urban Mazes," *New York Times*, November 26, 2004; Sean Naylor, *Not a Good Day to Die: The Untold Story of Operation Anaconda* (New York: Penguin, 2005), 162–63; Paul Hastert, "Spiral Development in Wartime" (PowerPoint presentation to National Defense Industrial Association, Tenth Annual Expeditionary Conference, Panama City, FL, October 27, 2005); T. J. Becker, "Not for Pilots Only," *Georgia Tech Research Horizons*, Spring/Summer 2004; Abby Vogel, "Mapping and Imagery Display: FalconView Goes Open Source for Corporate, Environmental, Government and Other Users," *Georgia Tech Research News*, August 12, 2009; Bailey, interview; Hastert, interview; Nathan Thornburgh, "The Invasion of the Chinese Cyberspies (and the Man Who Tried to Stop Them): An Exclusive Look at How the Hackers Called TITAN RAIN are Stealing U.S. Secrets," *Time*, September 5, 2005.

42. Thorn, interview; Naval Air Systems Command and Air Force Electronics Systems Center, "Joint Mission Planning Segment (JMPS) Statement of Objectives" (Microsoft Word document, 1998); Jake Thorn, "JMPS: Industry Day, a Roadmap to Success" (PowerPoint presentation, May 4, 1998).

43. For example, Mark A. Gillott: "This project holds great promise to break with the problems of the past." Gillott, "Breaking the Mission Planning Bottleneck," 24. And Mike Bartgis: "Since JMPS is based on PFPS [FalconView], there is no excuse for not meeting the user needs. . . . The user should be completely happy since the user will be on familiar ground." Bartgis, "AFMSS MPS and PFPS News," 8.

44. Bartgis, "AFMSS MPS and PFPS News," 8; Webster, interview; Thorn, interview.

45. Thorn, interview; Bennett, interview; Pyles, interview.

46. Robert Sandford and Paul Hastert, "PFPS Sustainment and Improvement: Giving the Warfighters What They Want Now" (slide presentation, 2006).

47. Thorn, interview; Bailey, interview; Toni Dineen, PFPS Technical Interchange Meeting minutes, August 26, 2009; Vogel, "Mapping and Imagery Display."

48. Antoine Henri Jomini, *The Art of War*, trans. William Price Craighill and George Henry Mendell (Philadelphia: J. B. Lippincott, 1862; Project Gutenberg, 2004), 69, http://www.gutenberg.org/ebooks/13549.

49. Stephen D. Biddle, *Military Power: Explaining Victory and Defeat in Modern Battle* (Princeton, NJ: Princeton University Press, 2004).

50. Alan Docauer, "Peeling the Onion: Why Centralized Control / Decentralized Execution Works," *Air & Space Power Journal*, April 2014, 25.

51. Cebrowski and Garstka, "Network-Centric Warfare," 29.

52. Office of the Chairman of the Joint Chiefs of Staff, *Joint Vision 2010* (Washington, DC: U.S. Department of Defense, 1996).

53. Spencer Wilkinson, *The Brain of an Army* (London: McMillan, 1891), 97.

54. On interservice tussles over command and control throughout the twentieth century, see C. Kenneth Allard, *Command, Control, and the Common Defense*, rev. ed. (Washington, DC: National Defense University Press, 1996). A case study of the first computerized system for managing joint conventional military operations traces a familiar pattern; see David E. Pearson, *The World Wide Military Command and Control System: Evolution and Effectiveness* (Maxwell, AL: Air University Press, 2000).

55. David N. Lucsko, *The Business of Speed: The Hot Rod Industry in America, 1915–1990* (Baltimore: Johns Hopkins University Press, 2008), 9; Christian Lüthje, Cornelius Herstat, and Eric von Hippel, "User-Innovators and 'Local' Information: The Case of Mountain Biking," *Research Policy* 34, no. 6 (2005): 951–65.

56. David L. Boslaugh, *When Computers Went to Sea: The Digitization of the United States Navy* (Los Alamitos, CA: IEEE Computer Society Press, 2003).

57. Donna J. Peuquet and Todd Bacastow, "Organizational Issues in the Development of Geographical Information Systems: A Case Study of U.S. Army Topographic Information Automation," *International Journal of Geographical Information Science* 5, no. 3 (1991): 303–19.

58. Daryl L. Fullerton, "Back to the Basics: Training Army Artillerymen to Grow Afghan National Army Artillerymen," *Air Land Sea Bulletin* 2008, no. 3 (2008): 4–7.

59. See, inter alia, Michael A. Raymond, "COP: Fusing Battalion Intelligence," *Fires Bulletin*, January–February 2008, 29.

## 5. Irregular Problems and Biased Solutions

1. John Ferris, "Netcentric Warfare, C4ISR and Information Operations: Towards a Revolution in Military Intelligence?," *Intelligence and National Security* 19, no. 2 (2004): 199–225; Stephen Biddle, "Speed Kills? Reassessing the Role of Speed, Precision, and Situation Awareness in the Fall of Saddam," *Journal of Strategic Studies* 30, no. 1 (2007): 3–46.

2. The U.S. Army's official two-volume history of the Iraq war is Joel D. Rayburn, Frank K. Sobchak, Jeanne F. Godfroy, Matthew D. Morton, James S. Powell, and Matthew M. Zais, *The U.S. Army in the Iraq War* (Carlisle Barracks, PA: U.S. Army War College, 2019). On the COIN

reform movement, see David H. Ucko, *The New Counterinsurgency Era: Transforming the U.S. Military for Modern Wars* (Washington, DC: Georgetown University Press, 2009); Michael R. Gordon and Bernard E. Trainor, *The Endgame: The Inside Story of the Struggle for Iraq, from George W. Bush to Barack Obama* (New York: Vintage Books, 2012).

3. General Rupert Smith describes COIN as "war amongst the people" in his book, *The Utility of Force: The Art of War in the Modern World* (New York: Vintage Books, 2008), 6.

4. Quoted in Noah Shachtman, "How Technology Almost Lost the War: In Iraq, the Critical Networks Are Social—Not Electronic," *Wired*, November 27, 2007.

5. Steve Niva, "Disappearing Violence: JSOC and the Pentagon's New Cartography of Networked Warfare," *Security Dialogue* 44, no. 3 (2013): 185–202; Jon R. Lindsay, "Reinventing the Revolution: Technological Visions, Counterinsurgent Criticism, and the Rise of Special Operations," *Journal of Strategic Studies* 36, no. 3 (2013): 422–53.

6. Stanley McChrystal, *My Share of the Task: A Memoir* (New York: Penguin, 2013), 148, 413.

7. Gary Luck and Mike Findlay, *Special Operations and Conventional Force Integration*, U.S. Joint Forces Command, Joint Warfighting Center, Focus Paper no. 5, 2008; *US Special Operations Command Pub 3–33: Conventional Forces and Special Operations Forces Integration and Interoperability Handbook and Checklist* (MacDill Air Force Base, FL: Special Operations Command, 2006).

8. Mark Urban, *Task Force Black: The Explosive True Story of the Secret Special Forces War in Iraq* (London: Little, Brown, 2010), 270–71.

9. See, for example, Raymond T. Odierno, Nichoel E. Brooks, and Francesco P. Mastracchio, "ISR Evolution in the Iraqi Theater," *Joint Forces Quarterly*, no. 50 (2008): 51–55.

10. These "teams" are very different organizations. A Tactical Intelligence Support Team is a detachment of a dozen or so Office of Naval Intelligence analysts, interrogators, case officers, and other support personnel, mostly reservists. A SEAL Team, by contrast, is a battalion-sized unit with six SEAL platoons, each of which has about twenty SEALs. "SEAL" is short for the "sea, air, and land" environments in which navy "frogmen" are trained to operate. A SEAL Team becomes a Naval Special Warfare Squadron (NSWRON) when it deploys "downrange" with all of its "enablers," such as intelligence, logistics, communications, and medical units. The Anbar SOTF (pronounced "so-tif") was built around a California-based NSWRON, but it also had additional SEAL platoons from a Virginia-based SEAL Team, a Kentucky-based Army Special Forces Operational Detachment A (ODA) Team (yet another "team," this one slightly smaller than a SEAL Platoon), and numerous "individual augmentees" from all over the place. The SOTF was a diverse little society in itself.

11. These numbers appear in an award citation I personally received. See the appendix for further discussion of my ethnographic methodology.

12. Risa Brooks, *Shaping Strategy: The Civil-Military Politics of Strategic Assessment* (Princeton, NJ: Princeton University Press, 2008), 226–55; Joshua Rovner, *Fixing the Facts: National Security and the Politics of Intelligence* (Ithaca, NY: Cornell University Press, 2011), 137–84.

13. "U.S. Central Command Slide Compilation, ca. August 15, 2002," tab K, slide 10, National Security Archive, http://nsarchive.gwu.edu/NSAEBB/NSAEBB214/Tab%20K.pdf.

14. Amatzia Baram, "Neo-tribalism in Iraq: Saddam Hussein's Tribal Policies 1991–96," *International Journal of Middle East Studies* 29, no. 1 (1997): 1–31; Hosham Dawood, "The Stateization of the Tribe and the Tribalization of the State: The Case of Iraq," in *Tribes and Power: Nationalism and Ethnicity in the Middle East*, ed. Faleh Jabar and Hosham Dawood (London: Saqi Books, 2003), 110–35.

15. Kevin M. Woods, Michael R. Pease, Mark E. Stout, Williamson Murray, and James G. Lacey, *Iraqi Perspectives Project: A View of Operation Iraqi Freedom from Saddam's Senior Leadership* (Norfolk, VA: U.S. Joint Forces Command, 2006), 89–95.

16. COIN classics include David Galula, *Counterinsurgency Warfare: Theory and Practice* (Westport, CT: Praeger Security International, 1964); Robert Thompson, *Defeating Communist Insurgency: The Lessons of Malaya and Vietnam* (New York: Praeger, 1966); Frank Kitson, *Low Intensity Operations: Subversion, Insurgency and Peacekeeping* (Harrisburg, PA: Stackpole, 1971).

Appropriations from the post-9/11 era include John A. Nagl, *Learning to Eat Soup with a Knife: Counterinsurgency Lessons from Malaya and Vietnam* (Chicago: University of Chicago Press, 2005); David Kilcullen, *Counterinsurgency* (New York: Oxford University Press, 2010).

17. Richard Dawkins, *The Blind Watchmaker: Why the Evidence of Evolution Reveals a Universe without Design* (New York: W. W. Norton, 1996), 191–92.

18. See, for example, Paul Aussaresses, *The Battle of the Casbah: Terrorism and Counterterrorism in Algeria, 1955–1957* (New York: Enigma Books, 2002), 17, 129.

19. Stathis N. Kalyvas, "Review of the New U.S. Army/Marine Corps Counterinsurgency Field Manual," *Perspectives on Politics* 6, no. 2 (2008): 351–53.

20. Stathis N. Kalyvas, *The Logic of Violence in Civil War* (New York: Cambridge University Press, 2006), 173–209.

21. For a striking example of one family acting in multiple, ostensibly incompatible roles, see Brian Feegan, "Entrepreneurs in Votes and Violence: Three Generations of a Peasant Political Family," in *An Anarchy of Families: State and Family in the Philippines*, ed. Alfred W. McCoy (Madison: University of Wisconsin Press, 1993), 33–108.

22. Roger Petersen, *Resistance and Rebellion: Lessons from Eastern Europe* (New York: Cambridge University Press, 2001), 32–79; Güneş Murat Tezcür, "Ordinary People, Extraordinary Risks: Participation in an Ethnic Rebellion," *American Political Science Review* 110, no. 2 (2016): 247–64.

23. Jeremy M. Weinstein, *Inside Rebellion: The Politics of Insurgent Violence* (New York: Cambridge University Press, 2006), 1–25.

24. Charles Tilly, "War Making and State Making as Organized Crime," in *Bringing the State Back In*, ed. Peter Evans, Dietrich Rueschemeyer, and Theda Skocpol (New York: Cambridge University Press, 1985), 169–87; Paul Staniland, "States, Insurgents, and Wartime Political Orders," *Perspectives on Politics* 10, no. 2 (2012): 243–64; Jacqueline L. Hazelton, "The 'Hearts and Minds' Fallacy: Violence, Coercion, and Success in Counterinsurgency Warfare," *International Security* 42, no. 1 (July 1, 2017): 80–113.

25. On Iraqi tribalism, see Baram, "Neo-tribalism in Iraq"; Lin Todd, *Iraq Tribal Study— al-Anbar Governorate: The Albu Fahd Tribe, the Albu Mahal Tribe and the Albu Issa Tribe*, Global Resources Group, Department of Defense, 2006, 3-1–4-9.

26. Gordon and Trainor, *Endgame*, 174–75, 240–63, 379–81.

27. Devlin's memo, "State of the Insurgency in al-Anbar," I MEF G-2, August 17, 2006, is reprinted in Thomas E. Ricks, *The Gamble: General David Petraeus and the American Military Adventure in Iraq, 2006–2008* (New York: Penguin, 2009), 339–43.

28. SIGACTS include direct fire attacks, indirect fire, IED detonations, and IED discoveries reported to or observed by U.S. forces. Figures from U.S. Multi-National Forces West figures in Anthony H. Cordesman, *Violence in Iraq: Reaching an "Irreducible Minimum"* (Washington, DC: Center for Strategic and International Studies, February 25, 2008), 32.

29. See, inter alia, Niel Smith and Sean MacFarland, "Anbar Awakens: The Tipping Point," *Military Review*, no. 2 (March–April 2008): 41–52; Ricks, *Gamble*; Bing West, *The Strongest Tribe: War, Politics, and the Endgame in Iraq* (New York: Random House, 2009); Mark F. Cancian, "What Turned the Tide in Anbar?," *Military Review*, no. 5 (October 2009): 118–121; Jim Michaels, *A Chance in Hell: The Men Who Triumphed over Iraq's Deadliest City and Turned the Tide of War* (New York: St. Martin's, 2010).

30. On U.S. military efforts to institutionalize COIN during this era, see Ucko, *New Counterinsurgency Era*, 103–40. On the tendency to misattribute the reasons for COIN success, see Joshua Rovner, "The Heroes of COIN," *Orbis* 56, no. 2 (2012): 215–32.

31. Ned Parker and Ali Hamdani, "How Violence Is Forging a Brutal Divide in Baghdad," *Times* (UK), December 14, 2006; Nir Rosen, "An Ugly Peace: What Changed in Iraq," *Boston Review*, December 2009.

32. Bob Woodward, "Why Did Violence Plummet? It Wasn't Just the Surge," *Washington Post*, September 8, 2008.

33. Austin Long, "The Anbar Awakening," *Survival* 50, no. 2 (2008): 67–94; John A. McCary, "The Anbar Awakening: An Alliance of Incentives," *Washington Quarterly* 32, no. 1 (2009): 43–59.

34. Stephen Biddle, Jeffrey A. Friedman, and Jacob N. Shapiro, "Testing the Surge: Why Did Violence Decline in Iraq in 2007?," *International Security* 37, no. 1 (2012): 7–40; John Hagan, Joshua Kaiser, Anna Hanson, Jon R. Lindsay, Austin G. Long, Stephen Biddle, Jeffrey A. Friedman, and Jacob N. Shapiro, "Correspondence: Assessing the Synergy Thesis in Iraq," *International Security* 37, no. 4 (2013): 173–98; Jon R. Lindsay and Roger Petersen, *Varieties of Insurgency and Counterinsurgency in Iraq, 2003–2009*, Center for Irregular Warfare and Armed Groups Case Study Series (Newport, RI: Naval War College, 2012).

35. Gary W. Montgomery and Timothy S. McWilliams, eds., *Al-Anbar Awakening: From Insurgency to Counterinsurgency in Iraq, 2004–2009*, vol. 2, *Iraqi Perspectives* (Quantico, VA: Marine Corps University Press, 2009); Marc Lynch, "Explaining the Awakening: Engagement, Publicity, and the Transformation of Iraqi Sunni Political Attitudes," *Security Studies* 20, no. 1 (March 21, 2011): 36–72, https://doi.org/10.1080/09636412.2011.549017; Mark Kukis, ed., *Voices from Iraq: A People's History, 2003–2009* (New York: Columbia University Press, 2011); Sterling Jensen, "Iraqi Narratives of the Anbar Awakening" (PhD diss., King's College London, 2014); Martha L. Cottam, Joe W. Huseby, and Bruno Baltodano, *Confronting Al Qaeda: The Sunni Awakening and American Strategy in Al Anbar* (Lanham, MD: Rowman and Littlefield, 2016).

36. See David Tucker and Christopher J. Lamb, *United States Special Operations Forces* (New York: Columbia University Press, 2007), 17–22. "Black" units operate under cover identities while "white" units operate overtly but discreetly.

37. Austin Long, *The Soul of Armies: Counterinsurgency Doctrine and Military Culture in the US and UK*, Cornell Studies in Security Affairs (Ithaca, NY: Cornell University Press, 2016), 1–34. See also Colin F. Jackson, "Defeat in Victory: Organizational Learning Dysfunction in Counterinsurgency" (PhD diss., Massachusetts Institute of Technology, 2008).

38. Richard H. Shultz Jr., *The Marines Take Anbar: The Four-Year Fight against Al Qaeda* (Annapolis, MD: Naval Institute Press, 2013), 186–89.

39. Quoted in Shultz, *Marines Take Anbar*, 209.

40. This general development was not limited to Iraq and is described in Hy S. Rothstein, *Afghanistan and the Troubled Future of Unconventional Warfare* (Annapolis, MD: Naval Institute Press, 2006), xiv.

41. Sean D. Naylor, "Support Grows for Standing Up an Unconventional Warfare Command," *Armed Forces Journal*, September 2007; Tucker and Lamb, *United States Special Operations Forces*, 179–204.

42. Dick Couch, *The Sheriff of Ramadi: Navy Seals and the Winning of Al-Anbar* (Annapolis, MD: Naval Institute Press, 2013), 166–69.

43. Sometimes this reputation can be helpful. On one operation during our deployment, the Iraqi target was at home watching *Under Siege*, a Steven Seagal action movie about a SEAL who singlehandedly defeats a band of terrorists. When the target realized that actual SEALs were outside, he surrendered without a fight. On Basic Underwater Demolition/SEAL training, see Dick Couch, *The Warrior Elite: The Forging of SEAL Class 228* (New York: Three Rivers, 2001). For a typically hagiographic overview of the NSW community, see Mir Bahmanyar and Chris Osman, *SEALs: The US Navy's Elite Fighting Force* (Oxford: Osprey, 2008).

44. On DEVGRU and the bin Laden raid, see Sean Naylor, *Relentless Strike: The Secret History of Joint Special Operations Command* (New York: St. Martin's, 2015), 42, 391–402.

45. William H. McRaven, *Spec Ops: Case Studies in Special Operations Warfare; Theory and Practice* (New York: Presidio, 1995), 3.

46. Scott R. Gourley, "NAVSPECWARCOM Year in Review," *The Year in Special Operations*, 2008 ed., 59–65, states that the NSW community in 2007 consisted of 6,700 personnel, including 2,300 SEALs and 600 Special Warfare Combat Crewmen, who are also "operators" but enjoy less prestige than SEALs.

47. Scott R. Gourley, "NAVSPECWARCOM Year in Review," *The Year in Special Operations*, 2007 ed., 148–55; Couch, *Sheriff of Ramadi*, 40–41, 219–22.

48. Michael T. Flynn, Rich Juergens, and Thomas L. Cantrell, "Employing ISR: SOF Best Practices," *Joint Forces Quarterly*, no. 50 (2008): 56–61; Christopher J. Lamb and Evan Munsing, *Secret Weapon: High-Value Target Teams as an Organizational Innovation*, Strategic Perspectives (Washington, DC: National Defense University, March 2011).

49. A. M. Turing, "On Computable Numbers, with an Application to the Entscheidungsproblem," *Proceedings of the London Mathematical Society* s2–42, no. 1 (1937): 230–65, https://doi.org/10.1112/plms/s2-42.1.230.

50. Minna Räsänen and James M. Nyce, "The Raw Is Cooked: Data in Intelligence Practice," *Science, Technology, & Human Values* 38, no. 5 (September 1, 2013): 655–77.

51. Thom Shanker and Matt Richtel, "In New Military, Data Overload Can Be Deadly," *New York Times*, January 16, 2011; Christopher Drew, "Military Is Awash in Data from Drones," *New York Times*, January 10, 2010.

52. On military HUMINT in Iraq, see Charles W. Innocenti, Ted L. Martens, and Daniel E. Soller, "Direct Support HUMINT in Operation Iraqi Freedom," *Military Review*, no. 3 (May–June 2009): 48–56.

53. For a more admiring portrayal of HUMINT in NSW Task Unit Ramadi, cf. Couch, *Sheriff of Ramadi*, 40–41, 219–22.

54. In this regard the SOTF recapitulated the general findings of Abigail J. Sellen and Richard H. R. Harper, *The Myth of the Paperless Office* (Cambridge, MA: MIT Press, 2002).

55. This is an example of by-products of report generation having more pragmatic importance than the report itself, as discussed in Martha S. Feldman, *Order without Design: Information Production and Policy Making* (Stanford, CA: Stanford University Press, 1989), 97–105.

56. SOF accidentally bombed an election convoy in Afghanistan when they confused an insurgent alias with a real personage, whom they tracked via SIGINT. Kate Clark, *The Takhar Attack: Targeted Killings and the Parallel Worlds of US Intelligence and Afghanistan* (Kabul: Afghanistan Analysts Network, May 2011).

57. This is an example of how prior plans, shared among distributed actors, structure real-time situated performances to produce predictable interactions, as described by Lucy A. Suchman, *Human-Machine Reconfigurations: Plans and Situated Actions*, 2nd ed. (New York: Cambridge University Press, 2007), 51–84.

58. Intelligence Science Board, *Educing Information: Interrogation; Science and Art* (Washington, DC: National Defense Intelligence College Press, 2006), 141–234.

59. Derek Jones, *Understanding the Form, Function, and Logic of Clandestine Cellular Networks: The First Step in Effective Counternetwork Operations* (Fort Leavenworth, KS: School of Advanced Military Studies, 2009).

60. Carter Malkasian, *Illusions of Victory: The Anbar Awakening and the Rise of the Islamic State* (New York: Oxford University Press, 2017).

61. "Transcript: DoD News Briefing—Secretary Rumsfeld and Gen. Myers," Department of Defense, February 12, 2002, http://archive.defense.gov/Transcripts/Transcript.aspx?TranscriptID=2636.

62. Naylor, *Relentless Strike*, 309.

63. Urban, *Task Force Black*, 82.

64. Naylor, *Relentless Strike*, 253, 256.

65. Urban, *Task Force Black*, 82.

66. Urban, *Task Force Black*, 84.

67. Naylor, *Relentless Strike*, 257–65.

68. Gordon and Trainor, *Endgame*, 205.

69. Cora Currier and Peter Maass, "Firing Blind: Flawed Intelligence and the Limits of Drone Technology," Intercept, October 15, 2015, https://theintercept.com/drone-papers/firing-blind/.

70. Quoted in Sean D. Naylor, "Petraeus Sounds Off on Afghanistan: General Says Killing or Capturing bin Laden Not Enough in Battle against Al-Qaida," *Army Times*, October 21, 2008.

71. Quoted in Woodward, "Why Did Violence Plummet?"

72. *The National Security Agency: Missions, Authorities, Oversight and Partnerships*, release no. PA-026–18 (Fort Meade, MD: National Security Agency, August 9, 2013), 6.

73. Gordon and Trainor, *Endgame*, 208.

74. Naylor, *Relentless Strike*, 291–310.

75. Naylor, *Relentless Strike*, 304.

76. Austin Long, "Whack-a-Mole or Coup de Grace? Institutionalization and Leadership Targeting in Iraq and Afghanistan," *Security Studies* 23, no. 3 (2014): 471–512.

77. Dana Priest and William M. Arkin, "Top Secret America: A Look at the Military's Joint Special Operations Command," *Washington Post*, September 2, 2011.

78. James A. Russell, *Innovation, Transformation, and War: Counterinsurgency Operations in Anbar and Ninewa Provinces, Iraq, 2005–2007* (Stanford, CA: Stanford University Press, 2011), 69–73.

79. Russell, *Innovation, Transformation, and War*, 88.

80. Shultz, *Marines Take Anbar*, 207.

81. Monte Morin, "Surveying the Situation in Volatile Ramadi," *Stars and Stripes*, August 16, 2006; Russell, *Innovation, Transformation, and War*, 119–22.

## 6. Increasing Complexity and Uneven Results

1. David Deptula, "Drones Best Weapons We've Got for Accuracy, Control, Oversight: Critics Don't Get It," *Breaking Defense*, February 15, 2013.

2. See, inter alia, James Cavallaro, Stephan Sonnenberg, and Sarah Knuckey, *Living under Drones: Death, Injury and Trauma to Civilians from US Drone Practices in Pakistan* (Stanford, CA: International Human Rights and Conflict Resolution Clinic, Stanford Law School; New York: Global Justice Clinic, NYU School of Law, 2012); Audrey Kurth Cronin, "Why Drones Fail: When Tactics Drive Strategy," *Foreign Affairs* 92, no. 4 (2013): 44–54; Sarah Kreps and John Kaag, *Drone Warfare* (Cambridge: Polity, 2014); Grégoire Chamayou, *A Theory of the Drone*, trans. Janet Lloyd (New York: New Press, 2015). For a variety of contending perspectives, pro and con, see Bradley Jay Strawser, ed., *Killing by Remote Control: The Ethics of an Unmanned Military* (New York: Oxford University Press, 2013); Matthew Evangelista and Henry Shue, eds., *The American Way of Bombing: Changing Ethical and Legal Norms, from Flying Fortresses to Drones* (Ithaca, NY: Cornell University Press, 2014); David Cortright, Rachel Fairhurst, and Kristen Wall, *Drones and the Future of Armed Conflict: Ethical, Legal, and Strategic Implications* (Chicago: University of Chicago Press, 2015).

3. See, inter alia, Bradley Jay Strawser, "Moral Predators: The Duty to Employ Uninhabited Aerial Vehicles," *Journal of Military Ethics* 9, no. 4 (December 1, 2010): 342–68; Daniel L. Byman, "Why Drones Work: The Case for Washington's Weapon of Choice," *Foreign Affairs*, August 2013; Charles J. Dunlap, "Clever or Clueless? Observations about Bombing Norm Debates," in Evangelista and Shue, *American Way of Bombing*, 109–30; Patrick B. Johnson, "Security Implications of Drones in Warfare," in Cortright, Fairhurst, and Wall, *Drones and the Future of Armed Conflict*, 121–41.

4. On the services' uneasy historical relationship with drones, see Thomas P. Ehrhard, "Unmanned Aerial Vehicles in the United States Armed Services: A Comparative Study of Weapon System Innovation" (PhD diss., Johns Hopkins University, 2000). On the continuity of themes, cf. Paul Dickson, *The Electronic Battlefield* (Bloomington: Indiana University Press, 1976); Peter W. Singer, *Wired for War: The Robotics Revolution and Conflict in the Twenty-First Century* (New York: Penguin, 2009).

5. Richard Whittle, *Predator: The Secret Origins of the Drone Revolution* (New York: Henry Holt, 2014), 7–18, 41–65.

6. Michael R. Thirtle, Robert V. Johnson, and John L. Birkler, *The Predator ACTD: A Case Study for Transition Planning to the Formal Acquisition Process* (Santa Monica, CA: RAND Corporation, 1997); Whittle, *Predator*, 79–106.

7. Whittle, *Predator*, 108–25; Andrew Cockburn, *Kill Chain: The Rise of the High-Tech Assassins* (New York: Henry Holt, 2015), 52–60.

8. Bill Grimes, *The History of Big Safari* (Bloomington, IN: Archway, 2014), 329–35; Whittle, *Predator*, 120–69.

9. Grimes, *History of Big Safari*, 332–33; Whittle, *Predator*, 150–228.

10. Bob Woodward, *Bush at War* (New York: Simon and Schuster, 2002), 76, 101; Whittle, *Predator*, 258.

11. Whittle, *Predator*, 247–300; William M. Arkin, *Unmanned: Drones, Data, and the Illusion of Perfect Warfare* (New York: Little, Brown, 2015), 90–98; Timothy M. Cullen, "The MQ-9 Reaper Remotely Piloted Aircraft: Humans and Machines in Action" (PhD diss., Massachusetts Institute of Technology, 2011), 246, 259–60; Cockburn, *Kill Chain*, 118–20; Grimes, *History of Big Safari*, 333–37.

12. Cullen, "MQ-9 Reaper," 197–265.

13. Cullen, "MQ-9 Reaper," 205–65; "Air Force Distributed Common Ground System Fact Sheet," U.S. Air Force, October 13, 2015, http://www.af.mil/AboutUs/FactSheets/Display/tabid/224/Article/104525/air-force-distributed-common-ground-system.aspx.

14. Paul Hastert, "Spiral Development in Wartime" (PowerPoint presentation to National Defense Industrial Association, Tenth Annual Expeditionary Conference, Panama City, FL, October 27, 2005), 38.

15. Cullen, "MQ-9 Reaper." Memoirs by aircrew are broadly consistent with, if less sophisticated than, Cullen's account—e.g., Matt J. Martin and Charles W. Sasser, *Predator: The Remote-Control Air War over Iraq and Afghanistan: A Pilot's Story* (Minneapolis: Zenith, 2010). Studies of teleoperation, generally including underwater robots and spacecraft, also corroborate Cullen's findings—e.g., David A. Mindell, *Our Robots, Ourselves: Robotics and the Myths of Autonomy* (New York: Viking, 2015).

16. Cullen, "MQ-9 Reaper," 273.

17. Cullen, "MQ-9 Reaper," 272.

18. Lambèr Royakkers and Rinie van Est, "The Cubicle Warrior: The Marionette of Digitalized Warfare," *Ethics and Information Technology* 12, no. 3 (2010): 289–96; Jamie Allinson, "The Necropolitics of Drones," *International Political Sociology* 9, no. 2 (2015): 113–27; Chamayou, *Theory of the Drone*; Cora Sol Goldstein, "Drones, Honor, and War," *Military Review*, December 2015; Lauren Wilcox, "Drone Warfare and the Making of Bodies Out of Place," *Critical Studies on Security* 3, no. 1 (2015): 127–31.

19. Dave Grossman, *On Killing: The Psychological Cost of Learning to Kill in War and Society*, rev. ed. (New York: Little, Brown, 2009).

20. Philip Alston, "Addendum: Study on Targeted Killings," in *Report of the Special Rapporteur on Extrajudicial, Summary or Arbitrary Executions* (New York: Office of the United Nations High Commissioner for Human Rights, 2010), 25. See also Royakkers and Est, "Cubicle Warrior"; Goldstein, "Drones, Honor, and War."

21. Jeff Schogol and Markeshia Ricks, "Demand Grows for UAV Pilots, Sensor Operators," *Air Force Times*, April 21, 2012.

22. Quoted in Whittle, *Predator*, 274. See also Scott Fitzsimmons and Karina Sangha, "Killing in High Definition: Combat Stress among Operators of Remotely Piloted Aircraft" (paper presentation, Eighty-Fifth Annual Conference of the Canadian Political Science Association, Victoria, 2013).

23. Cullen, "MQ-9 Reaper," 220n71, 235–36, 257, 277; Grimes, *History of Big Safari*, 335; Craig Whitlock, "When Drones Fall from the Sky," *Washington Post*, June 20, 2014.

24. On the Taliban's recovery during this period following its near defeat in 2002, see Antonio Giustozzi, *Koran, Kalashnikov, and Laptop: The Neo-Taliban Insurgency in Afghanistan* (New York: Columbia University Press, 2008).

25. Timothy P. McHale, *Executive Summary for AR 15–6 Investigation, 21 February 2010 CIVCAS Incident in Uruzgan Province*, U.S. Forces-Afghanistan, April 13, 2010.

26. Thom Shanker and Matt Richtel, "In New Military, Data Overload Can Be Deadly," *New York Times*, January 16, 2011.

27. *Classified Summary of the Command Investigation into the Friendly Fire Incident on 6 April 2011 in Regional Command-Southwest (RC-SW)*, International Security Assistance Force, April 25, 2011 (Freedom of Information Act).

28. Charles Perrow, *Normal Accidents: Living with High-Risk Technologies*, 2nd ed. (Princeton, NJ: Princeton University Press, 1999).

29. *Investigation Report of the Fixed Wing Close Air Support Airstrike in the Vicinity of Arghandab, Afghanistan June 2014*, U.S. Central Command, August 5, 2014 (Freedom of Information Act).

30. Julia Macdonald and Jacquelyn Schneider, "Battlefield Responses to New Technologies: Views from the Ground on Unmanned Aircraft," *Security Studies* 28, no. 2 (2019): 216–49.

31. Jamie Crawford, Barbara Starr, and Jason Hanna, "Human Error Led to MSF Strike, General Says," CNN, November 25, 2015, https://www.cnn.com/2015/11/25/politics/afghanistan-kunduz-doctors-without-borders-hospital/index.html; Matthieu Aikins, "Doctors with Enemies: Did Afghan Forces Target the M.S.F. Hospital?," *New York Times*, May 17, 2016.

32. Jason Lyall, Graeme Blair, and Kosuke Imai, "Explaining Support for Combatants during Wartime: A Survey Experiment in Afghanistan," *American Political Science Review* 107, no. 4 (November 2013): 679–705.

33. Michael T. Flynn, Matt Pottinger, and Paul D. Batchelor, *Fixing Intel: A Blueprint for Making Intelligence Relevant in Afghanistan* (Washington, DC: Center for a New American Security, January 2010), 7.

34. Kate Clark, *The Takhar Attack: Targeted Killings and the Parallel Worlds of US Intelligence and Afghanistan* (Kabul: Afghanistan Analysts Network, May 2011); Cockburn, *Kill Chain*, 191–200.

35. David Zucchino, "U.S. Report Faults Air Force Drone Crew, Ground Commanders in Afghan Civilian Deaths," *Los Angeles Times*, May 29, 2010.

36. Alex S. Wilner, "Targeted Killings in Afghanistan: Measuring Coercion and Deterrence in Counterterrorism and Counterinsurgency," *Studies in Conflict & Terrorism* 33, no. 4 (March 15, 2010): 307–29.

37. David E. Sanger, *Confront and Conceal: Obama's Secret Wars and Surprising Use of American Power* (New York: Crown, 2012), 68–113.

38. On September 18, 2001, Congress passed Public Law 107–40, which states, "The President is authorized to use all necessary and appropriate force against those nations, organizations, or persons he determines planned, authorized, committed, or aided the terrorist attacks that occurred on September 11, 2001, or harbored such organizations or persons, in order to prevent any future acts of international terrorism against the United States by such nations, organizations or persons."

39. James Risen and Marc Santora, "Man Believed Slain in Yemen Tied by U.S. to Buffalo Cell," *New York Times*, November 10, 2002.

40. Scott Shane, *Objective Troy: A Terrorist, a President, and the Rise of the Drone* (New York: Crown, 2015), 205–9, 245; *Yemen: Reported US Covert Actions 2001–2011* (London: Bureau of Investigative Journalism, March 29, 2012).

41. Shane, *Objective Troy*, 224.

42. Office of Legal Counsel, "Memorandum for the Attorney General re: Applicability of Federal Criminal Laws and the Constitution to Contemplated Lethal Operations against Shaykh Anwar al-Aulaqi," U.S. Department of Justice, July 16, 2010.

43. Shane, *Objective Troy*, 230; Jo Becker and Scott Shane, "Secret 'Kill List' Tests Obama's Principles," *New York Times*, May 29, 2012.

44. Greg Miller, "Under Obama, an Emerging Global Apparatus for Drone Killing," *Washington Post*, December 27, 2011.

45. "Yemen Strike Kills Mediator, Tribesmen Hit Pipeline," Reuters, May 25, 2010; Shane, *Objective Troy*, 227.

46. Shane, *Objective Troy*, 279–96; Mark Mazzetti, Charlie Savage, and Scott Shane, "How a U.S. Citizen Came to Be in America's Cross Hairs," *New York Times*, March 9, 2013. A daughter of al-Awlaki was also inadvertently killed in a SOF raid in Yemen in January 2017; Haroro J. Ingram and Craig Whiteside, "The Yemen Raid and the Ghost of Anwar al-Awlaki," *Atlantic*, February 9, 2017.

47. Richard Engel and Robert Windrem, "CIA Didn't Always Know Who It Was Killing in Drone Strikes, Classified Documents Show," NBC News, June 5, 2013, http://investigations.nbcnews.com/_news/2013/06/05/18781930-cia-didnt-always-know-who-it-was-killing-in-drone-strikes-classified-documents-show. JSOC received authorization in April 2012 to conduct terrorist attack disruption strikes in Yemen, reportedly with more stringent criteria than CIA signature strikes; Peter Bergen and Jennifer Rowland, "Drone Wars," *Washington Quarterly* 36, no. 3 (August 1, 2013): 7–26; Becker and Shane, "Secret 'Kill List.'" One congressional aide argued that CIA targeting was in fact more careful than JSOC: "The amount of time

that goes into a strike package at CIA is longer and more detailed than a strike package put together [at the Defense Department]. . . . Their standards of who is a combatant are different. Standards for collateral damage are different." Quoted in Ken Dilanian, "Debate Grows over Proposal for CIA to Turn Over Drones to Pentagon," *Los Angeles Times*, May 11, 2014.

48. Michael V. Hayden, "To Keep America Safe, Embrace Drone Warfare," *New York Times*, February 19, 2016.

49. Peter Baker, "Obama Apologizes after Drone Kills American and Italian Held by Al Qaeda," *New York Times*, April 23, 2015; Mark Mazzetti and Eric Schmitt, "First Evidence of a Blunder in Drone Strike: 2 Extra Bodies," *New York Times*, April 23, 2015; Eric Schmitt, "Adam Gadahn Was Propagandist for Al Qaeda Who Sold Terror in English," *New York Times*, April 23, 2015.

50. Baker, "Obama Apologizes"; Scott Shane and Eric Schmitt, "One Drone Victim's Trail from Raleigh to Pakistan," *New York Times*, May 22, 2013.

51. "Remarks by the President at the National Defense University," White House, Office of the Press Secretary, May 23, 2013, https://www.whitehouse.gov/the-press-office/2013/05/23/remarks-president-national-defense-university.

52. Klem Ryan, "What's Wrong with Drones? The Battlefield in International Humanitarian Law," in Evangelista and Shue, *American Way of Bombing*, 191–206.

53. John Mueller and Mark G. Stewart, "The Terrorism Delusion: America's Overwrought Response to September 11," *International Security* 37, no. 1 (July 1, 2012): 81–110; Benjamin H. Friedman and Jim Harper, *Terrorizing Ourselves: Why U.S. Counterterrorism Policy Is Failing and How to Fix It* (Washington, DC: Cato Institute, 2010).

54. Jennifer M. Welsh, "The Morality of 'Drone Warfare,'" in Cortright, Fairhurst, and Wall, *Drones and the Future of Armed Conflict*, 24–45.

55. Strawser, "Moral Predators."

56. Daniel Brunstetter and Megan Braun, "The Implications of Drones on the Just War Tradition," *Ethics & International Affairs* 25, no. 3 (2011): 337–58; Heather Roff, "Lethal Autonomous Weapons and Jus Ad Bellum Proportionality," *Case Western Reserve Journal of International Law* 47, no. 1 (2015): 37–52.

57. Hugh Gusterson, "Toward an Anthropology of Drones: Remaking Space, Time, and Valor in Combat," in Evangelista and Shue, *American Way of Bombing*, 191–206.

58. White House, "Procedures for Approving Direct Action against Terrorist Targets Located outside the United States and Areas of Active Hostilities," Presidential Policy Guidance, May 22, 2013. Prior reporting on the process includes Greg Miller, "Plan for Hunting Terrorists Signals U.S. Intends to Keep Adding Names to Kill Lists," *Washington Post*, October 23, 2012; Becker and Shane, "Secret 'Kill List'"; Greg Miller, Julie Tate, and Barton Gellman, "Documents Reveal NSA's Extensive Involvement in Targeted Killing Program," *Washington Post*, October 16, 2013. The most detailed reconstruction of U.S. targeted killing policy to date appears in Gregory S. McNeal, "Targeted Killing and Accountability," *Georgetown Law Journal* 102 (March 2014): 681–794.

59. John O. Brennan, "The Efficacy and Ethics of U.S. Counterterrorism Strategy" (speech at the Wilson Center, Washington, DC, April 30, 2012, https://www.wilsoncenter.org/event/the-efficacy-and-ethics-us-counterterrorism-strategy).

60. Becker and Shane, "Secret 'Kill List.'"

61. According to White House, "Procedures for Approving Direct Action," "Capture operations offer the best opportunity for meaningful intelligence gain from counterterrorism (CT) operations and the mitigation and disruption of terrorist threats. Consequently, the United States prioritizes, as a matter of policy, the capture of terrorist suspects as a preferred option over lethal action and will therefore require a feasibility assessment of capture options as a component of any proposal for lethal action."

62. Adam Goldman and Tim Craig, "American Citizen Linked to Al-Qaeda Is Captured, Flown Secretly to U.S.," *Washington Post*, April 2, 2015; Mark Mazzetti and Eric Schmitt, "Terrorism Case Renews Debate over Drone Hits," *New York Times*, April 12, 2015.

63. White House, "Procedures for Approving Direct Action"; McNeal, "Targeted Killing and Accountability," 730–59. See also Adam Klein, "A Response to the 'Drone Papers': AUMF

Targeting Is a Deliberate Process with Robust Political Accountability," *Lawfare* (blog), October 15, 2015, https://www.lawfareblog.com/response-drone-papers-aumf-targeting-deli berate-process-robust-political-accountability.

64. Philip Alston, "The CIA and Targeted Killings beyond Borders," *Harvard National Security Journal* 2 (February 2011): 283. See also Joshua Foust, "Unaccountable Killing Machines: The True Cost of U.S. Drones," *Atlantic*, December 30, 2011.

65. McNeal, "Targeted Killing and Accountability," 758–85.

66. Daniel Byman, "Understanding the Islamic State—A Review Essay," *International Security* 40, no. 4 (April 1, 2016): 127–65.

67. Eric Schmitt and Matthew Rosenberg, "C.I.A. Wants Authority to Conduct Drone Strikes in Afghanistan for the First Time," *New York Times*, September 15, 2017.

68. Charlie Savage and Eric Schmitt, "Trump Poised to Drop Some Limits on Drone Strikes and Commando Raids," *New York Times*, September 21, 2017.

69. W. J. Hennigan and Brian Bennett, "Trump Doesn't Micromanage the Military—but That Could Backfire," *Los Angeles Times*, July 7, 2017.

70. Savage and Schmitt, "Trump Poised to Drop Some Limits."

71. "Drone Wars: The Full Data," Bureau of Investigative Journalism, January 1, 2017, https://www.thebureauinvestigates.com/stories/2017-01-01/drone-wars-the-full-data.

72. On the controversies over drone casualty data, see Chris Woods, "Understanding the Gulf between Public and US Government Estimates of Civilian Casualties in Covert Drone Strikes," in Cortright, Fairhurst, and Wall, *Drones and the Future of Armed Conflict*, 180–98; Becker and Shane, "Secret 'Kill List.'"

73. Nelly Lahoud, Stuart Caudill, Liam Collins, Gabriel Koehler-Derrick, Don Rassler, and Muhammad al-Ubaydi, *Letters from Abbottabad: Bin Ladin Sidelined?* (West Point, NY: Combating Terrorism Center, May 2012), sec. SOCOM-2012–0000015-HT.

74. Rukimini Callimachi, "Al-Qaida Tipsheet on Avoiding Drones Found in Mali," Associated Press, February 21, 2013. The AP found, authenticated, and translated the document quoted here.

75. Callimachi, "Al-Qaida Tipsheet."

76. Patrick B. Johnston and Anoop K. Sarbahi, "The Impact of US Drone Strikes on Terrorism in Pakistan," *International Studies Quarterly* 60, no. 2 (2016): 203–19.

77. Erik Gartzke and James Igoe Walsh, "The Drawbacks of Drones: The Effects of UAVs on Militant Violence in Pakistan" (unpublished manuscript, January 2020), typescript; Erik Gartzke, "Blood and Robots: How Remotely Piloted Vehicles and Related Technologies Affect the Politics of Violence," *Journal of Strategic Studies* (forthcoming).

78. Stephanie Carvin, "The Trouble with Targeted Killing," *Security Studies* 21, no. 3 (2012): 529–55; Eric van Um and Daniela Pisoiu, "Dealing with Uncertainty: The Illusion of Knowledge in the Study of Counterterrorism Effectiveness," *Critical Studies on Terrorism* 8, no. 2 (2015): 229–45.

79. Michael C. Horowitz, Sarah E. Kreps, and Matthew Fuhrmann, "Separating Fact from Fiction in the Debate over Drone Proliferation," *International Security* 41, no. 2 (October 1, 2016): 7–42.

80. Patrick B. Johnston, "Does Decapitation Work? Assessing the Effectiveness of Leadership Targeting in Counterinsurgency Campaigns," *International Security* 36, no. 4 (Spring 2012): 47–79; Bryan C. Price, "Targeting Top Terrorists: How Leadership Decapitation Contributes to Counterterrorism," *International Security* 36, no. 4 (2012): 9–46.

81. Javier Jordan, "The Effectiveness of the Drone Campaign against Al Qaeda Central: A Case Study," *Journal of Strategic Studies* 37, no. 1 (2014): 4–29; Declan Walsh, "Drone Strikes on Al Qaeda Are Said to Take Toll on Leadership in Pakistan," *New York Times*, April 24, 2015.

82. Jenna Jordan, "Attacking the Leader, Missing the Mark," *International Security* 38, no. 4 (2014): 7–38; Austin Long, "Whack-a-Mole or Coup de Grace? Institutionalization and Leadership Targeting in Iraq and Afghanistan," *Security Studies* 23, no. 3 (2014): 471–512; Keith Patrick Dear, "Beheading the Hydra? Does Killing Terrorist or Insurgent Leaders Work?," *Defence Studies* 13, no. 3 (2013): 293–337.

83. Cronin, "Why Drones Fail," 52.

84. Aqil Shah, "Do U.S. Drone Strikes Cause Blowback? Evidence from Pakistan and Beyond," *International Security* 42, no. 4 (May 1, 2018): 6.

85. On the increasing legalization of combat, see Charles J. Dunlap Jr., "Lawfare Today: A Perspective," *Yale Journal of International Affairs* 3, no. 1 (2008): 146–53.

86. Ernesto Londoño, "Pentagon Cancels Divisive Distinguished Warfare Medal for Cyber Ops, Drone Strikes," *Washington Post*, April 15, 2013.

87. Phil Haun and Colin Jackson, "Breaker of Armies: Air Power in the Easter Offensive and the Myth of Linebacker I and II in the Vietnam War," *International Security* 40, no. 3 (January 1, 2016): 139–78.

88. Janina Dill, *Legitimate Targets? Social Construction, International Law and US Bombing* (New York: Cambridge University Press, 2014), 299–310.

89. Charles Garraway, "The Law Applies, but Which Law? A Consumer Guide to the Laws of War," in Evangelista and Shue, *American Way of Bombing*, 87–105.

90. Ryder McKeown, "Legal Asymmetries in Asymmetric War," *Review of International Studies* 41, no. 1 (2015): 117–38.

## 7. Practical Implications of Information Practice

1. Atul Gawande, "Why Doctors Hate Their Computers," *New Yorker*, November 12, 2018.

2. Carl von Clausewitz, *On War*, trans. Michael Howard and Peter Paret (Princeton, NJ: Princeton University Press, 1976), 153. Emphasis in the original.

3. Clausewitz, *On War*, 140. Emphasis in the original.

4. Jon Tetsuro Sumida, *Decoding Clausewitz: A New Approach to "On War"* (Lawrence: University Press of Kansas, 2008), 135–53.

5. U.S. Marine Corps, *MCDP 1: Warfighting* (Washington, DC: U.S. Government Printing Office, 1997); Stephen D. Biddle, *Military Power: Explaining Victory and Defeat in Modern Battle* (Princeton, NJ: Princeton University Press, 2004); Meir Finkel, *On Flexibility: Recovery from Technological and Doctrinal Surprise on the Battlefield*, Stanford Security Studies (Stanford, CA: Stanford University Press, 2011); Caitlin Talmadge, *The Dictator's Army: Battlefield Effectiveness in Authoritarian Regimes* (Ithaca, NY: Cornell University Press, 2015); Ryan Grauer, *Commanding Military Power* (New York: Cambridge University Press, 2016).

6. Some artificial intelligence methods, such as genetic programming, do enable a program to modify its own code. There are important reasons to be skeptical, however, that current approaches to AI will ever manage to generate autonomous intentionality, rather than just acting on the goals and values of human designers and users. Deficiencies in common sense and creative judgment have posed major problems for AI research for decades. Classic critiques include Harry Collins, *Artificial Experts: Social Knowledge and Intelligent Machines* (Cambridge, MA: MIT Press, 1990); Hubert L. Dreyfus, *What Computers Still Can't Do: A Critique of Artificial Reason*, rev. ed. (Cambridge, MA: MIT Press, 1992). The new wave of AI benefits from far greater computational capacity, but is still nowhere close to realizing pragmatic intentionality: Meredith Broussard, *Artificial Unintelligence: How Computers Misunderstand the World* (Cambridge, MA: MIT Press, 2018); Brian Cantwell Smith, *The Promise of Artificial Intelligence: Reckoning and Judgment* (Cambridge, MA: MIT Press, 2019).

7. *Dwight D. Eisenhower: Containing the Public Messages, Speeches, and Statements of the President, January 1 to December 31, 1957* (Washington, DC: Government Printing Office, 1958), 818.

8. Clausewitz, *On War*, 119.

9. Clausewitz, *On War*, 88.

10. In a classic work on leadership, Chester Barnard argues that the primary role of the leader is to debug the relationship between means and ends. See Chester I. Barnard, *The Functions of the Executive*, 30th anniversary ed. (Cambridge, MA: Harvard University Press, 1968). On the negotiation of ends and means throughout the military chain of command, see Eliot A. Cohen, *Supreme Command: Soldiers, Statesmen, and Leadership in Wartime* (New York: Simon and Schuster, 2002); Anthony King, *Command: The Twenty-First-Century General* (New York: Cambridge University Press, 2019).

11. Richard Stallman, "On Hacking," accessed March 19, 2019, http://stallman.org/articles/on-hacking.html. See also Pekka Himanen, *The Hacker Ethic: A Radical Approach to the Philosophy of Business* (New York: Random House, 2001). On the tropes of internet historiography, see Roy Rosenzweig, "Wizards, Bureaucrats, Warriors, and Hackers: Writing the History of the Internet," *American Historical Review* 103, no. 5 (1998): 1530–52.

12. On the interaction between user innovation and managed firms in the civilian economy, see Eric von Hippel, *Free Innovation* (Cambridge, MA: MIT Press, 2017).

13. Eric von Hippel, "'Sticky Information' and the Locus of Problem Solving: Implications for Innovation," *Management Science* 40, no. 4 (1994): 429–39.

14. Eric von Hippel, *Democratizing Innovation* (Cambridge, MA: MIT Press, 2005).

15. See the appendix for an application of the book's framework to itself.

16. M. Taylor Fravel, "Shifts in Warfare and Party Unity: Explaining China's Changes in Military Strategy," *International Security* 42, no. 3 (January 1, 2018): 37–83.

17. Academy of Military Science, *Science of Military Strategy* [in Chinese] (Beijing: Military Science Press, 2013), 93–97, 130. I am grateful to Mingda Qiu for translation assistance.

18. State Council Information Office, *China's Military Strategy* (Beijing: State Council Information Office of the People's Republic of China, May 2015), http://www.chinadaily.com.cn/china/2015-05/26/content_20820628.htm.

19. For a thorough net assessment of the US-China balance, see Eric Heginbotham, Michael Nixon, Forrest E. Morgan, Jacob L. Heim, Jeff Hagen, Sheng Li, Jeffrey Engstrom et al., *The U.S.-China Military Scorecard: Forces, Geography, and the Evolving Balance of Power, 1996–2017* (Santa Monica, CA: RAND Corporation, 2015), http://www.rand.org/pubs/research_reports/RR392.html. On the limits of Chinese capabilities, see Stephen Biddle and Ivan Oelrich, "Future Warfare in the Western Pacific: Chinese Antiaccess/Area Denial, U.S. AirSea Battle, and Command of the Commons in East Asia," *International Security* 41, no. 1 (July 1, 2016): 7–48. For contrasting views, see Evan Braden Montgomery, "Contested Primacy in the Western Pacific: China's Rise and the Future of U.S. Power Projection," *International Security* 38, no. 4 (2014): 115–49; Andrew S. Erickson, Evan Braden Montgomery, Craig Neuman, Stephen Biddle, and Ivan Oelrich, "Correspondence: How Good Are China's Antiaccess/Area-Denial Capabilities?," *International Security* 41, no. 4 (April 1, 2017): 202–13.

20. Owen R. Cote Jr., "Assessing the Undersea Balance between the U.S. and China" (working paper, MIT Security Studies Program, Cambridge, MA, 2011).

21. James C. Mulvenon and Andrew N. D. Yang, eds., *The People's Liberation Army as Organization* (Santa Monica, CA: RAND Corporation, 2002); Kevin Pollpeter and Kenneth W. Allen, eds., *The PLA as Organization v2.0* (Washington, DC: Defense Group, 2015).

22. James T. Quinlivan, "Coup-Proofing: Its Practice and Consequences in the Middle East," *International Security* 24, no. 2 (1999): 131–65; Talmadge, *Dictator's Army*, 12–40.

23. On the folly of culturalist arguments about innovation, see Thomas G. Mahnken, *Uncovering Ways of War: U.S. Intelligence and Foreign Military Innovation, 1918–1941* (Ithaca, NY: Cornell University Press, 2002).

24. Mandiant, *APT1: Exposing One of China's Cyber Espionage Units*, white paper, February 2013, https://www.fireeye.com/content/dam/fireeye-www/services/pdfs/mandiant-apt1-report.pdf.

25. Jon R. Lindsay and Tai Ming Cheung, "From Exploitation to Innovation: Acquisition, Absorption, and Application," in *China and Cybersecurity: Espionage, Strategy, and Politics in the Digital Domain*, ed. Jon R. Lindsay, Tai Ming Cheung, and Derek S. Reveron (New York: Oxford University Press, 2015), 51–86; Andrea Gilli and Mauro Gilli, "Why China Hasn't Caught Up Yet: Military-Technological Superiority and the Limits of Imitation, Reverse Engineering, and Cyber-Espionage," *International Security* 43, no. 3 (2018): 141–89.

26. David C. Gompert and Martin Libicki, "Cyber Warfare and Sino-American Crisis Instability," *Survival* 56, no. 4 (2014): 7–22; Joshua Rovner, "Two Kinds of Catastrophe: Nuclear Escalation and Protracted War in Asia," *Journal of Strategic Studies* 40, no. 5 (July 29, 2017): 696–730.

27. Robert Martinage, *Toward a New Offset Strategy: Exploiting U.S. Long-Term Advantages to Restore U.S. Global Power Projection Capability* (Washington, DC: Center for Strategic and

Budgetary Assessments, October 27, 2014), http://csbaonline.org/research/publications/toward-a-new-offset-strategy-exploiting-u-s-long-term-advantages-to-restore; Robert O. Work, "The Third U.S. Offset Strategy and Its Implications for Partners and Allies" (speech, Washington, DC, January 28, 2015), https://www.defense.gov/News/Speeches/Speech-View/Article/606641/the-third-us-offset-strategy-and-its-implications-for-partners-and-allies/.

28. Jon R. Lindsay and Erik Gartzke, "Introduction: Cross-Domain Deterrence, from Practice to Theory," in *Cross-Domain Deterrence: Strategy in an Era of Complexity*, ed. Jon R. Lindsay and Erik Gartzke (New York: Oxford University Press, 2019), 1–26.

29. Bernard Brodie, *A Guide to Naval Strategy* (Princeton, NJ: Princeton University Press, 1944), 277.

30. Karl E. Weick and Karlene H. Roberts, "Collective Mind in Organizations: Heedful Interrelating on Flight Decks," *Administrative Science Quarterly* 38, no. 3 (1993): 357–81.

31. Counterforce advocates include Brad Roberts, *The Case for U.S. Nuclear Weapons in the 21st Century*, Stanford Security Studies (Stanford, CA: Stanford University Press, 2015); Matthew Kroenig, *The Logic of American Nuclear Strategy: Why Strategic Superiority Matters* (New York: Oxford University Press, 2018). Counterforce skeptics include Robert Jervis, *The Illogic Of American Nuclear Strategy* (Ithaca, NY: Cornell University Press, 1984); Charles L. Glaser, *Analyzing Strategic Nuclear Policy* (Princeton, NJ: Princeton University Press, 1990).

32. Keir A. Lieber and Daryl G. Press, "The New Era of Counterforce: Technological Change and the Future of Nuclear Deterrence," *International Security* 41, no. 4 (April 1, 2017): 10.

33. Lieber and Press, "New Era of Counterforce," 33–34.

34. Austin Long and Brendan Rittenhouse Green, "Stalking the Secure Second Strike: Intelligence, Counterforce, and Nuclear Strategy," *Journal of Strategic Studies* 38, no. 1–2 (2014): 38–73, speculate that intelligence collection and fusion systems employed in recent U.S. conventional campaigns might also be used for counterforce targeting.

35. Brendan Rittenhouse Green, Austin Long, Matthew Kroenig, Charles L. Glaser, and Steve Fetter, "Correspondence: The Limits of Damage Limitation," *International Security* 42, no. 1 (2017): 193–207.

36. Alexander P. Butterfield, "The Accuracy of Intelligence Assessment: Bias, Perception, and Judgment in Analysis and Decision" (student paper, Naval War College, Newport, RI, March 1993), 5.

37. Ron Rosenbaum, *How the End Begins: The Road to a Nuclear World War III* (Simon and Schuster, 2011), 62.

38. Government Accountability Office, *Strategic Weapons: Changes in the Nuclear Weapons Targeting Process since 1991*, GAO-12–786R, July 31, 2012.

39. William Burr, ed., "The Creation of SIOP-62: More Evidence on the Origins of Overkill," Electronic Briefing Book No. 130, National Security Archive, July 13, 2004, https://nsarchive2.gwu.edu/NSAEBB/NSAEBB130/.

40. On the risks of accident and misperception in nuclear command and control systems during the Cold War, see Bruce Blair, *Strategic Command and Control* (Washington, DC: Brookings Institution Press, 1985); Scott D. Sagan, *The Limits of Safety: Organizations, Accidents, and Nuclear Weapons* (Princeton, NJ: Princeton University Press, 1995).

41. "Jimmy Carter's Controversial Nuclear Targeting Directive PD-59 Declassified," Electronic Briefing Book No. 390, National Security Archive, September 14, 2012, https://nsarchive2.gwu.edu/nukevault/ebb390/.

42. *Nuclear Posture Review* (Washington, DC: U.S. Department of Defense, February 2018), 25.

43. Avery Goldstein, "First Things First: The Pressing Danger of Crisis Instability in U.S.-China Relations," *International Security* 37, no. 4 (2013): 49–89; Caitlin Talmadge, "Would China Go Nuclear? Assessing the Risk of Chinese Nuclear Escalation in a Conventional War with the United States," *International Security* 41, no. 4 (April 1, 2017): 50–92.

44. I refer, of course, to Robert Jervis, *The Illogic of American Nuclear Strategy* (Ithaca, NY: Cornell University Press, 1984).

45. See, for instance, Scott Borg, "Economically Complex Cyberattacks," *IEEE Security and Privacy Magazine* 3, no. 6 (2005): 64–67; Mike McConnell, "Cyberwar Is the New Atomic Age,"

*New Perspectives Quarterly* 26, no. 3 (2009): 72–77; Richard A. Clarke and Robert K. Knake, *Cyber War: The Next Threat to National Security and What to Do about It* (New York: Ecco, 2010); Lucas Kello, "The Meaning of the Cyber Revolution: Perils to Theory and Statecraft," *International Security* 38, no. 2 (2013): 7–40.

46. Academy of Military Science, *Science of Military Strategy*, 193.

47. Thomas Rid, *Cyber War Will Not Take Place* (London: Hurst, 2013); Jason Healey, ed., *A Fierce Domain: Conflict in Cyberspace, 1986 to 2012* (Washington, DC: Cyber Conflict Studies Association, 2013); Jon R. Lindsay, "Stuxnet and the Limits of Cyber Warfare," *Security Studies* 22, no. 3 (2013): 365–404; Brandon Valeriano and Ryan C. Maness, *Cyber War versus Cyber Realities: Cyber Conflict in the International System* (New York: Oxford University Press, 2015).

48. Jordan Branch, "What's in a Name? Metaphors and Cybersecurity" (unpublished manuscript, August 2019), typescript.

49. Daniel W. Drezner, "The Global Governance of the Internet: Bringing the State Back In," *Political Science Quarterly* 119, no. 3 (2004): 477–98; Milton L. Mueller, *Networks and States: The Global Politics of Internet Governance* (Cambridge, MA: MIT Press, 2010); Sean Starrs, "American Economic Power Hasn't Declined—It Globalized! Summoning the Data and Taking Globalization Seriously," *International Studies Quarterly* 57, no. 4 (2013): 817–30; Peter F. Cowhey and Jonathan D. Aronson, *Digital DNA: Disruption and the Challenges for Global Governance* (New York: Oxford University Press, 2017).

50. Jon R. Lindsay, "Restrained by Design: The Political Economy of Cybersecurity," *Digital Policy, Regulation and Governance* 19, no. 6 (2017): 493–514; Brandon Valeriano, Benjamin M. Jensen, and Ryan C. Maness, *Cyber Strategy: The Evolving Character of Power and Coercion* (New York: Oxford University Press, 2018).

51. Erik Gartzke and Jon R. Lindsay, "Weaving Tangled Webs: Offense, Defense, and Deception in Cyberspace," *Security Studies* 24, no. 2 (2015): 316–48.

52. Jon R. Lindsay, "Tipping the Scales: The Attribution Problem and the Feasibility of Deterrence against Cyber Attack," *Journal of Cybersecurity* 1, no. 1 (2015): 53–67; Jon R. Lindsay and Erik Gartzke, "Coercion through Cyberspace: The Stability-Instability Paradox Revisited," in *Coercion: The Power to Hurt in International Politics*, ed. Kelly M. Greenhill and Peter Krause (New York: Oxford University Press, 2018), 179–203.

53. Erik Gartzke and Jon R. Lindsay, "Thermonuclear Cyberwar," *Journal of Cybersecurity* 3, no. 1 (February 2017): 37–48.

54. On the emergence of this concept and concerns about strategic instability, see Jason Healey, "The Implications of Persistent (and Permanent) Engagement in Cyberspace," *Journal of Cybersecurity* 5, no. 1 (January 1, 2019): 1–15.

## Appendix

1. John Gerring, "What Is a Case Study and What Is It Good For?," *American Political Science Review* 98, no. 2 (2004): 349.

2. Kenneth N. Waltz, *Theory of International Politics* (Reading, MA: Addison-Wesley, 1979), 9.

3. Waltz, *Theory of International Politics*, 8.

4. Steven Shapin, "Here and Everywhere: Sociology of Scientific Knowledge," *Annual Review of Sociology* 21 (1995): 304.

5. Carl von Clausewitz, *On War*, trans. Michael Howard and Peter Paret (Princeton, NJ: Princeton University Press, 1976), 120.

6. K. T. Fann, *Peirce's Theory of Abduction* (The Hague: Martinus Nijhoff, 1970); Jörg Friedrichs and Friedrich Kratochwil, "On Acting and Knowing: How Pragmatism Can Advance International Relations Research and Methodology," *International Organization* 63, no. 4 (2009): 701–31.

7. The hermeneutic circle is a notion from the phenomenological tradition of philosophy that describes the interdependence of subjective interpretation and objective reality. Pierce's notion of abduction has much in common with phenomenology, which informs the concept of intentionality in this book.

8. On ethnographic methods, see Barney G. Glaser and Anselm L. Strauss, *The Discovery of Grounded Theory: Strategies for Qualitative Research* (Chicago: Aldine, 1967); Michael Agar, *The Professional Stranger: An Informal Introduction to Ethnography*, 2nd ed. (Bingley, UK: Emerald Group, 2008).

9. Susan Leigh Star, "The Ethnography of Infrastructure," *American Behavioral Scientist* 43, no. 3 (November 1, 1999): 377–91.

10. Diana Forsythe, *Studying Those Who Study Us: An Anthropologist in the World of Artificial Intelligence* (Stanford, CA: Stanford University Press, 2001), 146–62.

11. Ulrike Schultze, "A Confessional Account of an Ethnography about Knowledge Work," *MIS Quarterly* 24, no. 1 (2000): 3–41.

12. Leon Anderson, "Analytic Autoethnography," *Journal of Contemporary Ethnography* 35, no. 4 (2006): 373–95.

13. Pierre Bourdieu, *Outline of a Theory of Practice* (New York: Cambridge University Press, 1977), 72.

14. Barry Posen, *The Sources of Military Doctrine: France, Britain, and Germany between the World Wars* (Ithaca, NY: Cornell University Press, 1984), chap. 5.

15. Timothy M. Cullen, "The MQ-9 Reaper Remotely Piloted Aircraft: Humans and Machines in Action" (PhD diss., Massachusetts Institute of Technology, 2011).

16. A chapter initially drafted for this book was revised as a stand-alone article: Jon R. Lindsay, "Restrained by Design: The Political Economy of Cybersecurity," *Digital Policy, Regulation and Governance* 19, no. 6 (2017): 493–514.

17. Alexander Wendt, *Social Theory of International Politics* (New York: Cambridge University Press, 1999); Emanuel Adler, *World Ordering: A Social Theory of Cognitive Evolution* (New York: Cambridge University Press, 2019).

18. Waltz, *Theory of International Politics*.

19. Robert O. Keohane, *After Hegemony: Cooperation and Discord in the World Political Economy* (Princeton, NJ: Princeton University Press, 1984).

20. Gideon Rose, "Neoclassical Realism and Theories of Foreign Policy," *World Politics* 51, no. 1 (October 1, 1998): 144–72; Hans J. Morgenthau, *Politics among Nations*, ed. Kenneth W. Thompson and David Clinton, 7th ed. (Boston: McGraw-Hill Education, 2005).

21. Peter J. Katzenstein and Nobuo Okawara, "Japan, Asian-Pacific Security, and the Case for Analytical Eclecticism," *International Security* 26, no. 3 (January 1, 2002): 153–85.

# Index

A-10 aircraft, 115
Abdulmutalab, Umar Farouk, 195
Abu Ghraib prison, 168
AC-130 gunships, 186, 192
adaptive information practice, 6, 8–12, 54,
    56, 62–64, 68–70, 74, 84, 97–99, 107,
    112–13, 118, 130–38, 176–78, 182,
    186–90, 203, 211–15, 220–27, 232
adaptive management, 11–12, 213, 220–27,
    231–32, 240
Aegis fire control system, 34–35
Afghanistan, 18–19, 26, 110, 126, 128–29,
    143, 155, 181, 187–88, 191–94, 202
AGM-84E Standoff Land Attack Missile
    (SLAM), 3, 123–24
Aidid, Mohammad, 166
Air Defence of Great Britain (ADGB), 73–76
Air Force Electronic Systems Center, 127
Air Force Mission Support System
    (AFMSS), 8, 10, 117–18, 120–22, 127–28,
    130, 217
Air Force Tactical Air Command, 115
airborne intercept (AI) radar, 105–6
Al-Asad, 152
Al-Awlaki, Abdulrahman, 196, 198
Al-Awlaki, Anwar, 194–98
Al-Awlaki, Nasser, 195–96
Al-Farekh, Mohanad Mahmoud, 200–201
Allen, John, 148–49, 176
Allied Powers, 26, 34, 192
Al-Maliki, Nouri, 175
Al-Masri, Abu Ayyub, 175
Al-Qaeda, 137, 143–45, 149, 173, 175–77,
    182, 185, 194–200, 202, 205–7

Al-Sadr, Moqtada, 144
Al-Shabwani, Jabir, 196
Alston, Philip, 201
Al-Zarqawi, Abu Musab, 143, 175–76
Amanullah, Zabet, 193
Amber drones, 183
*American Sniper*, 149
Amin, Muhammad, 193
Anbar Awakening (Sawaha al-Anbar),
    144–45, 148, 173, 176, 233
Anbar Province. *See* Special
    Operations Task Force (SOTF) Anbar
    Province
Andreasen, John, 112
antiaccess/area denial (A2/AD) network,
    20, 228–30
*Army Field Manual 3-24: Counterinsurgency*,
    137
Arquilla, John, 138
Askariyya shrine, 143
Authorization for Use of Military Force,
    194–95

B-1B aircraft, 192
B-2 aircraft, 4, 59
Baath Party, 136, 139, 142–43
Balad air base, 138, 147
Bateson, Gregory, 47
Battle of Barking Creek, 93
Battle of Britain
    adaptive practice patterns, 10, 84, 97–99,
    107
    archive accessibility, 71
    battle phases, 101

Battle of Britain *(continued)*
  command and control (C2) performance, 8, 71–72, 78, 82, 94–97, 106–7
  constrained external problems, 8, 71–72, 74, 79–83, 96–97, 104–6
  geographical factors, 72, 74, 79–80, 83, 104, 106
  German bombing challenges, 81–83
  information system coordination, 78–79
  institutionalized internal solutions, 8, 71–79, 82–83, 85–88, 107–8
  insulated practice patterns, 104–6, 108
  managed practice patterns, 8–10, 72, 97
  nighttime fighting challenges, 104–7
  Observer Corps role, 88–89, 95–96
  Operation Sea Lion plans, 81, 101–2, 106
  radar importance, 71–76, 78, 80, 82–88, 91–97, 104–6
  relative power assessments, 80, 106–8
  serviceable aircraft patterns, 102–4
  *See also* Royal Air Force (RAF)
Battle of Jutland, 35
Battle of New Orleans, 52
Battle of Passchendaele, 60
Battle of the Atlantic, 34, 108
Bay of Pigs, 66
Belgium, 71
Bennett, John, 123–24
Biddle, Stephen, 23, 131
Big Safari, 184–86, 190
Biggin Hill, 75
Bin Laden, Osama, 150, 185, 196, 205
Bin Mohammed, Abdullah, 205–7
*Black Hawk Down*, 166
Black Hawk helicopters, 70, 166
Bletchley Park, 84, 89
Blue, Neal and Linden, 183
Bomber Command. *See* Royal Air Force (RAF)
Bosnia, 183–84, 186
Boyd, John, 17
Bremer, Paul, 139
Brennan, John, 195, 200
Brentano, Franz, 44–45
Bristol Beaufighter aircraft, 104, 106
Bristol Blenheim aircraft, 104
Brodie, Bernard, 13, 231
Brooks, Frederick, 61, 119
Brown, Ron, 121
Bush, George W., 18, 136, 139, 144, 175, 185, 194
*Business @ the Speed of Thought*, 122

C-130 aircraft, 116, 120, 191
cathode ray tubes (CRTs), 86
Cebrowski, Arthur, 17–18, 110–11, 131
Central Intelligence Agency (CIA), 3, 147, 183, 185, 194–97, 200, 202–3, 208, 225

Chain Home radar system, 71–76, 78, 80, 82–84, 86–88, 91–97, 105. *See also* Royal Air Force (RAF)
Chamberlain, Neville, 77
Channel Gap, 74, 104
China
  antiaccess/area denial (A2/AD) network, 20, 228–30
  Belgrade embassy bombing, 3–5, 9, 215, 240
  Communist Party of China, 20, 229
  FalconView usage, 127
  informatization development, 1, 4, 13, 19–20, 227–30
  nuclear capabilities, 236–37
  People's Liberation Army (PLA), 12, 20, 228–30, 236–37
  Redstone Arsenal, 127
  *Science of Military Strategy*, 227–28, 238
  security breaches, 129
  Strategic Support Force, 229
  technical data theft, 230
Churchill, Winston, 9, 72, 75–76, 79, 108
Ciborra, Claudio, 26
Clark, Kate, 193
Clark, Wesley, 3
Clarke, Richard, 184–85
Clausewitz, Carl von, 11, 17, 21, 35–36, 49, 53, 79–80, 217–21, 227
Clinton, Bill, 4, 17–18
Coalition Provisional Authority, 139–40
Cold War, 1, 8, 16, 18–19, 22, 56, 113–14, 117, 130, 133, 149, 182, 185, 228, 230, 235, 240
Colombia, 127
Combat Flight Planning Software (CFPS), 121
Combined Air Operations Center (CAOC), 109–12, 126, 138, 154, 184, 186–88, 190
Combined Joint Special Operations Task Force (CJSOTF), 146–48, 155, 164–66, 176
command, control, communications, computers, intelligence, surveillance, and reconnaissance (C4ISR) systems, 28, 132, 231
command and control (C2) systems, 1, 25, 28, 66, 71–72, 82, 89, 94–97, 106–7, 109, 129, 132, 166, 180, 182, 219–20, 228, 237
Committee for the Scientific Survey of Air Defense, 74
Communism, 16, 20, 183, 229, 236
concept of operations (CONOP), 163, 165–66, 171
constrained external problems, 7–10, 34, 38, 48–57, 65, 69, 71, 96–97, 104, 130–31, 142, 148, 153, 186, 191, 201, 212, 232. *See also* unconstrained external problems

coordination as sociotechnical institution, 42–50, 54–55, 66, 69, 79, 108, 131, 153–54, 158–64, 178, 208, 222, 241
COPLINK, 176–77
Couch, Dick, 149
Creech, Wilbur, 115
Creech Air Force Base, 188
Croatia, 121, 133
Cromemco System 2 (CS-2) computers, 115–16
Cronin, Audrey Kurth, 207–8
cruise missiles, 34, 123, 182–84, 195
Cullen, Timothy, 188
cybersecurity, 9, 12, 30, 50, 64, 129, 213, 229–30, 237–40

Davis, Joshua, 212
Defense Advanced Research Projects Agency, 183
Defense Science Board, 61
Delta Force, 175
Denmark, 71
*Department of Defense Artificial Intelligence Strategy*, 20
Deptula, David, 19, 180
Derwish, Kamal, 195, 198
Devlin, Peter, 143–44
Distinguished Warfare Medal, 209
Distributed Common Ground System (DCGS), 187, 190–91
Docauer, Alan, 131
Donaldson, Lex, 56
Dover radar station, 83
Dowding, Hugh, 71, 73, 75–80, 94, 96–97, 99–101, 104–8
drones
    adaptive practice patterns, 182, 186–90, 203, 211
    Afghanistan deployment, 155, 181, 187–88, 191–94, 202
    Al-Qaeda targeting, 182, 185, 194–200, 202, 205–7
    Amber model, 183
    Barack Obama oversight, 10, 194–202, 216
    Big Safari unit, 184–86, 190
    cataloging challenges, 155–56
    civilian deaths, 10, 107, 182, 191, 193, 196–98, 202–5
    command and control (C2) systems, 180, 182
    constrained external problems, 186, 191, 201
    counterterrorism benefits, 12, 194, 202, 209–10
    disposition matrix database, 200
    distributed information systems, 185–90, 211

Donald Trump oversight, 10, 202–3, 205, 216
FalconView imagery, 126, 187–88
find, fix, finish, exploit, analyze (F3EA) cycles, 180–81, 194
Firebee model, 182
fratricide, 10, 182, 191–93, 213
Gnat model, 183
ground forces relationship, 180–81, 186, 194
Hellfire missiles, 184, 187, 191–92
high-value individual (HVI) identification, 180, 216
institutionalized internal solutions, 186, 193
insulated practice patterns, 182, 192–93, 211
intelligence data amounts, 155–56, 180
Iraq deployment, 155, 181, 187, 202
Kettering Bug prototype, 182
Lightning Bug model, 182
live video capabilities, 155, 186–88
loitering abilities, 180–81
managed practice patterns, 182, 193, 205–7, 211
organic internal solutions, 186
origins, 10, 182–83, 216
personality strikes, 197–98, 201
Predator model, 10, 126, 180, 182–91, 195–96, 206
problematic practice patterns, 182, 190, 203, 205, 211
Reaper model, 180, 182, 187–89, 191, 196
remote control morality concerns, 181–82, 189, 198, 208–12
signature strikes, 197–98, 201
sniper capabilities, 185–86
Special Operations Task Force (SOTF) Anbar Province usage, 152, 155, 166, 175
targeting accuracy, 10, 107, 181–82, 191–94, 197–98, 202–5, 209–10, 216

E-3 aircraft, 69–70
Eden, Lynn, 25
Edwards, Paul, 36
Eglin Air Force Base, 115–17, 121–22, 124
Eisenhower, Dwight, 220
Eleventh Reconnaissance Squadron, 183–84
enforcement as sociotechnical institution, 42–53, 59, 65, 98–101, 104, 108, 131, 153–54, 164–67, 172, 178, 208–10, 222, 241
Epic, 216–17
Execution Planner (XPlan), 113

F-117 aircraft, 122
F-14 aircraft, 34
F-15 aircraft, 69–70, 189

F-16 aircraft, 113, 117, 119–20, 122, 206, 226
F-22 aircraft, 129
F-35 aircraft, 129
F/A-18 aircraft, 123, 129
FalconView mapping system
  adaptive practice patterns, 9, 113, 133–35, 215
  beta version completion, 119
  *Business @ the Speed of Thought* chapter, 122
  Carnegie Mellon study, 127–28
  Combat Flight Planning Software (CFPS) interface, 121
  costs, 113, 129
  diffusion across military services, 113, 119, 123–24, 127
  Digital Terrain Elevation Data, 122
  drone imagery, 126, 187–88
  ease of use, 120, 126
  functional evolution, 113, 120–29, 133
  funding, 120–24, 127–28, 134
  Georgia Tech Research Institute (GTRI) development, 113, 118–19, 121–22, 124, 126, 128, 215
  innovative features, 114–15, 118, 215
  institutionalized internal solutions, 114, 125, 135
  insulated practice patterns, 113
  managed practice patterns, 132, 135, 215
  Mission Planning User Conference, 124
  *Mythical Man-Month* basis, 119
  open-source code, 129
  organic internal solutions, 114, 125, 135
  origins, 9–10, 113, 118
  Portable Flight Planning Software (PFPS) bundle, 113, 117
  Special Operations Task Force (SOTF) Anbar Province implementation, 159
  TaskView application, 125
  unclassified status, 120–21, 126
  user extensions, 125–27, 129, 133–34, 217, 225
  visual formatting options, 113, 121, 125–26
Fallujah, 5, 142–43, 145, 148, 152
Farouq, Ahmed, 198
Fedayeen Saddam, 139
Federal Bureau of Investigations (FBI), 200, 225
Federal Directorate of Supply and Procurement (FDSP), 3–4
Federal Intelligence Surveillance Court, 200
Federally Administered Tribal Areas (FATA), 194, 197
Ferris, John, 73, 77
Fighter Command. *See* Royal Air Force (RAF)

find, fix, finish, exploit, analyze (F3EA), 5, 154, 164, 167–71, 175–76, 180–81, 194
find, fix, track, target, engage, assess (F2T2EA), 110–12, 154
Firebee drones, 182
Fleming, Jerry, 115–16, 121
Flynn, Michael, 192
Fogleman, Ronald, 110, 183–84
Ford, Gerald, 185
Fort Hood, 195
Fort Huachuca, 183
FPLAN, 115–18, 121
France, 71, 73, 77–81, 86, 93
fratricide, 6, 10, 22, 65–66, 70, 90, 178, 182, 191–93, 213, 229
Fuller, J. F. C., 60

Gadahn, Adam, 198
Garstka, John, 111
Gaskin, Walter, 148
Gates, Bill, 119, 122
Gates, Robert, 193
Gawande, Atul, 216–17
General Atomics, 183, 187, 190, 195
Georgia Tech Research Institute (GTRI), 113, 118–19, 121–22, 124, 126, 128, 215
Ghadiya, Abu, 175
Gnat drones, 183
Goldberg, Rube, 64
Goldwater-Nichols reforms, 56
Göring, Hermann, 81
*Graf Zeppelin*, 82
Grauer, Ryan, 23
Grossman, Dave, 189
ground control intercept (GCI), 105–6
Gulf War, 16, 117, 133, 139, 184, 227

Haig, Douglas, 60
Haqqani, 202
Hayden, Michael, 197
Hayek, Friedrich, 63
Hellfire missiles, 184, 187, 191–92
high-value individuals (HVIs), 161–65, 167, 173, 175, 180, 216
Hill Air Force Base, 117
Hippel, Eric von, 63
Hitler, Adolf, 71, 79, 81, 101–2, 106
Holder, Eric, 200
human intelligence (HUMINT), 67, 141, 151, 156–57, 164, 168, 171
Hurricane aircraft, 72, 93, 102
Hussein, Saddam, 16, 139, 142
Hutchins, Edward, 26–27, 38

identification friend-or-foe (IFF), 87, 93–94
improvised explosive devices (IEDs), 5, 9, 26, 68, 136, 176
informatization. *See* China

*Inspire*, 196
institutionalized internal solutions, 7–10, 12, 17, 23–24, 54–56, 60, 71–73, 107, 112, 114, 125, 131–32, 135–36, 145, 147, 152–54, 160–61, 186, 193, 212, 216–22, 231–36. *See also* organic internal solutions
insulated information practice, 8, 10–12, 28, 54, 56, 59–62, 65, 68–70, 104–6, 113, 118, 138, 147, 153–54, 160, 170–73, 178, 182, 192–93, 211–13, 216, 222, 232–37
intelligence, surveillance, and reconnaissance (ISR), 109–11, 155–56, 165–66, 175
intentionality as guiding preference, 42–45, 50, 55, 60, 79, 97, 152–53, 158, 209, 216, 220–23, 227
International Security Assistance Force (ISAF), 192, 194
Iran, 18, 66, 139–40, 234, 240
Iraq
    Al-Qaeda in Iraq (AQI), 137, 143–45, 149–50, 173, 175–77, 202
    Askariyya shrine destruction, 143
    Baath Party, 136, 139, 142–43
    Balad air base, 138, 147
    Bedouin tribes, 142
    counterinsurgency efforts, 10, 19, 23, 137–42
    drone deployment, 155, 180, 187, 202
    FalconView usage, 126
    Fallujah Forward Operating Base (FOB), 5
    Green Zone, 140
    improvised explosive device (IED) presence, 5, 9, 26, 136, 176
    Iran, conflict with, 139–40
    offset strategy, 16
    Operation Iraqi Freedom, 112, 143
    Royal Ulster Constabulary, 177
    Saddam Hussein leadership, 16, 139, 142
    Shia people, 137, 140, 142–45, 175
    suicide attacks, 138
    Sunni people, 137, 139–40, 142–45, 175
    tooth to tail ratio, 29–30
    weapons of mass destruction, 18, 67, 139, 173
    Western Euphrates River Valley (WERV), 142–43, 149, 152, 158, 176, 181
    *See also* Special Operations Task Force (SOTF) Anbar Province
Irvine, Dallas, 29
Islamic State in Iraq and Syria (ISIS), 173, 202
Israel, 182

James, William, 45, 47–48, 67
Janowitz, Morris, 30

Johnson, Lyndon, 199
Joint Chiefs of Staff, 17, 56
Joint Committee on Anti-Aircraft Defense, 76
Joint Forces Command (JFCOM), 18
Joint Mission Planning System (JMPS), 127–30, 134–35
Joint Operational Tactical System, 64
joint operations center (JOC), 138, 154, 175
Joint Special Operations Command (JSOC), 8, 10, 137–41, 143–44, 146–50, 154, 168, 170, 173, 175, 194, 196–97, 200, 202, 225
*Joint Vision 2010*, 17–18, 110, 131
Jomini, Antoine Henri, 129–30
Jordan, 148
Jumper, John, 110–11, 184–85

Karem, Abraham, 182–83
Kelly, Thomas, 112
Kesselring, Albert, 81
Kettering Bug, 182
Khan, Samir, 196
Kirsh, David, 38
Kometer, Michael, 109–10, 112
Kosovo, 2–3, 6, 110, 128, 184–87, 189

Latif, Muhammad Mahmoud, 143
Latour, Bruno, 36–37, 45, 109
Law, John, 42, 75
Leigh-Mallory, Trafford, 79
Liber, Keir, 234
Libya, 194, 202
*Lifting the Fog of War*, 18
Lightning Bug drones, 182
Lindemann, Frederick, 75–76
Lo Porto, Giovanni, 197–98
London Air Defense Area (LADA), 73, 75
Long, Austin, 148
Luftwaffe
    aircraft exchange ratios, 102–4
    Battle of Britain Day losses, 101
    blitzkrieg tactics, 81, 101
    bombing challenges, 81–83
    external airfields, 80, 86
    geographical challenges, 72, 80, 104, 106
    ground support goals, 72, 81, 101
    initial British bombings, 77
    institutionalized internal solutions, 8
    insulated practice patterns, 10, 82–83, 108
    intelligence issues, 72, 80–82
    Knickebein beacon, 89
    night sorties, 104–6
    North region avoidance, 78, 80
    Operational Trials Wing 210 (OTW-210) success, 83
    organizational routines, 89–90
    pilot supply levels, 103–4
    political conflicts, 81

Luftwaffe *(continued)*
  radar assumptions, 82–83
  relative power assessments, 80
  target lists, 81
  unconstrained external problems, 8
  vertical stacking tactics, 94
  *See also* Battle of Britain

MacFarland, Sean, 144, 177
Maglio, Paul, 38
managed information practice, 8, 10–11, 54,
  56–59, 63, 68–70, 132, 135, 164, 173–76,
  178, 182, 193, 205–7, 211–15, 220, 236,
  239–40
Marine Expeditionary Force (MEF), 10, 143,
  147–48, 163, 166, 171, 173, 176–78. *See
  also* Special Operations Task Force
  (SOTF) Anbar Province
Marshall, Andrew, 16–17
Maruf, Abu, 143
Mattis, James, 18
Mayaguez incident, 66
McChrystal, Stanley, 138, 150, 175–76
McMaster, H. R., 21
McNeal, Gregory, 201
McRaven, William H., 150
measurement as sociotechnical institution,
  42–45, 47–55, 59, 84–90, 108, 131,
  153–57, 172, 177–78, 208–10, 222, 241
mediated experience, 38–41, 48, 58, 65, 84.
  *See also* reflexive experience
Messerschmitt 109 (Me-109) aircraft, 78, 81,
  83, 93, 101–4
Messerschmitt 110 (Me-110) aircraft, 83,
  102–3
Milosevic, Slobodan, 3
Mindell, David, 41–42
Minuteman missiles, 235
Mission Planning User Conference, 124
Mission Support System (MSS), 116–19
Modernized Integrated Database (MIDB), 4
Mohammad, Jude Kenan, 198
Momsen, Charles, 26
Monsoor, Michael, 149
Morehouse, Jim, 111–12
Multi-National Forces West, 146–47
Musharraf, Pervez, 194
*Mythical Man-Month, The*, 119

Napoleon, 28, 35, 42, 130, 218
National Counterterrorism Center, 200
National Security Agency (NSA), 175,
  196–97, 200, 225
NATO, 2–4, 16, 133, 183, 192
Naval Special Warfare (NSW), 6, 135, 146,
  148–52, 154, 157, 164–67, 173, 176–79.
  *See also* Special Operations Task Force
  (SOTF) Anbar Province

Naval Tactical Data System, 64
Nazi Party, 26, 71, 79, 182
Netherlands, 71
Nicaragua, 183
Nidal, Hasan, 195
North, Douglass, 42
North Korea, 234
Northrop Grumman, 128–29
Norway, 71, 80
NSW Development Group (DEVGRU), 150.
  *See also* Naval Special Warfare (NSW)
nuclear weapons, 12, 19, 25, 30, 59, 184, 213,
  222, 230, 233–37, 240

Obama, Barack, 10, 194–202, 216
observe, orient, decide, act (OODA) loop,
  17, 20, 41, 43, 131
Observer Corps, 88–89, 95–96
Office of Force Transformation, 18
Office of Naval Intelligence (ONI), 138, 151
Ogarkov, Nikolai, 16
Ogden Air Logistics Center, 122
Ogden Data Device (ODD), 117–18
Olson, Mancur, 65
Omar, Mullah Mohammed, 186
*On War*, 35–36
Operation Allied Force, 110
Operation Anaconda, 126
Operation Barbarossa, 101
Operation Desert Storm, 109–10, 128
Operation Enduring Freedom, 110
Operation Iraqi Freedom, 112, 143
Operation Sea Lion, 81, 101–2, 106
Operational Intelligence Center (OIC), 34
organic internal solutions, 7–10, 54–56, 62,
  64–65, 112–14, 125, 131–32, 135, 145,
  147, 152–54, 178, 186, 212, 216, 220–22,
  226, 231–33. *See also* institutionalized
  internal solutions
Owens, William, 17, 203

Pakistan, 181, 191, 194, 196–97, 200–204,
  207–8
Paret, Peter, 36
Park, Keith, 78–79, 106
Pentagon, 16, 119, 122, 183, 193, 200, 202,
  230
People's Liberation Army (PLA), 12, 20,
  228–30, 236–37
Perrow, Charles, 21–22, 57, 192
Petraeus, David, 137, 144, 175
Pevensey radar station, 83
Phony War, 81
plan position indicator (PPI) radar displays,
  105
Poland, 71, 81
Polaris submarine, 37
Pompeo, Michael, 202

Portable Flight Planning Software (PFPS), 113, 117. *See also* FalconView
*Predator*, 183
Predator drones, 10, 126, 180, 182–91, 195–96, 206. *See also* drones
Presidential Policy Guidance (PPG), 199–200, 202–3
Press, Daryl, 234
Principles, Standards, and Procedures (PSP), 203
problematic information practice, 8–12, 24, 54, 56, 65–70, 131, 136, 154, 182, 190, 203, 205, 211–13, 216, 222, 229, 236
Program Evaluation and Review Technique (PERT), 37
Pyles, John, 118–20, 122–23

Quine, W. V. O., 51
Quiver targeting application, 126, 134–35

Raeder, Erich, 79
Ramadi, 142–45, 152, 175, 177
Raymond, Eric, 63
Reagan, Ronald, 185
Reaper drones, 180, 182, 187–89, 191, 196. *See also* drones
Redstone Arsenal, 127
referential integrity, 40, 46–49, 59–60, 66, 87, 96, 142, 158, 193
reflexive experience, 40–41, 48, 58, 60, 63, 65, 192. *See also* mediated experience
remotely piloted aircraft (RPA). *See* drones
revolution in military affairs (RMA), 1–6, 11–13, 16–25, 110–11, 130–31, 136–38, 141, 212–13, 228, 230–31, 234–35
Roberts, Karlene, 41
Rochlin, Gene, 62
Rosenbaum, Ron, 235–36
Royal Air Force (RAF)
    adaptive practice patterns, 10, 84, 97–99, 107
    air defense role, 8, 71–73, 76–77, 79–80
    aircraft exchange ratios, 102–4
    Chain Home radar deployment, 71–76, 78, 80, 82–83, 86–88, 91–97, 105
    constrained external problems, 8, 71–72, 74, 79–83, 96–97, 104–6
    coordination structure, 91–97
    Filter Room, 74, 88, 91–97, 99
    friendly fighter tracking, 89, 95
    geographical benefits, 72, 74, 104, 106
    Hugh Dowding leadership, 71, 73, 75–80, 94, 96–97, 99–101, 104–8
    identification friend-or-foe (IFF) readings, 87, 93–94
    institutionalized internal solutions, 8, 71–79, 83–84, 107–8
    insulated practice patterns, 104–6
    managed practice patterns, 8–10, 72, 97, 215
    nighttime vulnerability, 104–7
    North region training sanctuary, 78, 80
    Ops Room, 40, 75, 78, 82, 84, 91–100, 111, 158
    origins, 73
    pilot supply levels, 103–4, 106
    political disagreements, 76–78
    radar usage, 80, 82–84, 104–6
    range and direction finding (RDF), 75, 84–86
    standardized data formats, 96–97
    strategic bombing ideology, 71–72, 77
    U.S. Air Force influence, 8, 71
    *See also* Battle of Britain
Royal Navy, 34, 78–81, 107–8
Rumsfeld, Donald, 18, 137, 173, 212
Russia, 17, 20, 129
Rye radar station, 83

Sandford, Robert, 117–21, 127
Sarajevo, 183
Saudi Arabia, 186, 195
S-boats, 83
Schlieffen, Alfred von, 14–16, 18, 20
*Science of Military Strategy*, 227–28, 238
SEAL (Sea, Air, and Land) platoons, 6, 138, 146, 149–52, 156, 159, 161, 163–69, 173, 179, 196, 203. *See also* Naval Special Warfare (NSW)
Secure Internet Protocol Router Network (SIPRNET), 155, 158, 166, 171, 188
sensitive site exploitation (SSE), 167–69
September 11 terrorist attacks, 28, 185–86, 190, 194–95
Serbia, 2–3, 240
Shah, Aqil, 208
Shahzad, Faisal, 195
*Sheriff of Ramadi, The*, 149
Sherman, William T., 18
Shia people, 137, 140, 142–45, 175
signals intelligence (SIGINT), 84, 89–90, 141, 155, 162–64, 175
Single Integrated Operational Plan (SIOP), 236
Snook, Scott, 69–70
Snow, C. P., 72
Sobczyk, Walter, 117
Somalia, 194, 200, 202–5
Soviet Union, 13, 16–17, 101, 106, 133, 182, 230, 235
Space and Naval Warfare Systems Command (SPAWAR), 134–35
Special Operations Command (SOCOM), 123, 127–28, 135, 146, 152, 234–36

Special Operations Task Force (SOTF)
Anbar Province
adaptive practice patterns, 6, 138
administrative structure, 145–47, 153, 158,
170
Al-Qaeda in Iraq (AQI) conflicts, 149–50,
173
classified environment, 158
Combined Joint Special Operations Task
Force-Arabian Peninsula (CJSOTF)
headquarters, 146–48, 155, 164–66,
176
command responsibilities, 147
concept of operations (CONOP), 163,
165–66, 171
constrained external problems, 142, 148,
153
control cycle phases, 153, 155–64, 166,
172, 178
counterinsurgency efforts, 8, 139, 148–49,
165, 173, 178
counterterrorism preferences, 138–39,
148, 153, 165, 169–70, 173, 176
direct action bias, 149–54, 161, 164–65,
170, 173–74, 178–79, 181, 216, 223
drone surveillance, 152, 155, 166, 175
FalconView implementation, 159
find, fix, finish, exploit, analyze (F3EA)
cycles, 5, 154, 164, 170–71, 194
fragmented data locations, 158–60,
170–71, 181
geographic dispersion, 152–53
high-value individual (HVI) targets,
161–65, 167, 173, 175
human intelligence (HUMINT) usage,
151, 156–57, 164, 168, 171
institutionalized internal solutions, 145,
147, 152–54, 160–61, 216
insulated practice patterns, 10, 138, 147,
153–54, 160, 170–73, 178, 216
Iraqi interpreters, reliance on, 157
managed practice patterns, 164, 173–76,
178
Marine Expeditionary Force (MEF) role,
10, 143, 147–48, 163, 166, 171, 173,
176–78
Multi-National Forces West hosting,
146–47
Naval Special Warfare (NSW) role, 6, 138,
146, 148–52, 154, 157, 164–67, 173,
176–79
organic internal solutions, 138–39, 145,
147, 152–54, 178, 216
problematic practice patterns, 154, 216
Secure Internet Protocol Routing
Network (SIPRNET) portals, 155, 158,
166, 171

sensitive compartmented information
facility (SCIF), 155
sensitive site exploitation (SSE), 167–69
signals intelligence (SIGINT) receptions,
155, 162–64
tactical local area networks (TACLANs),
158–60
target intelligence package (TIP), 163–65,
167, 171
technical intelligence, surveillance, and
reconnaissance (ISR) usage, 155–56,
165–66
training programs, 151, 173, 178
tribal engagement, 138, 164, 178
Sperrle, Hugo, 81
Spitfire aircraft, 72, 93, 102, 108
Stallman, Richard, 223
Sterelny, Kim, 38
Strategic Air Command, 236
Strategic Command (STRATCOM), 236–37
Stuka aircraft, 81
submarines, 26, 34, 37, 91, 107–8, 184–85,
228, 235
Sunni people, 137, 139–40, 142–45, 175
Swinton, Lord, 77
Syria, 142, 175, 202

Tactical Air Command, 116
Tactical Air Mission Planning System
(TAMPS), 123–24
tactical local area networks (TACLANs),
158–60
Taliban, 185–86, 192–94, 199, 202
target intelligence package (TIP), 163–65,
167, 171
Theater Battle Management Core System,
126
third offset strategy, 1, 12–13, 19–20,
230–32
Thorn, Jake, 115–17, 120–21, 128–29
Ticonderoga cruiser, 34
Tizard, Henry, 72, 74–76, 78–79, 98, 105
Tomahawk cruise missiles, 123, 195
Treaty of Ghent, 52
Trenchard, Hugh, 76–77
Trident badge, 150
Trump, Donald, 10, 202–3, 205, 216
Turing, Alan, 154
Tuttle, Jerry O., 64

unconstrained external problems, 7–10, 51,
56–57, 59–60, 63, 212. See also
constrained external problems
unmanned aerial vehicle (UAV). See drones
USS Cole, 195
USS Eisenhower, 64
USS Vincennes, 8–9, 34

V-1/V-2 guided weapons, 107, 182
Vaughan, Diane, 25
Ventnor radar station, 83
Vietnam War, 16, 30, 149, 183

War of 1812, 52
Wartime Integrated Laser Demonstration
(WILD), 184
Watson-Watt, Robert, 72–76, 105
Weber, Max, 62
Webster, Joe, 119–20, 123
Wehrmacht, 101, 107
Weick, Karl, 41
Weinstein, Warren, 197–98

Western Euphrates River Valley
(WERV), 142–43, 149, 152, 158, 176,
181
Westmoreland, William, 15–16, 20
Wilkinson, Spencer, 132
Williamson, Oliver, 36
Winn, Rodger, 34
Women's Auxiliary Air Force, 86, 91

Xi Jinping, 20, 229

Yemen, 181, 194–96, 198, 200, 202–7, 240

Zimmerman, David, 73–74, 77